KU-345-696

STRATHCLYDE UNIVERSITY LIBRARY
30125 00380262 5

HANDBOOK OF GEOPHYSICAL EXPLORATION

SECTION I: SEISMIC EXPLORATION

VOLUME 21

SUPERCOMPUTERS IN SEISMIC EXPLORATION

HANDBOOK OF GEOPHYSICAL EXPLORATION

I. SEISMIC

II. ELECTRICAL

III. GRAVITY

IV. MAGNETIC

V. WELL-LOGGING

VI. RADIOMETRIC

VII. REMOTE SENSING

VIII. GEOTHERMAL

SECTION I. SEISMIC EXPLORATION

Editors: Klaus Helbig and Sven Treitel

Volume 1. Basic Seismic Theory
2. Seismic Instrumentation
3. Seismic Field Techniques
4. Seismic Inversion and Deconvolution
5. Seismic Migration
6. Seismic Velocity Analysis
7. Seismic Noise Attenuation
8. Structural Interpretation
9. Seismic Stratigraphy
10. Production Seismology
11. 3-D Seismic Exploration
12. Seismic Resolution
13. Refraction Seismics
14. Vertical Seismic Profiling
15. Seismic Shear Waves
16. Seismic Coal Exploration
17. Mathematical Aspects of Seismology
18. Physical Properties of Rocks
19. Engineering Seismics
20. Pattern Recognition and Image Processing
21. Supercomputers in Seismic Exploration

SEISMIC EXPLORATION

Volume 21

SUPERCOMPUTERS IN SEISMIC EXPLORATION

by
E. EISNER
Texaco
Houston, Texas, USA

PERGAMON PRESS
OXFORD · NEW YORK · BEIJING · FRANKFURT
SÃO PAULO · SYDNEY · TOKYO · TORONTO

U.K.	Pergamon Press plc, Headington Hill Hall, Oxford OX3 0BW, England
U.S.A.	Pergamon Press, Inc., Maxwell House, Fairview Park, Elmsford, New York 10523, U.S.A.
PEOPLE'S REPUBLIC OF CHINA	Pergamon Press, Room 4037, Qianmen Hotel, Beijing, People's Republic of China
FEDERAL REPUBLIC OF GERMANY	Pergamon Press GmbH, Hammerweg 6, D-6242 Kronberg, Federal Republic of Germany
BRAZIL	Pergamon Editora Ltda, Rua Eça de Queiros, 346, CEP 04011, Paraiso, São Paulo, Brazil
AUSTRALIA	Pergamon Press Australia Pty Ltd., P.O. Box 544, Potts Point, N.S.W. 2011, Australia
JAPAN	Pergamon Press, 5th Floor, Matsuoka Central Building, 1-7-1 Nishishinjuku, Shinjuku-ku, Tokyo 160, Japan
CANADA	Pergamon Press Canada Ltd., Suite No. 271, 253 College Street, Toronto, Ontario, Canada M5T 1R5

First edition 1989

Library of Congress Cataloging in Publication Data

Supercomputers in seismic exploration/[edited] by
E. Eisner—1st ed.
p. cm. — (Handbook of geophysical exploration.
Section I. Seismic exploration; v. 21)
Includes bibliographies.
1. Seismic prospecting—Data processing.
2. Supercomputers.
I. Eisner, E. (Elmer) II. Series.
TN269.S364 1984 vol. 21 622'. 159 s—dc19
[622'. 159] 88—25410

British Library Cataloguing in Publication Data

Supercomputers in seismic exploration.
Seismology. Applications of computer systems
I. Eisner, E. (Elmer) II. Series
551.2'2'0285
ISBN 0-08-037018-7

Printed in Great Britain by A. Wheaton & Co. Ltd., Exeter

PREFACE

This volume is a direct outgrowth of a workshop held in conjunction with the November 1986 meeting of the Society of Exploration Geophysicists. The question explored there was whether existing computational capability was the primary hindrance to complete yet practical inversion of geophysical data for routine interpretation purposes.

L. Baker considers a typical seismic model and determines the computational requirements associated with several standard numerical modeling techniques.

For this volume, M. Edwards and M. Reshef supplement this discussion with a detailed nuts and bolts discussion of the current status of such work using CRAY surpercomputers.

O. Holberg, paying great attention to the tradeoffs involved in the choice of algorithm and architecture, gives a design analysis for constructing a specialized computer for use on this problem.

O. Johnson and E. Leiss present an up to the minute survey of the many diverse approaches to new computer designs, any one of which may turn out to provide the right choice for use on the inverse problems of concern in geophysical exploration and/or production.

A. McAulay discusses his work on inversion, and also reviews new hardware design considerations which may be of particular use in this area.

S. Ronen and R. Schreiber elaborate on the possibilities inherent in the use of a systolic computer for these problems.

W. Moorhead expands on the hypercube architecture and provides guidance on its effective use in this area.

P. Mora and A. Tarantola present an optimistic discussion indicating that existing hardware and algorithmic developments may get us a long way in the direction we want to go.

The following three authors address the kernel of seismic inversion, iterative modeling.

K. Marfurt and C. Shin explain how to take conventional finite difference and finite element solvers and make them more efficient for the special iterative modeling example. They provide some algorithmic developments which should promote tractability by reducing the apparent computational load requirement.

I. Mufti shows how finite difference algorithms can be improved to give more affordable results. Impatient to wait for perfection and fearing that the best is the enemy of the good, he shows us what he is able to do here and now. He presents the results obtained from a large-scale 3-D finite-difference seismic model and discusses their interpretive significance.

G. Schuster illustrates what can currently be accomplished with the boundary integral technique, an alternative that may aid in efficient 3-D modeling.

T. F. Russell presents a survey of reservoir modeling and the opportunities which improved computational capacity would offer. Since geophysical exploration in support of reservoir development is of rapidly increasing concern, this material should be of wide interest.

Readers are also alerted to follow the developments by D. Rothman at MIT who is using cellular automata to model fluid flow in porous media. Unfortunately it was not possible to include an account here because of time constraints.

The individual chapters represent the diverse points of view of their authors, and no attempt has been made to establish a consensus. While industrial security considerations have operated to some extent, every effort has been made to present a current perspective on this problem as of January 1988. I believe that large scale cooperation is the key to rapid progress and hope that this volume will serve that end.

I wish to express my sincere thanks to the chapter authors for their cooperation in making this volume timely, and to Texaco for logistic support and encouragement.

Elmer Eisner, Editor
Texaco – Sr. Scientist Emeritus
Houston, Texas
March 1988

TABLE OF CONTENTS

CHAPTER 1

IS 3-D WAVE-EQUATION MODELING FEASIBLE IN THE NEXT TEN YEARS?

by
L. J. BAKER
Exxon Production Research Company

Can a massively-parallel computer be built in the next ten years capable of performing realistic 3-D-seismic modeling within a few hours? To answer this question, we first review the uses of seismic forward modeling and describe a seismic model of typical size. Next, we summarize the mathematical methods and corresponding computational requirements for such a model. Finally, based on the current performance of a distributed-memory parallel computer and projections of increased computer power in ten years, we conclude that realistic 3-D acoustic modeling is possible in ten years.

BACKGROUND

Prior to the introduction of supercomputers, seismic modeling was essentially confined to 1-D and 2-D ray tracing. Ray trace modeling is still the most widely used seismic modeling technique. In addition to its historical incumbency, ray tracing is usually much faster than wave-equation methods. Furthermore, ray tracing mirrors the way a geophysicist thinks about how a seismic section is formed. However, in instances where there is complex geology or complex wave phenomena, geophysicists do turn to acoustic or elastic wave-equation modeling. Some of the specific advantages of wave-equation modeling over ray tracing are:

1. automatic generation of diffractions, critical refractions, and multiples,
2. more accurate amplitudes and waveforms, especially in the presence of small structures and thin beds,

3. no missing seismic events, regardless of complexity.

The use of seismic modeling programs in the petroleum industry varies from company to company. Initially, with limited computing power, wave-equation modeling was primarily used for fundamental research studies. Now, however, wave-equation modeling is used more often to test seismic interpretations, to plan data acquisition, and to calibrate the effectiveness of seismic processing schemes (and other modeling programs). We also use forward modeling schemes in inversion, but a typical model-driven inversion program requires ten or more forward simulations, thus requiring even greater computational power.

Most seismic modeling is currently conducted in 2-D, and some ray tracing is performed in 3-D. Clearly, 3-D modeling capability is becoming increasingly important as we collect more 3-D data and search for smaller reservoirs. For modeling the full seismic experiment and especially for modeling geology with structure, it is essential to model the actual seismic data-gathering process, including multiple shots. A limited amount of 3-D wave-equation modeling has been reported, primarily to demonstrate supercomputers' capabilities (Adams, 1984; Edwards, Hsiung, Kosloff, and Reshef, 1985). More recently, Mufti (Mufti, 1987) has discussed low resolution 3-D acoustic modeling primarily based on exploding reflector modeling. Thus one might say that 3-D wave-equation modeling, in a limited sense, is possible even today. In the remainder of this chapter, we will consider 3-D modeling for models whose 2-D cross-sections are comparable in size to today's 2-D models, penetrated by medium-resolution seismic waves.

A TYPICAL SEISMIC MODEL

Now let us estimate the size of a reasonable seismic model. Of course, there is considerable variance in geological models, but we attempt to describe a medium-sized model from the point of view of a seismic interpreter. A typical seismic model may have dimensions 3 km $\times$ 3 km $\times$ 4.5 km with velocities ranging from 1.5 km/s to 4.5 km/s. A typical source amplitude spectrum may contain energy from 10 Hz–50 Hz with peak energy at 25 Hz, so that the effective wavelength in the slowest medium is about 60 m. Thus the problem size, in wavelengths, is $50 \times 50 \times 75$. We would commonly listen long enough to detect reflections from the bottom, say 4 seconds.

Besides deciding upon the parameters for the geologic model itself, other modeling decisions must be made. In particular, there are three general classes of wave-equation models: exploding reflector, acoustic, and elastic. With an exploding reflector model, we attempt to model directly a stacked seismic section by simply propagating waves upward from seismic interfaces laden with explosive charges. While an exploding reflector model is a valuable first approximation, it contains several limitations; most notably it fails to generate certain raypaths, yields wrong times for multiples, and has the wrong polarity for waves reflected from both sides of an interface (Claerbout, 1985). In acoustic modeling, we treat the earth as a fluid, with only compressional waves. An acoustic model is a good first approximation to seismic response in the real world, and, in the remainder of this chapter, we will focus on acoustic modeling.

The obvious failing of the acoustic approximation is that it does not contain elastic effects, resulting from the existence of shear waves. Elastic modeling is much more computationally intensive than acoustic modeling. For 3-D modeling, we can expect that elastic modeling would require approximately 32 times more floating point operations and approximately 16 times more memory. Using the rule of thumb that computer speed generally increases by a factor of 10 every 5 years, we expect elastic capabilities to lag behind acoustic capabilities by about 7 years.

Yet another choice must be made. Do we want to see the response from a single source, a line of sources, or an areal grid of sources? Here we again take the middle road and concern ourselves with modeling a single seismic line from a 3-D model.

NUMERICAL METHODS

Four general methods are traditionally used to model wave propagation: finite differences (Kelly, Ward, Treitel, and Alford, 1976; Dablain, 1986; Bayliss, Jordan, LeMesurier, and Turkel, 1986; Cohen, 1986; Shubin and Bell, 1987), finite elements (Marfurt, 1984), pseudospectral (Kosloff, Reshef, and Lowenthal, 1984; Fornberg, 1987), and boundary integrals (Rokhlin, 1983; Apsel, 1984; Schuster, 1985). The computation required for the first three methods depends upon the problem size (spatial and temporal) and is usually independent of model complexity, while boundary integral methods depend upon both complexity and size. In the remaining discussion, we assume that the problem under consideration has sufficient structure that boundary integral methods are not appropriate. Given the massive number of gridpoints in a 3-D grid, we will rule out consideration of implicit frequency-domain approaches and thus will assume that an explicit time-stepping method is used.

For concreteness, let us consider the computational and memory requirements for two such methods: a fourth-order explicit finite-difference algorithm and the pseudospectral method. First, let us make the analysis for a finite-difference approach. For our model problem, if we assume 12 points/wavelength and a CFL condition (Courant, Friedrichs, and Lewy, 1928; Richtmyer and Morton, 1967) of 0.5, we would require a $600 \times 600 \times 900$ grid and 7200 timesteps (note that the range of velocities requires a small timestep and hence a large number of timesteps). Thus we would have a total of 2.3×10^{12} gridpoint-timesteps (ignoring domain trimming) and we would require approximately

1.5 GWords of core memory.

Assuming 25 calculations per gridpoint-timestep, our calculation requires 5.8×10^{13} floating point operations for a single source. Further assuming 50 source locations, we then need a total of approximately

3×10^{15} floating point operations.

Next, let us repeat the analysis for a pseudospectral method. In theory, pseudospectral modeling requires fewer points/wavelength than finite differences so we assume 6 points/wavelength. Based on the work of Kosloff (Edwards, Hsiung, Kosloff, and Reshef, 1985), we assume a $\text{CFL} = 0.2$. We would thus require a $300 \times 300 \times 450$ grid and 9000 timesteps. Thus we would require approximately

0.2 GWords of core memory,

about 1/8 as much as finite differences. Note that this memory requirement is already met by some of today's supercomputers. Assuming that the number of real floating point operations to calculate an n-long fast Fourier transform (FFT) is approximately $10\,n\,\log(n)$ and that 4 FFTs are required in each direction (2 forward and 2 inverse) yields the estimate of approximately 1.5×10^{10} floating point operations per timestep for a total of 1.4×10^{14} floating point operations per source. Again assuming 50 source locations, we then need a total of approximately

7×10^{15} floating point operations,

which is twice as many as with finite differences.

A number of caveats should be made concerning the above estimates. First, the size of a seismic model certainly depends upon the specific application. One can easily imagine geologic models considerably smaller or larger than the one described. Second, critical parameters such as source frequency and formation velocities can make a tremendous difference in the computational and memory requirements. For instance, if the typical wavelength is twice as long as in the above analysis, the amount of computation decreases by a factor of 16 and the memory required decreases by a factor of 8. Thus these computational and memory estimates should only be considered as ballpark figures, valid only to one or two orders of magnitude.

PARALLEL COMPUTERS

As conventional sequential computers approach inherent physical limitations, it is becoming apparent that the supercomputers of the future will be parallel machines. By exploiting a collection of processors to concurrently perform computations, parallel computers promise increased speed and improved cost/performance. For all but relatively trivial problems, effective parallelism requires that the multiple processing units of a parallel computer communicate their results with each other. Two general methods exist for arranging this communication: shared memory and distributed memory with message passing. Technology does not currently permit shared memory computer with massive parallelism. For this reason, we focus our attention on distributed memory computers which allow

DIMENSION	NODES	CHANNELS
0	1	0
1	2	1
2	4	4
3	8	12
4	16	32

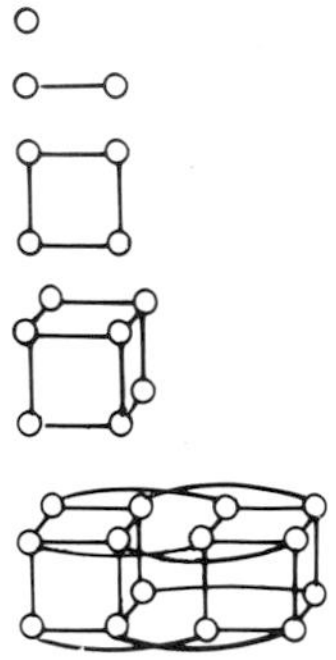

Fig. 1 The hypercube topology.

massive parallelism. The canonical example of a distributed memory computer is a hypercube. Hypercubes are so named because the communication channels between the multiple processors form an N-dimensional cube, as shown in Figure 1. An N-dimensional cube thus has 2^N processors and $N2^{N-1}$ communication channels.

FINITE DIFFERENCES ON A HYPERCUBE

In order to estimate better the efficiency of 3-D acoustic modeling on a massively-parallel computer, we implemented a 2-D acoustic modeling program, ACOUS2D, on a 16-processor Intel vector hypercube computer. ACOUS2D is an explicit finite-difference code, and, as such, is easy to parallelize. Figure 2 shows a simple ACOUS2D model.

The only complication to parallelization is ACOUS2D's use of domain trimming. ACOUS2D uses a fixed-size computational grid, but only makes calculations at gridpoints as absolutely required. More specifically, exploiting the hyperbolic behavior of the wave equation, ACOUS2D only calculates near the source in the early stages of the computation and only calculates near receivers in the final stages of the computation. If the computational grid is simply divided into horizontal strips assigned to nearest-neighbor processors, domain trimming results in the computational load not being evenly distributed among processors at the early and late stages of the computation. To improve load balancing in the face of domain

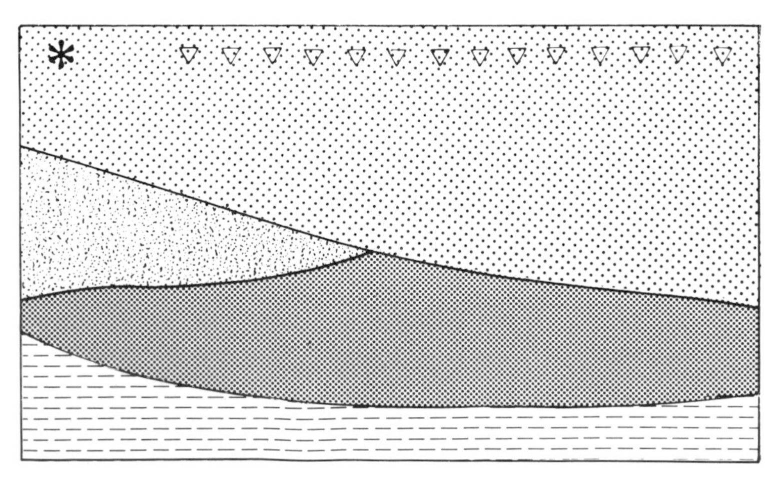

Fig. 2 A simple ACOUS2D model.

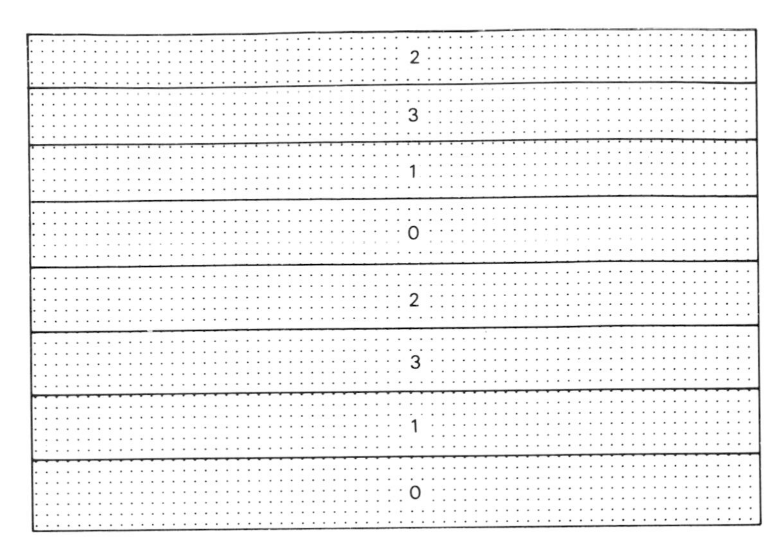

Fig. 3 Decomposition with 4 processors and 2 strips/processor.

trimming, the finite-difference grid is decomposed into strips with multiple strips assigned to each processor. Figure 3 shows a decomposition with 4 processors and 2 strips per processor. To achieve high efficiency requires choosing the appropriate number of strips per processor (typically 2-4) to trade off load balancing and communication overhead. Even in the absence of domain trimming, load balancing can be a significant factor because of the computational and communication overhead incurred by processor whose domain contains sources and receivers.

Figure 4 plots timing data from an ACOUS2D run with 3 strips per processor. We can see from the graph of the time for the interior loop that the load is not perfectly balanced. The nodes on the right of the graph, corresponding to the top of the model, do more work (interior loop) and less waiting (communication and synchronization). Additional testing shows that the bottleneck is primarily synchronization, and that communication overhead is minimal. Noting that processor "0" spent a little more than 500 seconds waiting and that the total run time was approximately 2400 seconds, we see that the efficiency was close to 80%. Without domain trimming, we expect efficiencies above 95%.

IS REALISTIC 3-D MODELING POSSIBLE IN TEN YEARS?

Currently, our general-purpose supercomputers (e.g., CRAY-XMP48, Amdahl 1400E) run at approximately 1-2 GFLOPS, while our fastest special-purpose

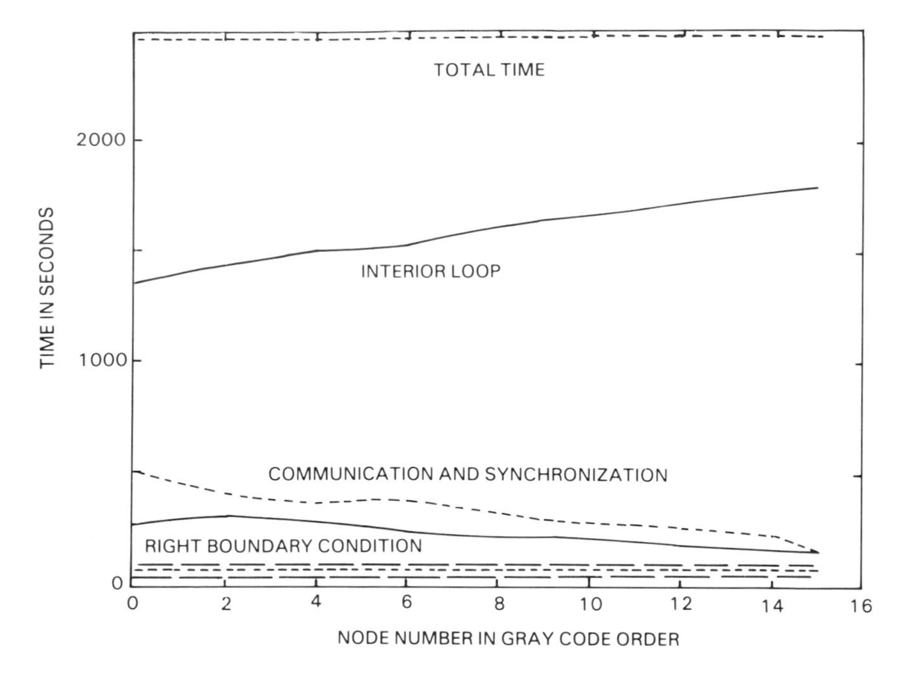

Fig. 4 Timing data for ACOUS2D on 16-processor Intel hypercube computer.

numerical computers run at approximately 10 GFLOPS (e.g., IBM's GF11). Given that the speedup in computers is leveling off and that we want to retain enough flexibility in our parallel computer to allow variants of a given algorithm, it is reasonable (and probably conservative) to assume that, in ten years time, we can build a massively-parallel machine which runs at 500 GFLOPS. One possible scenario is given by the following table for the individual processors in a distributed memory machine (as seen from our 2D modeling results on a hypercube, seismic modeling only requires a ring topology for interconnection network).

HYPERCUBES		
	Present	+10 Years?
Speed of node	5 MFLOPS	100 MFLOPS
Memory/node	0.5 MBytes	5 MBytes
Inter-node Communication	0.5 MBytes/sec	5 MBytes/sec
No. of Processors	1,000	5,000

With such a scenario, sufficient memory would not be a problem for either finite difference or pseudospectral modeling. In ten years, we can expect 50 GWords of memory while both the numerical methods require only about a GWord of memory. The situation is not so favorable in terms of computational power. Based on our supercomputer and hypercube experiences with 2-D modeling, we expect that a 3-D modeling program will perform at approximately half of a computer's advertised peak speed. Thus, in ten years, we can realistically expect a sustained speed of 250 GFLOPS for a 3-D modeling program. Our earlier analysis estimated that $3\text{-}7 \times 10^{15}$ floating point operations would be required to model a line with 50 shotpoints. These calculations indicate that, in ten years, a 3-D acoustic modeling program will be able to model data collection over a seismic line with a turnaround of 3-8 hours.

ACKNOWLEDGMENTS

The analysis reported here benefited from many discussions with my colleagues. In particular, I would like to acknowledge W. Y. Crutchfield, M. A. Dablain, and A. Weiser who made many helpful comments. The program ACOUS2D was developed jointly with J. B. Bell and G. R. Shubin and was initially implemented on an Intel hypercube by G. Chessire and C. B. Moler.

REFERENCES

Adams, N., 1984, 3-D elastic modeling on a vector computer: Presented at the 54th Ann. SEG Mtg., Atlanta.

Apsel, R. J., 1984, Exact synthetic seismograms for three-dimensional irregularly layered media using the boundary integral equation algorithm, Binteq: Presented at the 54th Ann. SEG Mtg., Atlanta.

Bayliss, A., Jordan, K. E., LeMesurier, B. J., and Turkel, E., 1986, A fourth-order accurate finite-difference scheme for the computation of elastic waves: Bull. Seis. Soc. Am., *76*, 1115–1132.

Claerbout, J. F., 1985, Imaging the earth's interior: Blackwell Scientific Publ.

Cohen, G., 1986, Fourth-order schemes for the 2-D wave equation in a homogeneous medium: Presented at the 56th Ann. SEG Mtg., Houston.

Courant, R., Friedrichs, K. O., and Lewy, H., 1928, Uber die partiellen differenzengleichungen der mathematischen physik: Math. Ann., *100*, 32.

Dablain, M. A., 1986, The application of high-order differencing to the scalar wave equation: Geophysics, *51*, 54–66.

Edwards, M., Hsiung, C., Kosloff, D., and Reshef, M., 1985, Elastic 3-D forward modeling by the Fourier Method: Presented at the 55th Ann. SEG Mtg., Washington, D.C.

Fornberg, B., 1987, The pseudospectral method: comparisons with finite differences for the elastic wave equation: Geophysics, *52*, 483–501.

Kelly, K. R., Ward, R. W., Treitel S., and Alford, R. M., 1976, Synthetic seismograms: a finite difference approach: Geophysics, *41*, 2–27.

Kosloff, D. D., Reshef, M., and Lowenthal, D., 1984, Elastic wave calculations by the Fourier method: Bull. Seis. Soc. Am., *74*, 875–891.

Marfurt, K. J., 1984, Accuracy of finite-difference and finite-element modeling of the scalar and elastic wave equations: Geophysics, *49*, 533–549.

Mufti, I. R., 1987, Interpretive lessons from three-dimensional finite-difference seismic models: Presented at the 57th Ann. SEG Mtg., New Orleans.

Richtmyer, R. D. and Morton, K. W., 1967, Difference methods for initial-value problems: Interscience Publ.

Rokhlin, V., 1983, Solution of acoustic scattering problems by meansof second kind integral equations: Wave Motion, *5*, 257–272.

Schuster, G. T., 1985, Modeling structural traps by a hybrid boundary integral equation and Born series method: Presented at the 55th Ann. SEG Mtg, Washington, D.C.

Shubin, G. R., and Bell, J. B., 1987, A modified equation approach to constructing fourth order methods for acoustic wave propagation: SIAM J. Sci Stat. Comp., 135–151.

CHAPTER 2

THREE-DIMENSIONAL SEISMIC PROCESSING, MIGRATION, AND MODELING USING PARALLEL PROCESSING ON CRAY SUPERCOMPUTERS

by
MOSHE RESHEF and MICKEY EDWARDS
Cray Research, Inc.

1.0 INTRODUCTION

The use of parallel processing and large memory storage will enable the seismic processing community to achieve a wide range of 3-D processing capabilities which have been denied or severely restricted by computer systems of the past. Parallel processing is the use of multiple processor units on the same problem or application. This chapter represents a snapshot of 3-D processing results which have been measured using existing CRAY X-MP computer systems. Some of the discussed algorithms have also been tested with CRAY-2 systems. Due to on-going CRAY-2 hardware enhancements and evolving operating system and Fortran compiler software, CRAY-2 results are construed as preliminary and only CRAY X-MP timings will be given. This chapter will not include mathematical development and will not attempt to identify or reference applicable literature publications. It is believed that discussed algorithms are familiar to the geophysical community.

1.1 CRAY X-MP HARDWARE AND SOFTWARE

This section provides a brief overview of CRAY X-MP hardware and software. Discussed are central memory, large secondary memory Solid-state Storage Device, I/O Subsystem, Central Processing Units, and parallel processing.

1.1.1 *CRAY X-MP Central Memory*

At writing time, central memory on publicly announced CRAY X-MP systems consists of 2 to 16 million 64-bit words and is interleaved in 16 to 64 memory banks. Interleaved memory is arranged so that consecutive addresses are in consecutive memory banks. The size of central memory is currently limited by a 24-bit instruction address field. Future follow-on compatible systems will contain both 24- and 32-bit address fields to allow larger central memories and execution of existing software. Memory for four processor systems is composed of ECL bipolar circuits. Single and two processor systems utilize static MOS memory. On multiple processor systems, all of central memory can be accessed by all processors. This is an important difference to so-called massive parallel processing architectures which typically necessitate the movement of arrays to and from local memory of different processors.

Many computer architectures require that vectors occupy consecutive locations with a memory increment or stride of one. Some architectures impose the penalty of slower access time for vectors stored with memory increments other than one. CRAY X-MP architecture allows vectors to be formed with uniform positive or negative memory increments and with random indexing. ECL and MOS memory banks can be accessed every four and eight clock periods (34 and 68 nanoseconds), respectively. No delay will occur if vector operations access enough other memory banks before needing to access the original memory bank. If a reference is made to a bank while it is busy with a previous reference, the conflicting reference must wait for the busy condition to clear. The wait time will be one to three clock periods with ECL memory or one to seven clock periods for MOS memory. In the multiple processor case, once initial memory bank conflicts are resolved, each processor will typically stream or receive/store operand/resultant elements at each clock period. The severity of overall system performance due to memory bank conflicts depends on the application and to some degree on programming skill. Experience to date for intensive computational seismic algorithms indicates that a four processor system degradation of less than 10 percent is normal. As the number of processors and the size of central memory increase, memory contention is limited by increasing the number of memory banks. The maximum number of memory banks is dependent on physical packaging and wiring. There are four memory ports for each processor. These ports provide concurrent loading of two vectors from memory, storing of one vector to memory, and input · output operations. The original CRAY-1 utilized a single memory port or path for the above operations.

1.1.2 *CRAY X-MP Solid-state Storage Device*

Optional large secondary memory with CRAY X-MP systems is provided by the non-rotating Solid-state Storage Device or SSD. The SSD can be configured with 32, 64, 128, 256 or 512 million 64-bit words of MOS memory storage. When the first SSD was introduced in 1982, 64-kilobit memory modules were used with a maximum size of 32 million words. Since then, 256 kilobit and one megabit memory has been used to increase memory capacity within the same physical space. The use of denser memory packaging has allowed the housing of the 32 million word SSD within the I/O Subsystem (IOS) eliminating the need for a stand-alone hardware unit. As more dense MOS memory modules become reliable and plentiful, SSD memory capacity will increase. For example, the next increase will provide two billion words of storage using four megabit MOS memory.

On a four processor CRAY X-MP system, the four largest SSDs are connected to central memory using two 1000 million bytes per second channels. All other configurations utilize a single 1000 Mbytes/sec channel. Originally, the maximum transfer rate was obtained only if maximum memory capacity was configured. The transfer rate was reduced by a factor of 1/2, 1/4, 1/8, etc. if the memory capacity was halved, quartered, etc. Today, above transfer rates of 1000 Mbytes/sec and 2000 Mbytes/sec are realized regardless of the memory capacity. Currently, Cray Research operating systems use the SSD as a disk device with standard I/O routines. Depending on the operating system, approximately 330 to 540 microseconds are required to access the SSD. Typically, 3-D algorithms need to access 3-D datasets in two ways. For example, constant-x and constant-y planes could be required. For this example, assume that 3-D datasets are initially stored in the SSD with z being the most rapid increasing storage index followed by x and then y. Constant-y or xz planes represent contiguous SSD storage and can be accessed with a single I/O request. Constant-x or yz planes represent scattered SSD storage and are accessed with multiple I/O requests. Asynchronous queued I/O (AQIO) routines can be used to efficiently access such 3-D datasets. With AQIO, a single call to the operating system passes a list of I/O addresses. Table 1 illustrates the efficiency of using AQIO routines with a four processor CRAY X-MP and 128 million word SSD. For timings, 50 read/write operations are performed for different record sizes. Record sizes vary from one sector (512 words or 4096 bytes) to 256 sectors. The second column gives the aggregate transmission rate when a single I/O request or operating system call is issued for each record. The third column gives transmission rate when a list of 50 addresses is passed with a single operating system call. If several I/O operations can be defined by adding a constant increment to SSD and central memory addresses, a single compound AQIO request can be issued. The fourth column gives transmission rate when a single AQIO request generates 50 I/O operations. Table 1 shows transmission rates greater than two billion bytes per second are achieved. Dividing transmission rates in half approximates SSD performance for usage with one and two processor CRAY X-MPs.

TABLE 1.

TRANSFER RATE IN MWORDS/SEC

SSD record size in sectors	Single requests	AQIO	Compound AQIO
1	1.22	13.60	33.50
2	2.43	14.28	36.35
4	4.76	28.51	72.28
8	9.42	56.66	133.46
16	18.98	104.59	177.25
32	33.36	152.89	231.90
64	62.75	208.88	263.97
128	75.28	254.01	285.72
256	131.33	265.67	291.11

Transfer rates for 50 requests using several I/O techniques and record sizes.

The SSD allows timely computation of many out-of-memory problems. Included are recursive algorithms which must retrieve and restore datasets for each of several thousand computational steps. The feasibility of many previously unpractical algorithms has been demonstrated by using the high bandwidth of the SSD to obtain computational or CPU-bound programs. To date experience shows that solution of the data motion problem is the most difficult challenge facing implementation of 3-D algorithms. After solving the data motion problem, parallel processing can be used to reduce computational times.

1.1.3 *CRAY X-MP I/O Subsystem*

The original CRAY-1 introduced in 1976 performed I/O operations using conventional I/O channels attached to central memory. The use of CRAY supercomputers in seismic processing began in 1981 with the introduction of the CRAY-1/S system which featured a detached I/O Subsystem (IOS) to perform I/O operations. All CRAY X-MP systems feature IOS systems. The IOS contains up to eight million 64-bit words of MOS memory primarily used for I/O buffering. For a system with on-line magnetic tapes, the IOS contains three or four I/O processors (IOPs). The IOS is connected to central memory with one or two 100 MBytes/sec channels. IOS buffer memory is used for "read ahead" and "write behind" I/O operations with disk and tape. Filling and flushing of buffers is transparent to the application program. Depending on the IOS model, 8 to 10 6250 bpi magnetic tapes can be streamed at maximum tape speed before degradation occurs. Approximately *xxx* dual density IBM 3480 tape cartridges can be streamed. Using two 100 MBytes/sec channels between the IOS and central memory, an aggregate transmission rate of *xxx* MBytes/sec can be sustained with disk operations. The IOS is not used with SSD transfers to and from central memory. An additional 100 MBytes/sec channel can be attached to the IOS to provide so-called back door SSD transfers of tape and disk data bypassing central memory. For detailed descriptions of IOS architecture and peripheral devices, appropriate Cray Research, Inc. manuals can be referenced.

1.1.4 *CRAY X-MP Central Processing Units*

CRAY X-MP systems contain one, two, or four CPUs. Each CPU or processor is identical. Each CPU has 14 functional units. Functional are fully segmented and can initiate a new operation every clock period. Functional units are independent and can perform operations concurrently. Each CPU has eight vector registers containing 64 64-bit words or elements. Most vector processing architectures are characterized by "memory-to-memory" operations. Each vector operation loads one or two operand arrays from memory and stores a resultant array to memory. CRAY architecture uses "register-to-register" vector operations. Input operand and output resultant arrays reside in vector registers. In many algorithms, the same variable is used more than once in computations and does not have to be re-loaded from memory. "Memory-to-memory" operations can be realized on CRAY X-MPs by using multiple memory ports. Output results from one functional unit can be used as the input to another functional unit. If the latter functional unit is not busy, "chaining" occurs as each individual result is generated. For vector lengths greater than 64, computations are performed in 64-element segments. Because of its single path to memory, long vector lengths produced "saw-toothed" performance curves with the CRAY-1. The increased number of CRAY X-MP memory ports removes long vector performance degradation associated with vector length multiples of 64.

For scalar operations, each CPU has eight 64-bit general purpose registers and 64 64-bit intermediate or backup registers. Contents of an intermediate register can be transferred to and from a general purpose scalar register in one clock period. Each CPU also contains eight 24-bit address registers and 64 24-bit intermediate address registers. Contents of an intermediate address register can be transferred to and from an address register in one clock period. CRAY Fortran compilers use above intermediate registers for storage of scalar variables and addresses. Scalar and vector instructions are 16-bit and 32-bit parcels and can be interspersed. Instructions are executed out of four instruction buffers, each containing 128 16-bit parcels. Instruction buffers are loaded from central memory at the rate of 32 16-bit parcels per clock period. Typically, mathematical library subroutines require less than 512 16-bit parcels and do not require additional instruction fetching from memory after initial loading of instruction buffers.

For more detailed description of central processing units, appropriate Cray Research, Inc. manuals can be referenced.

1.1.5 *CRAY X-MP Parallel Processing*

For the Fortran programmer, parallel processing can be achieved using multitasking, microtasking, macrotasking, and autotasking. Multitasking was the first developed capability and embraces tasking concepts existing for a number of years. Multitasking library utilities allow initiation of tasks, event posting, and synchronization. The Fortran programmer identifies parallel sections of the program and must subroutinize those sections. The programmer must also allocate global and local variables and arrays. Microtasking employs a pre-compiler which interprets compiler directives and produces a second source containing appropriate library utility calls. Microtasking can be used with subroutines or outer do-loops. Provided processors are available, each outer loop pass uses a different processor. Microtasking attempts to provide optimal performance in a batch environment by dynamically using available processors. Typically, the system overhead associated with microtasking is less than that of multitasking. Microtasking can be used with a smaller granularity. Macrotasking also uses a pre-compiler and combines both multitasking and microtasking. In the CRAY user community, multitasking is sometimes called macrotasking. Autotasking attempts to automatically parallelize a Fortran program. Autotasking represents the most difficult challenge and will undergo refinements for some time. Based on experience to date, selection of the parallel processing tool depends on the application.

CRAY X-MP hardware is designed to facilitate parallel processing. Inter-CPU communication features shared address registers, shared scalar registers, and

TABLE 2.

MAXIMUM THEORETICAL SPEEDUP

Fraction of time parallelizable f	Number of processors							
	$p=1$	$p=2$	$p=4$	$p=8$	$p=16$	$p=32$	$p=64$	$p=$ infinity
1.00	1.00	2.00	4.00	8.00	16.00	32.00	64.00	infinity
0.99	1.00	1.98	3.88	7.48	13.91	24.43	39.26	100.00
0.98	1.00	1.96	3.77	7.02	12.31	19.75	28.32	50.00
0.97	1.00	1.94	3.67	6.61	11.03	16.58	22.14	33.33
0.96	1.00	1.92	3.57	6.25	10.00	14.29	18.18	25.00
0.95	1.00	1.90	3.48	5.93	9.14	12.55	15.42	20.00
0.94	1.00	1.89	3.39	5.63	8.42	11.19	13.39	16.67
0.93	1.00	1.87	3.31	5.37	7.80	10.09	11.83	14.28
0.92	1.00	1.85	3.23	5.13	7.27	9.19	10.60	12.50
0.91	1.00	1.83	3.15	4.91	6.81	8.44	9.59	11.11
0.90	1.00	1.82	3.08	4.71	6.40	7.80	8.77	10.00
0.75	1.00	1.60	2.28	2.91	3.37	3.66	3.82	4.00
0.50	1.00	1.33	1.60	1.78	1.88	1.94	1.97	2.00
0.25	1.00	1.14	1.23	1.28	1.31	1.32	1.33	1.33
0.10	1.00	1.05	1.08	1.09	1.10	1.11	1.11	1.11
0.00	1.00	1.00	1.00	1.00	1.00	1.00	1.00	1.00

semaphore registers. Shared registers are used for passing addresses and scalar information from one CPU to another. Semaphore registers are used for control between CPUs. CPU deadlock is determined by hardware.

Compared to many parallel processing architectures, the CRAY X-MP uses a small number of "fast" processors opposed to a large number of "slow" processors. As shown by well-known Table 2, successful employment of multiple processors requires a high fraction of all computational time to be parallelizable. For example, if 4 per cent of computations must be serially executed by a single processor, the employment of 16 processors would provide a maximum theoretical speedup of 10 times that of a single processor. Maximum theoretical speedup does not include system overhead or delays associated with memory conflicts or data I/O. To date experience with computation of recursive out-of-memory 3-D algorithms indicates that less than 100 per cent parallelizable execution will be realized with CRAY X-MP systems.

1.2 THREE-DIMENSIONAL STANDARD PROCESSING

The initial problem in processing 3-D surveys is the size of the input data. In previous years, a typical 3-D marine survey covered a surface area of 3 by 5 miles generating 500 to 600 reels of 6250 bpi tape. Today, the number of tape reels comprising a marine survey can be increased by an order of magnitude. High resolution land surveys using large areal arrays can also result in a large number of field tapes.

The following is an illustrative example. Consider a survey of 300 lines with 1000 shots per line. Each shot is recorded at 120 receivers for a total of 36 million traces. For 1500 samples per trace, the total amount of 3-D data to be processed is 54 billion words or amplitudes. Storing this amount of data on 6250 bpi tapes in SEG-Y format will require approximately 1800 tapes. The time required to extract the entire data from one tape is approximately 2.5 minutes or, for this example, an accumulative I/O time of 75 hours. Optimization of data management procedures is crucial for the entire processing sequence. This includes fundamental concepts such as parallel tape streaming and minimizing the number of times input tapes are loaded into the computer system.

The usage of new cartridge tapes, for example, the dual density IBM 3480, as a replacement for the conventional 6250 bpi media in the field or during preprocessing, will ease data volume and data streaming problems. New emerging technology such as optical disks will greatly diminish logistics associated with physically mounting tapes. Optical disks will also allow more flexibility in algorithm design. For example, many memory bound or limited imaging algorithms could be re-visited. For algorithms which accumulate or superimpose results, optical disks could be remounted for computations with a subset of depths or for trace subsets of the input data. In other words, we do not have to perform all computations with a single input data pass.

Assuming we can find an efficient procedure of feeding input data tapes into the computer, the second major problem is determining if the entire input dataset can reside within the computer system. That is, is the disk farm large enough to hold all input data traces? For large 3-D surveys, we must determine if meaningful sized subsets of the overall survey can reside on disk storage. Subsequent sections will address storage requirements for specific algorithms. If the cost of a large disk farm can be justified, we must determine if disk data can be accessed fast enough. This is a function of desired algorithms. It is anticipated that storing enormous volumes of data on disk storage and accessing that data will remain a problem for some time in the future.

The third major problem in processing 3-D surveys is determining if sufficient computational power exists to execute desired algorithms. Currently, the largest and most powerful CRAY computer systems cannot perform pre-stack migration using all field traces of large 3-D surveys.

Today large 3-D surveys require several months of processing with the typical CRAY system used by most oil companies and geophysical contractors. Depending on processing and quality control philosophies, the above time can vary from company to company but computer requirements remain enormous even with current algorithms. Currently, most 3-D seismic processing systems do not utilize parallel processing. Processing on multiple processor systems commonly features multiprocessing where each processor executes a different job. This section will address parallel 3-D standard processing and results which have been measured with existing CRAY X-MP systems. The following software has been developed to execute on future CRAY product lines.

1.2.1 *Parallel Processing Model (PPM)*

The use of more sophisticated processing techniques will significantly increase the number of computations. It is anticipated that newer algorithms will require all available computational power and will require efficient parallel processing. From the size of the input data volume, it is anticipated that I/O channels and storage resources will easily be saturated. In order to optimally utilize an increased number of CPUs, the entire job must be parallelized representing a non-trivial effort. Based on above constraining factors, design concepts include the following:

1. Optimal utilization of more CPUs, more I/O channels, more memory, and faster clock period;
2. Top-down approach to parallel processing;
3. Modular application routines;
4. Pre-processing and data organization already performed by other systems;
5. Avoidance of output tapes;
6. Use of large SSDs for capturing subsets of data for afterward velocity determination, inversion, etc.

Figure 1 describes the data flow. The number of parallel tape streams depends on the number of CPUs or processors, number of I/O channels, and number of tape drives. Utilization of central memory is the first problem addressed by the

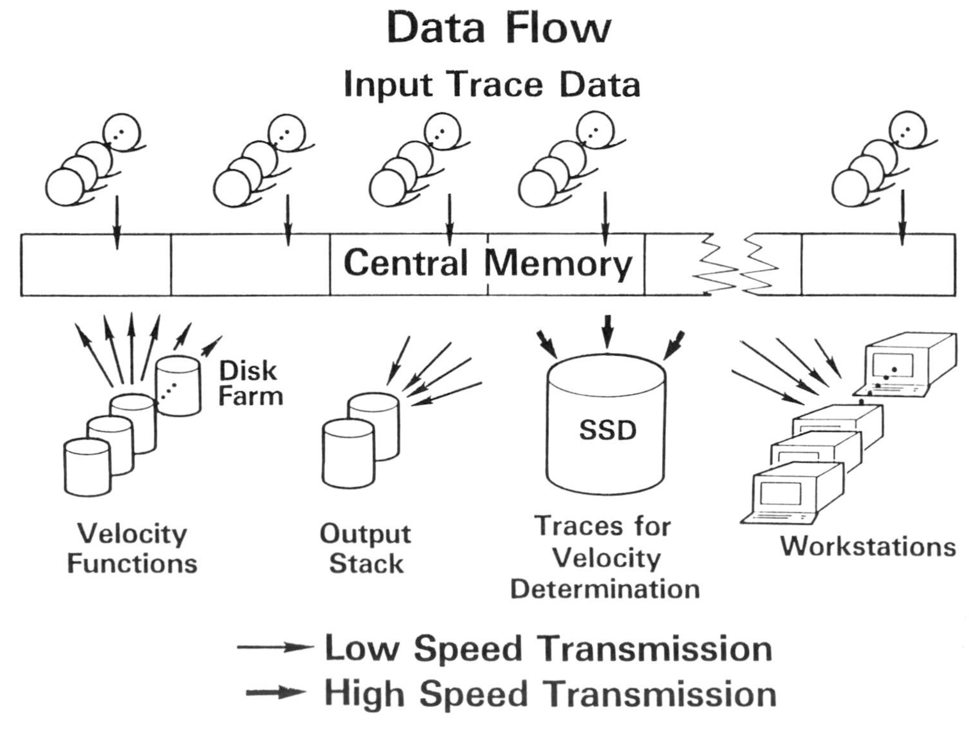

Fig. 1

PPM. Each data input task contains local array storage for double buffering in order to overlap I/O and computations. Size of local arrays is determined by central memory size, I/O bandwidth, I/O mechanism, depth of vector registers. CRAY X-MP implementation has used 64-trace buffers. The number of traces in a buffer or panel should be at least equal to the number of elements in a vector register to allow optional vectorization across the spatial axis instead of time. Panels of 64 traces are efficient with the Cray X-MP due to the detached I/O Subsystem (IOS). If the IOS or equivalent I/O mechanism is not employed (for example, CRAY-2 product line), local array storage will probably expand.

Global arrays, shared by all CPUs or computational tasks, contain subsets or partitions of output stack results and input velocity functions. For large 3-D surveys, only a portion or slice of the stack dataset can reside in central memory at a time. Stack subsets are overlapped. It is not necessary to assume ideal geometry or uniformly spaced input traces. As each input trace is processed, program determines which output CMP stack trace to be summed into. The width of stack subset overlapping depends on the degree of departure from ideal geometry and central memory size. Finished stack subsets are saved on disk storage prior to initializing new subset or retrieving previously accumulated results from disk. There is only one stack array in central memory. CRAY architectures allow all CPUs to share central memory. Existing 10 megabyte per second per disk stream capability is adequate.

The above on-the-fly in-memory stacking eliminates today's so-called binning process. Necessary position or navigation corrections are assumed to have been performed in pre-processing. It is also necessary to require some organization or sequence of processing input tape reels. A simplified example is shown by Figure 2. Field acquisition can be visualized as alternating south-to-north and north-to-south lines. Each line requires several tape reels. Processing is performed in the south-to-north direction with successive lines processed in a west-to-east manner. The above addressed marine surveys. Typically, 3-D land surveys feature some form of areal array coverage. In these cases, the above marine reel organization may not be feasible and another preliminary reel organization may be required.

In addition to stack subsets, global array storage contains stacking velocity function subsets. In central memory, the size of the velocity subset is equal to the stack subset. The input stacking velocity function can be defined on a coarser grid and interpolation used to supply a unique velocity function at each CMP. For disk storage, the velocity function dataset can be smaller than the stack dataset and access from disk is less than or equal to stack dataset access.

Global arrays have to be protected for "updating." As implemented on the

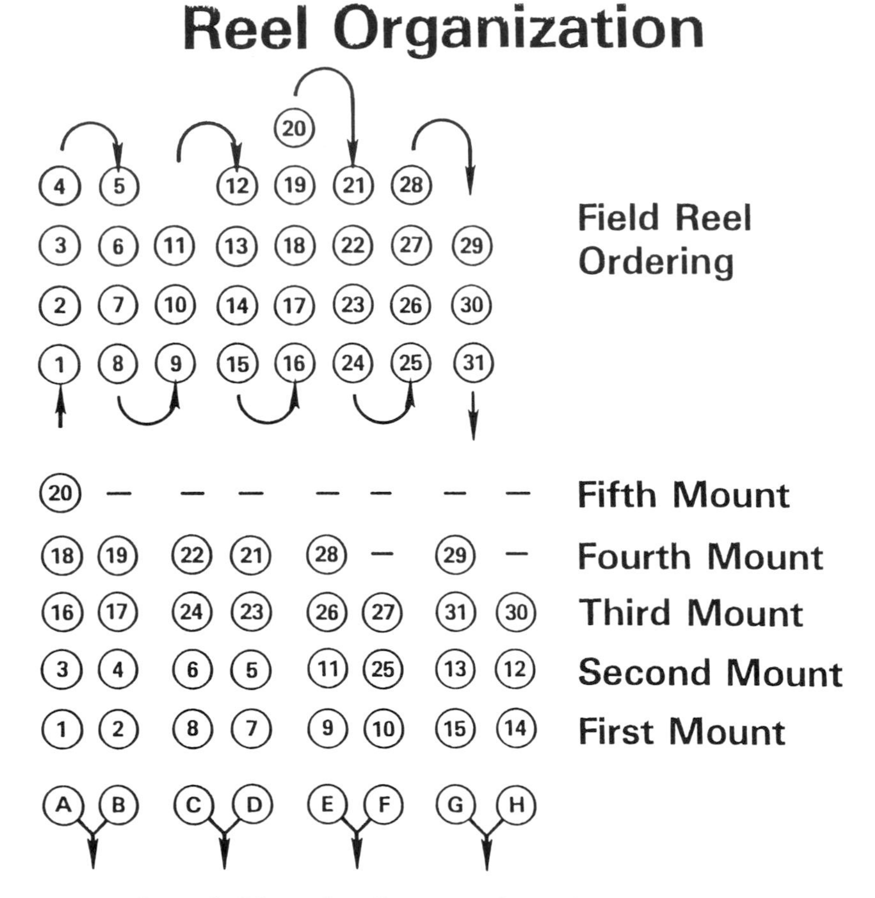

Example of Simple Streaming for 8 Tape Drives

Fig. 2

CRAY X-MP/4, macrotasking was implemented and software memory locks were used during addition into the stack array. There must be an easy access to global arrays for real-time monitoring. The program implementing the parallel processing model was fully automatic without human intervention or interference. Future operating systems must allow access of the stack data by another program in order to drive workstation monitoring graphics. The workstation user should have the capability of suspending and re-starting the processing.

Processing steps preceding the stack result can include all the operations that can be performed on any input trace independently. Processes like data conversion trace filtering, deconvolution, and NMO correction are examples of such operations. These processes are performed on the data immediately after it is extracted from the tapes and, since most of these processes are fast, they can be done with minor cost. Some form of DMO can be implemented in this scheme. Parallelism can be implemented at this stage in different ways. If for example the number of CPUs is less or equal to the number of I/O channels, the parallelism will be implemented in a higher level (the tape I/O stage). When the number of CPUs is greater than the number of I/O channels, more parallelism possibilities exist. Some of the trace processes may require computations of FFTs. Since very many transforms of the same length will be required, a low level parallelism can be achieved through vectorization of multiple transforms. Vectorization across spatial axis (or the number of traces in a panel) can be used.

1.2.2 *Performance Testing with Parallel Processing Model (PPM)*

We will now address existing CRAY X-MP performance using a small 3-D field survey. Benchmarked system is a CRAY X-MP/416 with 8.5 ns clock period, eight 6250 bpi tape drives, four I/O channels, four CPUs, and four million words of central memory running under operating system COS 1.16. The field data contains 877 sources with 192 traces per source or a total of 168,384 traces. Each trace contains 1500 samples or 6240 bytes. Processing includes SEG-Y data conversion, bandpass filtering, deconvolution, NMO, and stacking. Every third common source collection is saved on the SSD requiring approximately 85 million words. Three million words of central memory are used for subsets of velocity function and stack data. Parallel processing is achieved using macrotasking compiler directives. The above processing was I/O bound requiring approximately 2.5 minutes to input four tapes in parallel (0.83 megabytes per second per stream). All data was processed with an elapsed or wallclock time of 5.3 minutes. This is a rate of about 2 million traces per hour. Tape blocking of five traces would yield a transmission rate exceeding one megabyte per second and will increase processing rate. During the above 2.5 minutes, each CPU was busy 72.5% of the time. Approximately 5.4 ms of computational time was required for each input trace. 7.5 ms per trace (or 588 clock periods per amplitude) could have been spent before making the processing CPU bound. The above testing demonstrates expected model performance.

If an SSD with 512 million words is used, every third common source collection could be saved for an input dataset of more than 5,000 sources (or more than one million input traces). The above SSD would be filled in less than 30 minutes of wallclock time for processing in dedicated mode. If desired, the above processing could be CPU bound. If it is necessary to save CMP traces on tape, the processing rate would be at least halved, yielding approximately one million input traces per hour. In this case, approximately 15 ms of computations could be performed on each input trace before processing becomes CPU bound. Outputting CMP traces would require SSD Storage and would reduce the number of common source collections which could be saved.

1.3 THREE-DIMENSIONAL MIGRATION

This section will discuss parallel processed 3-D migration results using post- and pre-stack algorithms. Results indicate that parallel processing can be effectively utilized to reduce computational time.

1.3.1 *Post-stack Migration*

Discussed post-stack algorithms address wavenumber-frequency and spatial-frequency formulations.

1.3.1.1 *Wavenumber-frequency Domain Migration*

3-D post-stack migration in the wavenumber-frequency domain requires computation of 3-D Fourier transforms. Algorithms include so-called Stolt and phase-shift methods. This section will address large 3-D FFTs and optimal implementation of mapping or phase-shift operators.

The largest 3-D FFT computed to date used a $1024 \times 1024 \times 1024$ real-valued data set. Computer resources included eight million words of central memory, 40 million words of SSD storage, and eight DD-49 disk units. Using optimized assembler language routines which pack and unpack 32-bit samples at the approximate rate of $2\frac{1}{2}$ clock periods per element, 32-bit samples were used on disk storage. Approximately 90% capacity on four disk units is required for 1024^3 elements. A second set of four disk units was utilized to allow concurrent retrieval of four streams and storage of four streams. Successive retrieval/storage passes of the data ping-pong input/output data with the two sets of four disk units. A simple distributive storage scheme was employed such that constant-x or constant-y planes could be retrieved or stored by issuing four I/O requests. Disk positioning was minimized and transfer rates slightly less than 10 MBytes/sec/stream were realized. The SSD and asynchronous queued I/O (AQIO) routines were used to effect a transpose of the second and third (x, y or k_x, k_y) storage indices. Triple buffer arrays in central memory were employed. Three data passes were required to perform forward and inverse FFT computations. The first pass computed x and z (or t) transforms on all 1024 constant-y planes. For each constant-y plane, 1024 real-to-complex FFTs with respect to z (or t) were computed followed by 513 complex-to-complex FFTs with respect to x. The second pass computed y and k_y transforms for all 1024 constant-k_x planes. For each constant-k_x plane, each of the y and k_y transform steps computed 513 complex-to-complex FFTs. Finally, the third pass computed k_x and k_z (or ω) transforms for all 1024 constant-y planes. For each constant-y plane, 513 complex-to-complex FFTs with respect to k_x were computed followed by 1024 complex-to-real FFTs with respect to kz (or ω). An elapsed time of 7.7 minutes was required using a four processor CRAY X-MP system (9.5 nanosecond clock period) with a microtasked Fortran program. Program execution was CPU-bound. Both parallel and 1-D FFT computation was utilized. By using more efficient FFT routines, the elapsed time would be decreased.

Total disk and SSD data movement of 3.2 and 4.1 billion words, respectively, was performed. Dividing the total disk and SSD data movement by the elapsed time yields sustained transfer rates of 6.9 and 8.7 million words per second, respectively. The measured wallclock time did not include data staging or initialization of the input dataset. Removal of the SSD would have necessitated two additional data passes to perform transpose before and after y and k_y transforms. Disk storage can be eliminated by using a 512 million word SSD. That is, using 32-bit samples, 1024^3 real values or $1024 \times 1024 \times 512$ complex values can reside within a 512 million word SSD.

Many post-stack wavenumber-frequency domain migration algorithms can be readily implemented by inclusion of computations during the second pass of the data. For example, the so-called Stolt algorithm requires an interpolation of ω coordinates and scaling in the k_x, k_y, ω domain. Interpolation and scaling can be applied to each constant-kx plane prior to the k_y transform step and disk storage. If optimally implemented, above added computations are less than FFT computational time. Many post-stack phase-shift migration algorithms can also be readily implemented in a similar fashion. Prior to the k_y transform step with each constant-k_x plane, downward continuation or extrapolation operators for each output depth interval can be generated and applied and imaging can be performed. Imaging produces a complex-valued result in the k_x-k_y-z domain and the final ω transform step in the third data pass is not computed. The primary objection to this scheme is the storage overlay constraint which restricts the number of output depths to be less than or equal to one-half the number of unmigrated time samples in each trace. By increasing data storage, I/O data movement and FFT computations, the same number of output depths can be computed.

1.3.1.2 *Spatial-frequency Domain Migration*

3-D post-stack migration in the spatial-frequency domain typically requires more I/O operations and more computations than wavenumber-frequency domain algorithms. Many wavenumber-frequency domain techniques are two-pass algorithms. That is, the entire stacked dataset is retrieved from external storage only two times. The following two sections discuss recursive spatial-frequency domain algorithms which require SSD implementation to be feasible.

1.3.1.2.1 Generalized Phase-shift Migration

3-D post-stack migration using the generalized phase-shift method in the spatial-frequency domain has been implemented on the four processor CRAY X-MP system. This algorithm is based on the temporal transformed two-way nonreflecting acoustic wave equation. Deviating from conventional Taylor series or Runga-Kutta approximation, the exponential depth extrapolation operator is evaluated by an expansion with Bessel functions and modified Chebychev polynomials. Computation with modified Chebychev polynomials involves application of a coupled first-order system operator relating pressure and partial derivative of pressure with respect to depth. This algorithm is accurate for relatively large depth steps. For depth intervals where input velocities do not vary vertically for several output depth samples, migrated results for intermediate depths can be obtained by the incremental cost of computing additional Bessel functions without recomputation of Chebychev polynomial terms. The latter includes computation of spatial derivative terms which comprise the bulk of computational work. Spatial derivatives are computed in the Fourier domain.

The largest dataset migrated to date consisted of 360 lines with 375 stacked traces in each line. Each input trace contained 1500 amplitudes with 4 millisecond sampling rate. Mixed-radix FFT computation was performed on each trace

$(1500 = 2^2 \times 3x5^3)$. Discarding frequencies above 62.5 Hz requires SSD storage of $2 \times 360 \times 375 \times 400$ or 108 million words for pressure and pressure derivative. Computation of each depth step requires a total SSD data movement of 1.73 billion bytes. Using a four processor CRAY X-MP, a speed up factor of 3.75 over single processor execution was measured. With a 9.5 nanosecond clock period, an elapsed or wallclock time of 80 seconds was required for computation of each depth step. Specifying the input velocity function with vertical intervals of 50 meters, computation of $360 \times 375 \times 1000$ migrated results with depth sampling of 8 meters required an elapsed time of 3.5 hours. For the worst case, when a unique velocity is specified at each spatial grid point, an additional $360 \times 375 \times 1000$ or 135 million words of SSD storage for velocities will be required and the total elapsed time will exceed 22 hours. For this case, total data motion of 2.2×10^{11} words and 3.6×10^{13} floating point operations are required. Dividing the number of floating point operations by the wallclock time gives a sustained computational rate of approximately 450 MFLOPS.

1.3.1.2.2 ADI Crank-Nicolson Migration

We will now examine 3-D post-stack migration using an implicit method in the spatial-frequency domain. The algorithm is based on the one-way 30 degree wave equation using the moving coordinate system as advanced by Claerbout. Computer implementation uses the so-called splitting approach or an alternating direction implicit (ADI) Crank-Nicolson scheme. Each downward continuation or extrapolation step proceeds along x and then y coordinates. For each extrapolation, the ADI method requires the solution of $2N$ complex tridiagonal systems of dimension N where N is the number of input/output x and y coordinates. The algorithm is recursive where the result of the previous extrapolation step is required as input to the current extrapolation step. An arbitrary velocity function can be used.

The above algorithm has been demonstrated with two and four processor CRAY X-MP systems using an SSD. An input dataset of $200 \times 200 \times 1000$ migrated results. In order to achieve a high I/O streaming rate, input traces are distributed across 10 disk units. A Fourier transform with respect to time is initially performed using parallel FFT computation. 513 frequencies including the Nyquist are used with computations. To facilitate computations, the transformed dataset is transposed from (ω, x, y) to (x, y, ω) ordering.

Half of the frequencies for the 40,000 traces are initially stored in the SSD and processed. Afterwards, the remaining frequencies are moved from disk storage to the SSD and processed. Partial sum results for eight depths are accumulated for each retrieval and restorage of 3-D data subsets. The entire 3-D dataset is retrieved and restored 125 times. Using a multitasked FORTRAN program with an assembler language tridiagonal solver and queued asynchronous I/O, CPU bound execution is achieved. Using 32 million words of SSD storage and two million words of central memory, and elapsed or wallclock time of approximately two hours was required with a two processor CRAY X-MP system (9.5 nanosecond clock period). Execution with a single processor yielded an elapsed time of 3.8 hours.

For above two processor execution, 98% of the wallclock time is spent in the parallelizable part of the program. The I/O and generation of partial sums was performed with one processor. Memory contention accounted for approximately 2.6% of the wallclock time. Multitasking synchronization accounted for 0.9% of the time. I/O interrupts and contention accounted for 0.4% of the time. Miscellaneous operating system functions required 1.7% of the elapsed time.

The above program was reworked for a four processor CRAY X-MP system (9.5 nanosecond clock period). Using 40 million words of SSD storage and 3.67 million words of central memory, elapsed or wallclock times of 3.49, 1.85, and 1.01 hours were measured for execution with one, two, and four processors, respectively. 1.5×10^{12} floating point operations and 40 billion words of data movement were performed. Dividing these numbers by the smallest elapsed time gives sustained rates of 412 MFLOPS and 11 million words per second.

1.3.2 *Pre-stack Migration*

Pre-stack migration using Kirchhoff summation and an eikonal equation algorithm will be investigated.

1.3.2.1 *Kirchhoff Migration*

Kirchhoff summation in the spatial-time domain is the most economical of pre-stack algorithms. For brevity, immediate remarks address the simplest form of the algorithm. Each input trace will be summed into all output migration columns satisfying an aperture before processing the next input trace. Prior to summing operations, each input trace will be resampled or interpolated to a finer sample interval (for example, $\Delta t = 0.1$ ms) to allow selection of the nearest time sample. This precludes the need of interpolating each extracted amplitude. Excluding extremely small apertures, the computational cost of one-time resampling is small. Two square root computations (expressed as functions of distance from source and receiver points to the imaged point in space) are evaluated by vectorized table lookup using the random gather instruction. Input trace x and y coordinates do not have to belong to a uniform grid. The random gather instruction is also used to extract NMO corrected amplitudes. Using a microtasked FORTRAN program with a four processor CRAY X-MP system, the above operations plus multiplication by a weight and summation into output migration columns was performed at the approximate rate of 2.1 clock periods (17.85 ns) per summed amplitude. The partial derivation with respect to depth z computed on each completed output migrated column and resampling of each input trace was negligible. Tape buffering of unmigrated and migrated traces to and from central memory was hidden behind computations.

The primary objection to the above is the use of incorrect computational velocities (for example, stacking velocities). If we wish to perform computations with a more meaningful velocity term or compute more accurate travel times, the above algorithm must be augmented by ray tracing logic. It would be prohibitive to apply ray tracing for all propagation paths. The practical solution is that of applying ray tracing to a subset of propagation paths and using interpolation for most propagation paths. Linear interpolation can be implemented with high computational efficiency using concurrent or chained multiply-and-add computations. Another objection to the above "simplest" algorithm is usage of incorrect weighing factors. Once again, practical considerations lead to an interpolation scheme. With respect to computer implementation, the determining factor in selecting practical interpolation schemes is the size of central memory tables and arrays. Trade-offs between accuracy, computational speed, and memory requirements must be weighed for each specific computer configuration and processing environment. Central memory requirements can be reduced by using the SSD.

The overall speed of optimal implementation of Kirchhoff migration depends on many factors and parameters. Benchmark comparisons with conventional computer systems typically yield exceedingly high speedup factors. On the other hand, unprecedented processing rates are still insufficient to contemplate migration of all field traces of a large 3-D survey. For example, consider an aperture of ± 50 traces in the x and y directions. This means each input trace contributes to 101×101 or 10201 output migration columns. If each output column contains 1000 migrated results and if the "simplest" algorithm without interpolation is used, approximately 182 ns are required to migrate each input trace with a four processor CRAY X-MP system. At this processing rate, approximately 475,000 input traces could be migrated in a 24-hour period.

1.3.2.2 *Eikonal Migration*

We will now examine pre-stack migration using ray theory. The 3-D acoustic eikonal equation is directly solved using a Runga-Kutta method to find source and receiver travel times to all output migration coordinates in space. An arbitrary velocity function is used. This algorithm was used with a small 3-D land survey featuring a fixed or stationary areal array of 192 receivers with 320 sources. Each trace contains 1300 amplitudes with 4 millisecond sampling. Output imaged space contains $52 \times 58 \times 850$ points with 6.7 meter depth intervals and 20 meter x and y grid spacing.

Initially, all input traces are stored on disk storage. Using 64-bit amplitudes, 80 million words of storage are required. The first step of the algorithm solves the eikonal equation for all 192 receivers and stores times on the SSD. $192 \times 52 \times 58 \times 850$ or 493 million words of storage are required. The second step processes all common source files. If the source location coincides with a receiver, $52 \times 58 \times 850$ travel times are retrieved from the SSD. Otherwise, eikonal travel times are computed. Traces corresponding to the current source are retrieved from disk storage. Applying linear interpolation, amplitudes are extracted using the random gather instruction and are added to accumulative output dataset residing in central memory. Since travel times for all 192 receivers are required for the imaging of each source, the entire 493 million word SSD file is retrieved for each of 320 sources. This totals 1.26×10^{12} bytes of data movement. Using a four processor CRAY X-MP/416 system (8.5 nanosecond clock period), four common source files are processed in parallel and an elapsed time of 3.6 hours is required for the entire survey. The 61,440 input traces were migrated at the approximate rate of 210 milliseconds per trace or 82 nanoseconds (9.7 clock periods) per summed amplitude.

In the general case, with varying receiver arrays, SSD management becomes more complicated. The following discussion will assume uniformly spaced source and receiver locations. Let $N_x N_y N_z$ denote the number of output migration coordinates. The first step computes travel times for all source and receiver locations storing results on disk storage. Let N_r denote the maximum number of receiver locations such that $N_r N_x N_y N_z$ is less than 512×10^6. If N_r is sufficiently large, performance proportional to the previous example can be expected. Travel times for groups of N_r receiver locations will be transferred from disk to SSD storage. Travel times for each source having the current N_r receiver locations are retrieved from disk to central memory. Input traces corresponding to current source and receiver locations are retrieved from disk to central memory and summed into accumulative output dataset residing in central memory. When all sources for the current N_r receiver locations are processed, travel times for another set of N_r receiver locations are moved from disk to SSD storage and the above operations are repeated. Compared to the previous example, the above sketched solution differs in additional disk I/O. Travel times for receiver locations are only accessed twice (one write and one read). As a function of N_r, travel times for source locations are accessed more than twice. If necessary, striped disk storage and parallel streaming can be utilized.

The above 3-D pre-stack eikonal migration approach is more computationally intensive than most 3-D pre-stack Kirchhoff migration algorithms implemented by CRAY users.

1.4 THREE-DIMENSIONAL FORWARD MODELING

This section addresses forward modeling for the acoustic and elastic wave equations. Conventional finite-difference and Fourier or pseudo-spectral methods are discussed.

1.4.1 *Finite-Difference Forward Modeling*

The finite-difference method has been the most common implementation of forward modeling with the acoustic wave equation. We will now consider the conventional approach using explicit fourth-order differencing in space and time stepping or time integration of second-order differencing. An absorbing boundary is applied as a 15-point weighing function. A free surface case is not included. Assuming constant density, three variables or arrays are required at each spatial grid point. A macrotasked FORTRAN program featuring in-memory and out-of-memory SSD versions has been developed for the four processor CRAY X-MP system.

A model of $121 \times 121 \times 121$ grid points in the x, y, and z directions requires 5.3

million words of array storage and can be computed in memory. Using four processors (8.5 ns clock period), an elapsed time of 90 seconds is required for computation of 1000 time steps. A model of $250 \times 250 \times 250$ grid points requires 46.9 million words of SSD storage. Computation of 1000 time steps requires an elapsed or wallclock time of less than 15 minutes using four processors. Larger models (for example, $550 \times 550 \times 550$ grid points) can be computed with a 512 million word SSD. By using higher order schemes, accurate first partial derivatives can be computed and variable density modeling can be performed.

1.4.2 *Fourier Method Forward Modeling*

Forward modeling using the Fourier or pseudo-spectral method has been implemented for both acoustic and elastic wave equations. The Fourier method computes spatial derivatives in the Fourier domain. An absorbing boundary is applied as a 15-point weighing function. An optional free surface can be used. Arbitrary velocities and densities can be used.

Implementation of the acoustic case requires five variables or arrays at each grid point for the general case. A spatial grid of $256 \times 256 \times 256$ points requires SSD storage of 84 million words. For the constant density case, four variables or arrays are required. The use of a constant density halves the number of FFTs computed. Using a four processor CRAY X-MP (9.5 nanosecond clock period) with a multitasked FORTRAN program, approximately 2.49 hours of elapsed time are required for the computation of 1000 time steps with $256 \times 256 \times 256$ grid points. Inclusion of a free surface increases the wallclock time to 3.29 hours. The comparable constant density problem can be computed within 1.33 and 1.74 hours without and with free surface condition, respectively. For the constant density case without free surface, total data motion of 1.36×10^{11} words and 2.4×10^{12} floating point operations are required. Dividing the number of floating point operations by the wallclock time gives a sustained computational rate of approximately 500 MFLOPS. Conventional time integration of second-order differencing was utilized. Multitasking speedup factors exceeding 3.9 over execution with a single processor were observed. A CPU bound program was obtained by using asynchronous queued I/O for SSD data motion.

The elastic case is implemented using the equations of momentum conservation, strain-displacement and stress-strain relations for a linear isotropic solid undergoing infinitesimal deformation. Implementation requires 15 variables or arrays at each grid point. Optional free surface is achieved by an extended region of zero P- and S-wave velocities. Implementation requires six arrays in the extended region. For example, computation with a spatial grid of $225 \times 225 \times 225$ points without a free surface requires 171 million words of SSD storage. Computation with a free surface by the inclusion of 90 additional z coordinates requires 198 million words of SSD storage. Odd base FFTs are used to avoid computation with Nyquist frequencies. For a free surface-less model of $225 \times 125 \times 225$ grid points in the x, y, and z directions, respectively, SSD storage of 95 million words is required. Using a four processor CRAY X-MP (9.5 nanosecond clock period) with a FORTRAN multitasked program, computation of 1000 time step requires approximately 3.5 hours of elapsed time. Total data motion of 3.8×10^{11} words and 5.5×10^{12} floating point operations are required. Dividing the number of floating point operations by the wallclock time gives a sustained computational rate of approximately 435 MFLOPS. As in the acoustic case, time integration of second-order differencing and asynchronous queued I/O was used.

For both acoustic and elastic cases, time stepping by time integration of second-order differencing can be eliminated by using a new rapid expansion method (REM). The REM is based on a modified Chebychev expansion with the formal solution to the governing equations. Sine and cosine expressions are expanded in series using Bessel functions and modified Chebychev polynomials. All time dependency is contained in Bessel function terms. Spatial differentiation is identical to that used by previously discussed algorithms. Spatial differentiation is contained in Chebychev polynomials terms. Compared to time integration of second-order differencing, larger time steps or increments are permissible with REM implementation. For output time sections, results at intermediate times are obtained by resubstituting intermediate times in the computational equations. This involves computation with Bessel function terms and does not require recomputation of spatial derivatives. The REM method can be applied to acoustic and elastic wave propagation using formulations other than the Fourier method.

If relatively large time steps or increments are used, REM implementation is faster than temporal differencing. Ideally, we would prefer to compute one jump from zero time to the time of the last time step and then compute successive Δt time steps by performing series summations with additional sets of Bessel functions. Central memory storage is required for intermediate results. The size of central memory limits the maximum time step or increment. SSD storage of global variables is identical to previously discussed temporal differencing algorithms. Machine or computer accuracy can be efficiently obtained. An accurate absorbing boundary condition for large time steps has not yet been found.

The largest elastic model computed with the REM method contained $315 \times 315 \times 315$ spatial grid points with a spacing of 20 meters. The free surface condition was not included and absorbing boundaries were not applied. SSD storage of 469 million words and 14 million words of central memory were required. The maximum velocity in the model was a P-wave velocity of 4000 meters/second. Using $\Delta t = 1$ millisecond, second-order temporal differencing requires 1000 time steps or operations for model propagation to 1 second. Using the REM method, series summations required 628 terms for a jump or increment to one second. Each of the above terms requires a Chebychev recursion step which is computationally equivalent to one time step or operation with temporal differencing. The REM method required approximately 37% fewer operations. Using a four processor CRAY X-MP system (8.5 nanosecond clock), the above time jump or increment required more than 11 hours of elapsed or wallclock time. Using additional sets of Bessel functions, successive intermediate times with $\Delta t = 4$ ms were then computed. The total elapsed time for all computations was 11.3 hours. Since the majority of computations comprise FFT computation of spatial derivatives, the total elapsed time would essentially be the same if intermediate times are computed with $\Delta t = 1$ or 2 milliseconds.

1.4.3 *Forward Modeling Graphics*

Time history output for selective lines on the $z = 0$ surface or cross sectional snapshots for selective time steps can be obtained with minor degradation to elapsed times. On the other hand, output of 3-D snapshots presents a formidable problem. Consider an acoustic model of $250 \times 250 \times 250$ grid points. Disk storage of all snapshots is infeasible. If we output 32-bit pressure values, disk storage of 1000 time steps would require 62.5 billion bytes or more than 52 DD-49 disk units. One future solution could be parallel streaming to a number of optical disks. If an aggregate bandwidth of 60 million bytes per second could be achieved, approximately 18 minutes of I/O streaming time would be required. I/O time could be overlapped with computations. Another future solution could be the use of the HSX-1 high speed external channel which can download data at rates up to 100 million bytes per second. This solution would require less than 11 minutes of I/O time for the above 1000 time steps with 32-bit values. Future hardware, high speed graphics devices, or workstations would have to be designed to accept data at 100 million bytes per second rate.

CHAPTER 3

WAVE EQUATION COMPUTATIONS AND TRULY PARALLEL PROCESSING

by
OLAV HOLBERG
A/S Informasjonskontroll
P.O. Box 265
1371 Asker, Norway

Most computationally demanding problems of interest to the geophysical community can be formulated as recursive wavefield extrapolation in time or space. This includes depth migration and migration velocity analysis before or after stack by downward continuation of acoustic or elastic wavefields, and also acoustic or elastic forward modeling, reverse time depth migration and eventually modeling driven inversion of surface-seismic and well-seismic data by general high-order finite-difference (FD) techniques or by hybrid FD-pseudospectral techniques.

Numerical wave propagation and parallel implementation strategies are discussed briefly. Finite-difference algorithms for extrapolating wavefields in time or space can be mapped onto any parallel hardware structure supporting nearest neighbor communication in one, two, or three dimensions, but will run efficiently on a large number of processors only when the bandwidths of the communication channels are properly dimensioned relative to the arithmetic bandwidth of each processor. For efficient parallel implementation we propose to use a one-dimensional array of vector processors. By totally eliminating the need for physically long communication channels, this computational structure can be operated efficiently at a very high speed.

Such a machine, specifically designed to perform well on wave equation computations, can be realized with sufficient interprocessor communication and arithmetic capabilities to outperform current supercomputers by two orders of magnitude using state-of-the-art (1987) technology. When efficient numerical algorithms are used, this structure will be capable of completing realistic three-dimensional calculations with execution times of the order of minutes.

INTRODUCTION

Most computationally demanding problems of interest to the geophysical community can be formulated as a time evolution problem or a depth extrapolation problem where one or more wavefields are recursively extrapolated in time or space by numerical solution of an appropriate wave equation.

The forward modeling problem, i.e. the generation of synthesized seismic shot records from a given geological model, is most conveniently solved by recursively stepping the numerical representation of the wavefield resulting from a time-varying source function forward in time. The time histories of the wavefield can then be recorded at spatial locations of interest, thereby allowing the numerical creation of synthetic seismograms for any source-receiver configuration.

Depth imaging, i.e. depth migration and migration velocity analysis before or after stack, is similarly performed by recursive downward continuation of the wavefield recorded at the surface. At each computational depth level an imaging is performed. Individual shot records are typically imaged by a temporal correlation of upcoming and downgoing waves (Claerbout, 1971; Berkhout, 1985), while zero-offset (stacked) data are imaged by computing the pressure at time zero (Loewenthal, 1976; Schneider, 1978). Alternatively, when the recorded wavefield is treated as a time-dependent secondary source distribution rather than a boundary condition in space, depth migration can be accomplished by reverse time propagation (Levin, 1984).

The full seismic inverse problem, i.e. the iterative estimation of elastic earth parameters from observed wavefields generated by known sources, can be solved either by time extrapolation or by depth extrapolation. In the time evolution approach, each iteration involves a forward propagation of the actual sources in the current medium, a reverse time propagation of the current residuals, and a temporal correlation of the two wavefields thus obtained at each point of space (Tarantola, 1987). Posed as a depth extrapolation problem, each iteration involves a downward extrapolation of observed surface displacements and tractions, an upward extrapolation using the initial conditions of the previous field as sources, and a temporal correlation of the two wavefields thus obtained at each point of space (Tarantola, Jobert, Trezeguet and Denelle, 1978).

Today there is only one major obstacle to the introduction of such wave theoretical techniques in routine operations: availability of computational power. The recursive wave calculations that constitute the computationally demanding part of proper wave theoretical imaging are monumental tasks even for today's most powerful supercomputers. Proper 3-D acoustic depth migration of a single seismic shot record requires hours of computing time on current vector computers. That amounts to years of processing time for a fraction of a typical 3-D seismic survey. Full elastic inversion would be computationally more demanding by at least one order of magnitude.

Conventional supercomputer architecture is approaching fundamental limits in speed imposed by signal propagation and heat dissipation. Thus, some sort of massively parallel processing will be needed. The emergence of Very Large Scale Integration technology and automated design tools has made implementation of hardware for highly parallel computing feasible. However, attempts at producing highly parallel general purpose computers have not yet been successful. Systems originally designed to be completely general purpose turn out not to be so in that they are extremely inefficient in some applications. Numerous experiments have shown that the performance of parallel architectures is very algorithm dependent. The theoretical peak performance of such systems can only be approached in special cases.

It is therefore suggested that the way to approach the problem is not to look for a "general purpose" parallel architecture and figure out how to match ones algorithms to it, but rather to work out efficient parallel implementations of particular algorithms, and then design an architecture to execute them efficiently. At present, this is the only known way to meet extreme computational requirements. For such dedicated architectures it is essential to start with the problem and then attempt to discover what type of architecture is most appropriate. Because both the algorithm and the architecture can be varied, there is a substantially higher degree of freedom than in conventional systems design.

Properly matching parallel algorithm and architecture is a difficult process where it is of paramount importance to keep the global system complexity down. The most crucial decision is the choice of the underlying algorithms since the suitability of the algorithms largely determines both the design cost and the performance of the total system.

For this reason we shall limit our discussion to conceptually simple but powerful algorithms of finite-difference type. Such solution techniques replace the continuous partial differential equations by discrete difference approximations on a regular computational grid. These equations can be written consistently with the continuity conditions of continuum mechanics which require that both the tractions and the displacements remain continuous across all possible interfaces in a solid medium. The boundary conditions for each grid point are then satisfied implicitly for arbitrarily inhomogeneous media. This is, however, not strictly correct for a medium with a liquid-solid interface where there can be slip parallel to the interface, and shear stress components parallel to the interface are not required to be continuous. On the other hand, most do proceed as if it were correct, with no very apparent problems. An explicit specification of the boundary conditions at the interfaces is then not required and a single relatively simple numerical kernel can therefore be applied to the entire computational domain.

The present paper begins with an outline of the essential elements of efficient numerical wave propagation in time and space. This is followed by a discussion on parallel implementation strategies. Finally, some guidelines are given for what could be called "an algorithm-driven design of a parallel wave equation processor."

TIME EVOLUTION

Any time evolution process governed by linear partial differential equations can be formulated as follows:

$$\frac{\partial B}{\partial t} = AB + s \tag{1}$$

Here $B = B(x_1, x_2, x_3, t)$ is a wave vector and $A = A(x_1, x_2, x_3)$ is an operator matrix containing combinations of spatial differentiators weighted by the parameters characterizing the medium in which the waves propagate, and s is a source term. Examples of (1) for elastic and acoustic waves are given in Appendix I.

Stepping the wave vector B in (1) forward or backward in time requires the evaluation of spatial derivatives at all grid points within the computational domain. This is most efficiently done by general high-order differencing (Holberg, 1987), i.e. by evaluation of expressions of the type

$$\begin{aligned} d_j^+(u) &= (1/\Delta x_j) \sum_{l=1}^{L} \alpha_l [u(x_j + l\Delta x_j) - u(x_j - (l-1)\Delta x_j)] \\ &\approx \partial/\partial x_j (u(x_j + \Delta x_j/2)) \end{aligned} \tag{2a}$$

or

$$d_j^-(u) = (1/\Delta x_j) \sum_{l=1}^{L} \alpha_l [u(x_j + (l-1)\,\Delta x_j) - u(x_j - l\Delta x_j)]$$

$$\approx \partial/\partial x_j(u(x_j - \Delta x_j/2)) \tag{2b}$$

where $2L$ is the length of the differentiator.

Equation (1) is neutrally stable or stable when energy dissipation is incorporated. This does not guarantee that any numerical implementation of (1) is stable. However, the fact that (1) does not itself give rise to exponentially growing solutions, allows us to optimize the difference coefficients α_l, $l = 1, \ldots, L$, in (2) to control the corresponding numerical dispersion within a spatial frequency band without constraints. We can then produce curves for trade-off between mesh refinement and differentiator length as shown in figure 1 for different levels of maximum relative error in numerical group velocity.

Because the memory and arithmetic requirements increase geometrically with the mesh refinement, application of the shortest operators is very uneconomic and also technologically unfeasible for realistic 3-D models. On the other hand, as the grid is coarsened, the approximation of the interface conditions at material discontinuities might become less accurate. This latter aspect of discretization is, however, not yet fully understood. Whereas the numerical dispersion is determined completely by the numerical group velocity, the errors in numerical reflection and transmission at an interface are controlled both by the group velocity errors in the media separated by the interface, and by the error involved in the approximation of the interface conditions. However, the amplitude of the reflected and transmitted wavelets does not deteriorate in accuracy once an incoming wavelet has interacted with the interface. It is reasonable, therefore, to argue (Brown, 1984) that it is much more important to use accurate differentiators than it is to use accurate interface conditions, and to use this as a justification for coarse grid calculations.

However, too coarse grids definitely make accurate representation of curved interfaces difficult. When curved interfaces are aligned to a "staircase" in the grid, the artificially introduced edges can give rise to spurious diffractions in low-velocity regions where the grid may be coarse relative to the dominant wavelength. The results can be improved by averaging the interfaces over a grid block (Shubin, Baker and Bell, 1985). In our experience, about 3 grid points per shortest wavelength appears to be adequate in most situations. This is also roughly the optimum choice from a purely computational point of view (Holberg, 1987).

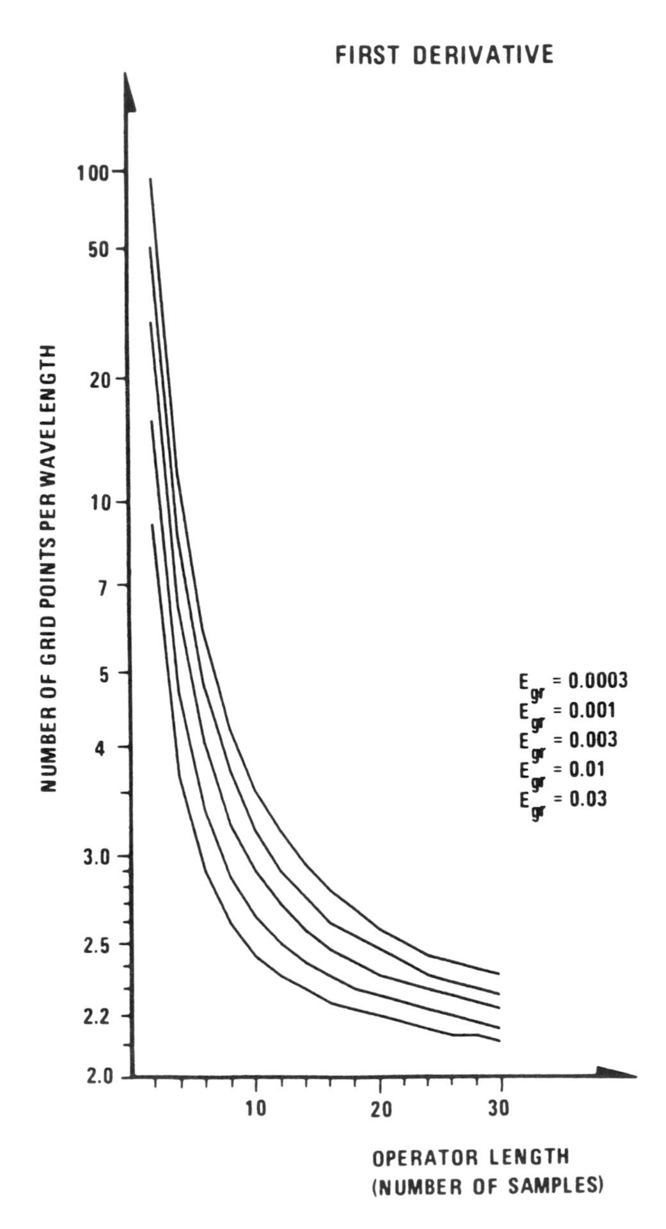

Fig. 1 The relation between differentiator length and required sampling interval for different levels of maximum relative error in numerical group velocity (from Holberg, 1987).

Generally the elastic earth parameters vary much more rapidly in the vertical direction than in the horizontal directions. It is therefore possible to increase the accuracy of the representation of curved interfaces by using a finer grid in depth while retaining a laterally coarse grid.

The simplest explicit time integration schemes are obtained by using standard second-order finite-difference substitution for the time derivative. The relation for stepping the wave vector forward or backward in time is then

$$B(t \pm \Delta t) = B(t) \pm \Delta t A B(t) \tag{3}$$

For such schemes the stability limit is roughly the Courant limit, i.e., $\Delta t \leqslant (\Delta x/c)_{\min}$ (Kosloff, Reshef and Loewenthal, 1984; Holberg, 1987) where c is the velocity of propagation and Δx is the spatial grid spacing. However, in order to keep the error in time differentiation comparable to the error in spatial differentiation, a finer temporal grid spacing Δt must often be used. Of the order of 20 grid points per shortest wavelength in time appears to be adequate for realistic geophysical simulations. The time step can be increased, for instance, by using higher order Taylor expansions to predict the wave vector at time $t + \Delta t$. This is potentially more efficient but would result in added algorithmic complexity because the operator A in (1) would have to be applied more than once for each computational time step.

A reasonably sized time evolution problem measures $60 \times 60 \times 100$ shortest wavelengths in the spatial dimensions. This corresponds to a subsurface volume of $1.8 \times 1.8 \times 3.0\ \text{km}^3$ when the source is assumed to have a high-cut frequency of 40 Hz, producing a shortest wavelength of 30 m when the lowest shear wave velocity is 1200 m/s. If we use a spatial sampling interval of 10 m (3.0 grid points per shortest wavelength) and tolerate a peak relative error in numerical group velocity of 0.5 %, then it is seen from figure 1 that a spatial differentiator length of 8 ($L = 4$ in equation 2) is required. The number of add-multiply-add operations is then roughly $18(L + 1) = 90$ per grid point per time step when (3) is used for time integration of the elastic wave equation. The number of spatial grid points is 10^7 and the number of time steps required to simulate a seismic record of length 4 sec is about 4000. For such a simulation the total number of arithmetic operations is thus 10^{13}. The storage requirement is of the order of one gigabyte.

DEPTH EXTRAPOLATION

This class of problems is most conveniently solved in the space-frequency domain because this domain allows independent treatment of all temporal frequencies.

Replacing time t by temporal frequency ω via Fourier transform involves no loss of generality under the assumption that the medium in which the waves propagate is time invariant. Depth extrapolation processes can then be formulated as follows:

$$\frac{\partial B}{\partial x_3} = AB \tag{4}$$

The wave vector is $B = B(x_1, x_2, x_3, \omega)$ and as before $A = A(x_1, x_2, x_3, \omega)$ is an operator matrix containing spatial differentiators. Examples of (4) for elastic and acoustic waves are given in Appendix II.

As it stands, equation (4) is inherently unstable because it produces paired exponentially growing and decaying non-propagating solutions, known as evanescent waves, in addition to the propagating solutions. By applying a double Fourier transform with respect to x_1 and x_2 it can be shown (Ursin, 1983) that the eigenvalues of the system matrix A, for the special case of a horizontally layered acoustic or elastic medium, are of the form

$$\lambda = \pm i[\omega^2/c^2 - (k_1^2 + k_2^2)]^{1/2} \tag{5}$$

where k_j denotes wavenumber or spatial frequency in the x_j direction and c is the relevant compressional or shear velocity. Suppressing evanescent energy corresponds to preventing the eigenvalues of A from becoming real. A necessary criterion for numerical stability is therefore to constrain the numerical differentiation such that

$$D_1^2 + D_2^2 \leqslant \omega^2/c^2 \tag{6}$$

where $D_j(k_j)$, $j = 1, 2$ represent the spatial frequency responses of the differentiators. This is illustrated for the two-dimensional case in figure 2.

If the derivatives are computed by multiplication in the spatial Fourier domain, the multiplication can be restricted such that (6) is fulfilled globally for the highest velocity at each depth level (Kosloff and Baysal, 1983). In the presence of lateral velocity variations this procedure causes undesired elimination also of waves propagating near the horizontal direction. A more promising scheme is therefore to use space-variant differentiators such that (6) can be fulfilled locally. This can be implemented by designing a number of optimum differentiators with different spatial cut-off frequencies. The precomputed difference coefficients can then be accessed by table look-up, i.e. by taking $\alpha_l = \alpha_l(\omega/c(x))$, $l = 1,\ldots, L$, in (2) such that the ratio between the temporal frequency and the local velocity is used as a key to

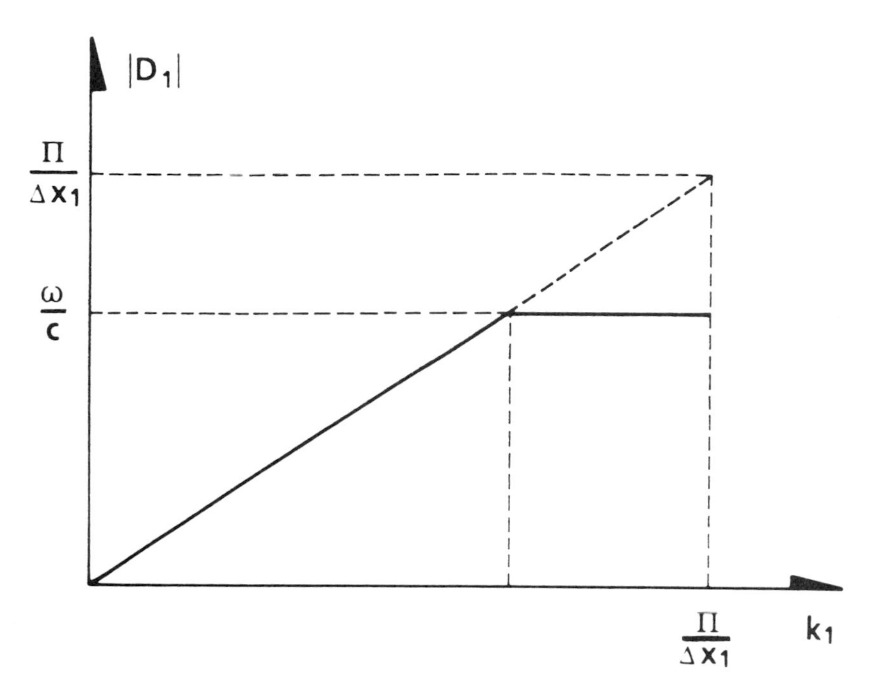

Fig. 2 Bandlimitation of lateral differentiator in order to suppress evanescent instabilities in depth extrapolation.

address the correct operator at each grid point during the downward continuation process. The feasibility of this approach is yet to be demonstrated for other than acoustic problems.

When relation (6) is satisfied, (4) can be integrated in depth with standard techniques for ordinary differential equations such as the Runge-Kutta method. For the scalar one-way wave equation all terms required to compute the wavefield at depth level $z + \Delta z$ can be optimally combined into a single space-variant convolutional operator (Holberg, 1988).

A full-scale one-way acoustic depth extrapolation problem for depth migration of individual seismic shot records would typically be performed for 300 temporal frequencies representing 5 s of data with a bandwidth of 60 Hz. This wavefield would be extrapolated 400 steps downwards on a 200×200 lateral mesh using a spatial sampling interval of 2.0 grid points per shortest wavelength in all directions. The corresponding subsurface model then measures $2.0 \times 2.0 \times 4.0\ \text{km}^3$ when the lowest compressional wave velocity is 1500 m/s. The number of complex add-multiply-add operations per gridpoint per depth step per frequency component is of the order of 50 for an accurate and efficient algorithm, making the total number of arithmetic operations for this problem roughly 2.5 10^{12}.

For full elastic depth extrapolation we lack sufficient practical experience to give reasonable accurate estimates. Preliminary investigations, however, indicate that the elastic depth extrapolation problem is about one order of magnitude more computationally intensive than the one-way acoustic problem.

PARALLEL IMPLEMENTATION

The depth extrapolation problems exhibit a high degree of inherent parallelism because all temporal frequencies can be processed independently, and in parallel. The local nature of the spatial finite-difference operators makes it possible to decompose the time evolution algorithms, and to further decompose the depth extrapolation algorithms, by dividing the spatial domain into subregions in such a way that the physical/computational space can be mapped directly onto a processor array of dimension one, two, or three.

When L_j is the operator half-length in the x_j direction, each processor needs access to data points within surface zones of "thickness" L_j and $L_j - 1$ in the neighboring spatial subdomains. The situation is illustrated for 3-D, 2-D and 1-D decomposition in figure 3. Before an x_j-differentiation of a specific variable can proceed in each subdomain, the extra perimeter points must be updated with copies of the latest values determined by the neighboring processors.

For a subdomain of dimension $n_1 n_2 n_3$ grid points the required interprocessor communication bandwidth relative to the arithmetic bandwitdh of each processor is therefore roughly

$$C_r = \frac{2[(2L_1 - 1)\, n_2 n_3 + (2L_2 - 1)\, n_1 n_3 + (2L_3 - 1)\, n_1 n_2]}{(L_1 + 1)\, n_1 n_2 n_3 + (L_2 + 1)\, n_1 n_2 n_3 + (L_3 + 1)\, n_1 n_2 n_3} \tag{7}$$

where terms containing L_3 vanish for depth extrapolation problems. In this relation one add-multiply-add operation has been counted as one arithmetic operation. This is reasonable since symmetric convolutions of the type (2) typically account for 80 % or more of the total arithmetic workload and each processor therefore should be equipped with two adders for each multiplier to avoid an add-bound implementation. If one multiply-add operation is counted as one arithmetic operation, $L + 1$

should be substituted by $2L+1$ in the denominator of (7). The "extra" $+1$ operation accounts for the work required to linearly combine the derivatives.

When $n_j < L_j$, i.e. when the dimension of the local data volume is smaller than the operator half-length, each processor needs to update copies of data points beyond the domains of the nearest neighbors in the array. However, relation (7) still holds because the data can be propagated recursively in such a way that the processors do not have to transfer data that they would not have accessed for their own local computations anyway.

From (7) it is seen that C_r is essentially proportional to the ratio between the surface area and the volume of each subdomain and virtually independent of the operator length. This indicates a strong need to minimize the surface area of each

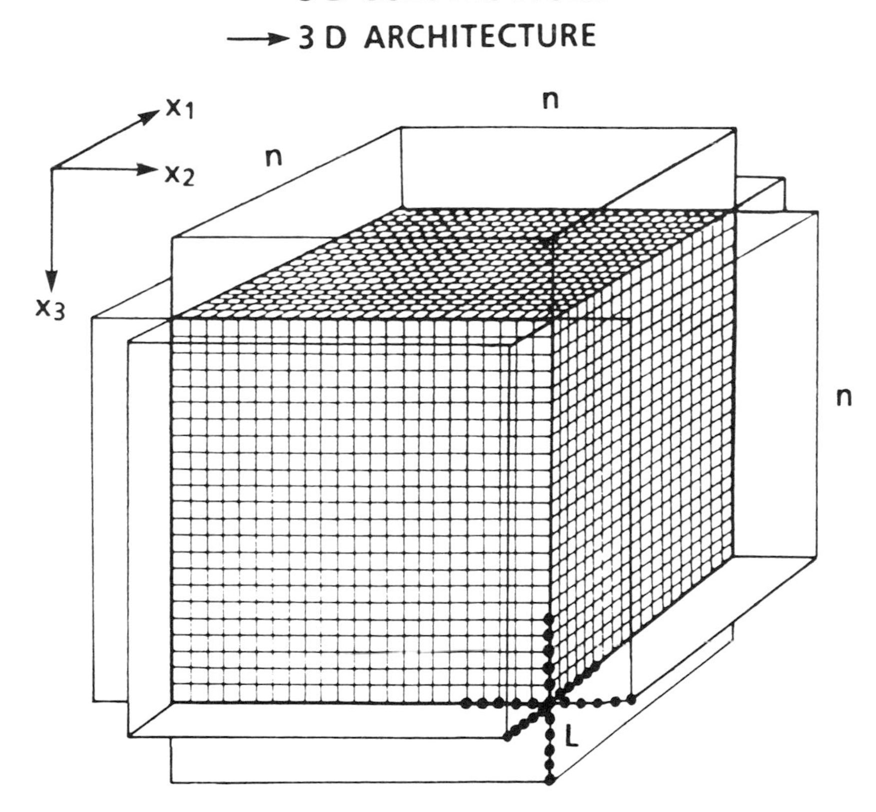

Fig. 3a Subdomain and range of interaction with neighboring subdomains for 3-D decomposition.

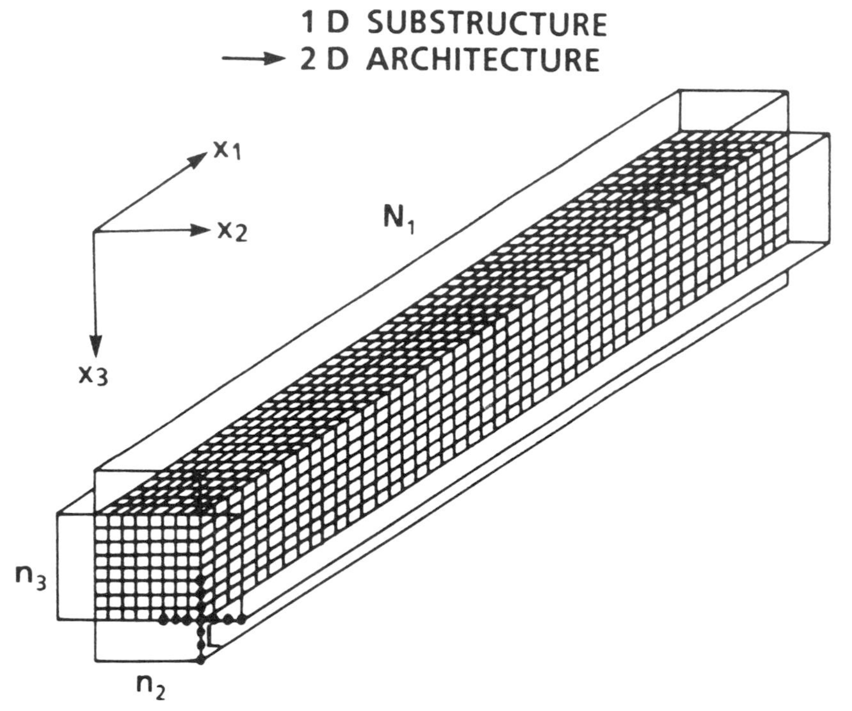

Fig. 3b Subdomain and range of interaction with neighboring subdomains for 2-D decomposition.

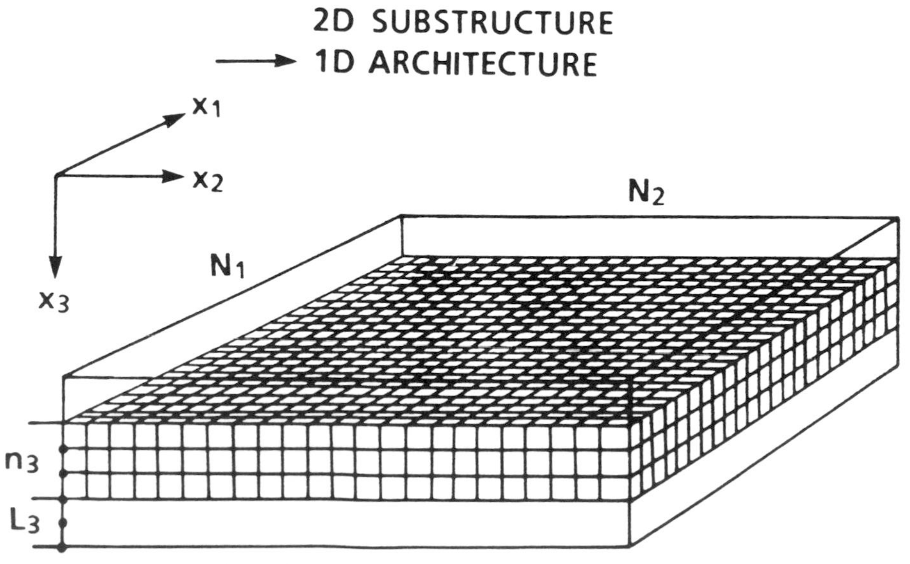

Fig. 3c Subdomain and range of interaction with neighboring subdomains for 1-D decomposition.

subdomain by making the proportions cubic. Accordingly, the ideal hardware structure should be a 3-D mesh of processors.

However, to ensure high performance and cost-effective design, each processor should be given a pipelined vector architecture (Kogge, 1981). Due to the start-up time required to fill the pipelines, the efficiency of such systems degrades as the vector length decreases. This argument is pointing in the direction of essentially 1-D pencil shaped subdomains and a 2-D hardware structure. The discussion can be stretched further towards 2-D plane subdomains and 1-D hardware structures because the local memory bandwidth requirements can be relaxed when it is possible to use vectors with the same spatial orientation as the differentiators.

The task of wiring the processors in three, two, and one dimension is difficult moderate, and easy, respectively. There is no practical way a high number of powerful processors can be interconnected in three, or even two, dimensions without using physically long communication channels that may strongly reduce the performance of the overall architecture and undercut the advantage of parallelism. Both the Floating Point Systems T-series and the Intel n-cube suffer from this limitation. These machines consist of 2^n vector processors interconnected as if they were on the corners of a cube in n-space (hypercube). This topology includes 3-D, 2-D, and 1-D nearest neighbor meshes. Each processor has a performance potential of roughly 10 million multiply-add operations per second. However, because of the complexity of the multidimensional wiring, the communication links have been made bit-serial with a net bandwidth of about 0.1 million 32-bits words or less per second in each direction. Therefore, from (7), the 3-D time evolution algorithms can not be run efficiently on these machines unless we use more than about 50^3 grid points per node. Current versions are not supplied with enough memory to do that.

Reliability is another critical issue in truly parallel systems design. Even if each processor module could be built with an exeptional MTBF (Mean Time Between Failure) of 20 years, a system incorporating a few hundred processors would get an expected MTBF of only a few weeks which is clearly not satisfactory. Incorporating redundancy or fault tolerance in multidimensional processor arrays is also a massive headache because entire processor rows or planes would have to be by-passed in order not to ruin the load balancing.

The difficulties above can be circumvented by using a one-dimensional architecture. For the time evolution algorithms, the interprocessor communication requirements are then tougher although they can be relaxed by using a smaller grid spacing (and a shorter differentiator) in depth while retaining a laterally coarse grid. However, because a one-dimensional interconnection scheme can be made strictly nearest neighbor also in a physical sense, we are able to use fast word-parallel channels for data transfer, effectively leaping the communication bandwidth by almost two orders of magnitude.

OUTLINE DESIGN OF AN EFFICIENT PARALLEL WAVE EQUATION PROCESSOR

The deal efficiently with large-scale wave equation computations a processor structure as shown in figure 4a is proposed. The system comprises an arbitrarily high number, say 300 for practical applications, of powerful reasonably general

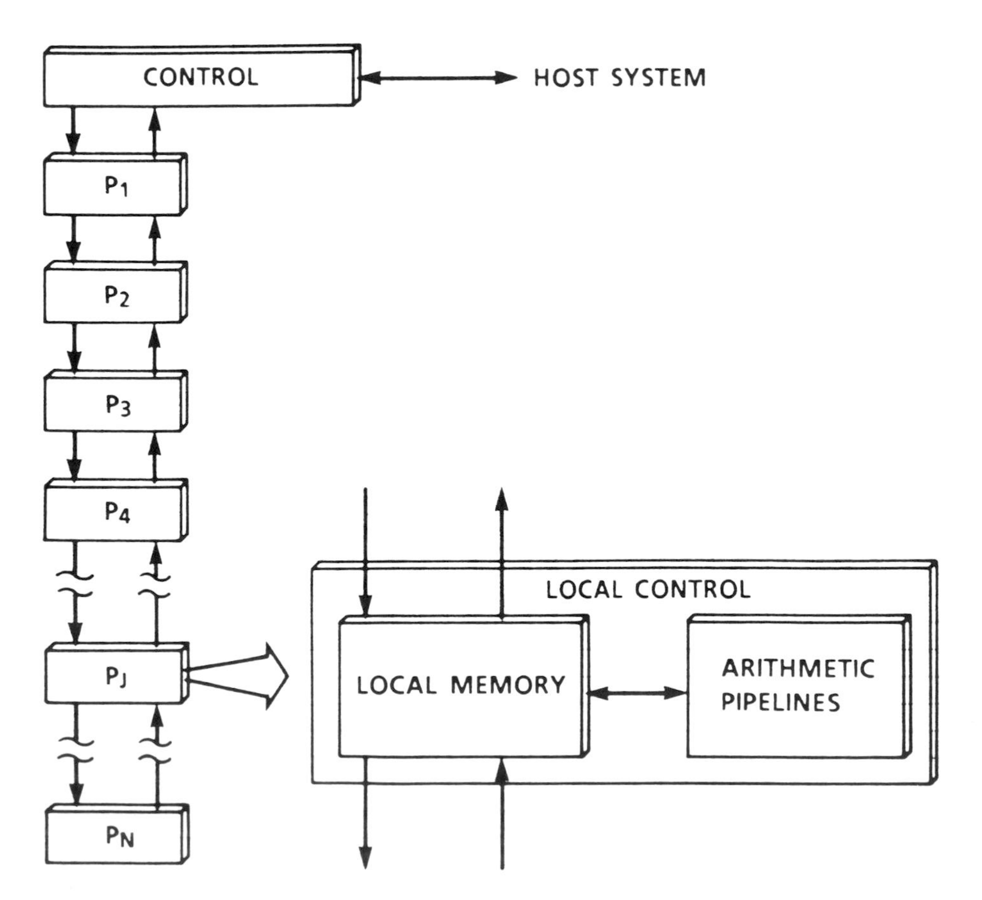

Fig. 4a Proposed architecture for efficient parallel wave equation computations.

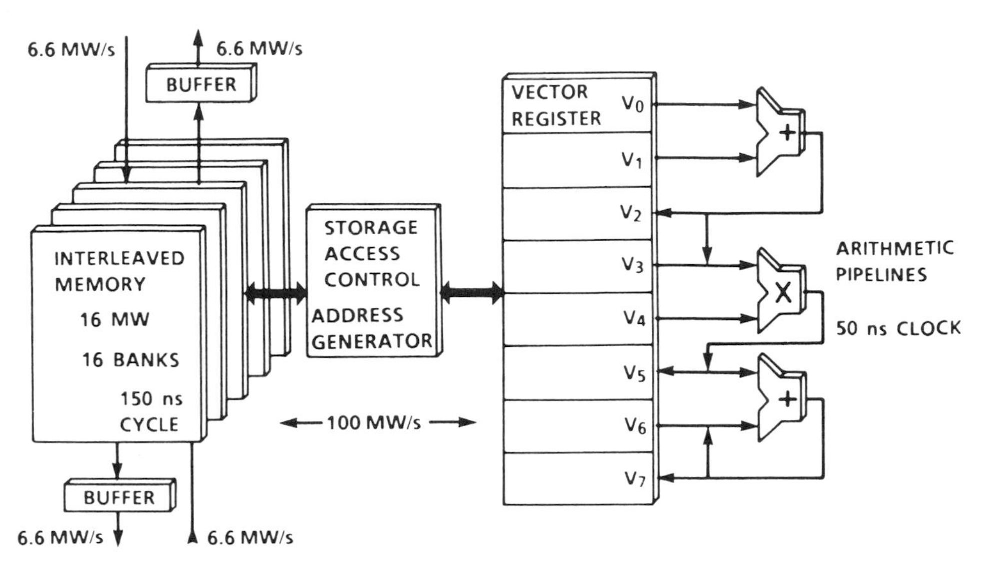

Fig. 4b Proposed vector architecture for each processing element in figure 4a.

purpose vector processors interconnected in one dimension, and a control unit interfacing the processor array to a host computer. Because the system is truly homogeneous, i.e. each module is identical and contains identical connections to other modules, programming is greatly simplified.

Each processor operates asynchronously. Both data and control information can be propagated recursively from either end of the array. Correct sequencing is ensured by triggering instructions at the availability of operands, i.e. the architecture is data driven and self-timed (Kung, 1984). No global communication is thus required to operate the system. However, a single bus would probably be included for service and diagnostic purposes.

Redundancy can be incorporated by equipping each processor with a self-testing system. If a fatal error is detected, a by-pass function can be enabled such that the defective processor acts as a pure communication link. A constant computational resource can then be kept available to the user by incorporating extra processors at the end of the array.

The floating point arithmetic hardware of each processor consists of three pipelined functional units, i.e. two adders and one multiplier. For the kind of applications outlined above, these arithmetic pipelines would be configured to perform chained add-multiply-add or add and multiply or multiply-add separately, as indicated in figure 4b. Rather than relying on novel and unproven technology, the processing hardware would be implemented using standard VLSI building blocks, resulting in a processor that can be economically replicated. In 1987 commercially available floating point arithmetic chips are capable of producing one result every 50 ns. This would give each processor a peak performance of 60 MFLOPS (Million Floating Point Operations per Second). To support this speed, the arithmetic unit would have access to 16 interleaved 1 MW (Million Words) local memory banks, allowing up to 5 memory references per add-multiply-add operation when a read/write cycle time of 150 ns is assumed.

Four word-parallel one-way communication channels link each processor to its two nearest neighbors in the array. These channels operate at the same speed as each memory bank. This gives an interprocessor communication bandwidth which is higher than required by relation (7) to match the arithmetic bandwidth for the worst possible case, i.e. time evolution with $n_3 = 1$ and $L_1 = L_2 = L_3$.

Because the recursive wave calculations are very much compute-bound when the complete wavefield can be kept within the memory of the machine, the bandwidth of the interface to the host computer is not critical. For the current generation of supermini computers we conservatively estimate that a sustained transfer rate of 1 MW/s can be achieved. Loading the elastic earth parameters for each grid point in the time evolution example above would then take roughly 30 s. If 300 processors were used the computations would be completed in about 12 minutes. The underlying computational performance is then close to 18 GFLOP. For the acoustic depth extrapolation example, the time required to load the data would be about 1 ms per recorder channel. The initial temporal Fourier Transform would be performed on-the-fly by accumulating weighted contributions from the data as they propagate down the structure. The raw number crunching would be completed in less than 3 minutes. During the recursive downward continuation the velocity information about the subsurface would have to be supplied at a rate less than 1 word per 100 add-multiply-add operation or less than 0.2 MW/s.

DISCUSSION

Recursive wavefield extrapolation in time or space represents the backbone of modern geophysical imaging techniques. Such algorithms of finite-difference type can be mapped onto any parallel hardware structure supporting nearest neighbor communication in one, two, or three dimensions, but will run efficiently on a large number of processors only when the bandwidths of the interprocessor communication channels are properly dimensioned relative to the arithmetic bandwidth of each processor. A truly powerful parallel computer properly designed to meet this requirement is yet to appear.

At a first glance a one-dimensional architecture would be the least attractive alternative because it requires the most interprocessor communication for a given number of processors. However, this drawback is more than offset by the ability to use fast strictly local interprocessor communication channels and long vectors for arithmetic pipelining. Using state of the art 1987 technology it is possible to produce a one-dimensional computational structure with sufficient interprocessor communication and arithmetic capabilities to outperform current supercomputers by two orders of magnitude on discrete wave calculations. Preliminary investigations indicate that such a machine, specifically designed to perform well on discrete wave calculations, can be developed for about 10 million USD.

Application of this architecture is not limited to finite-difference algorithms. True believers of pseudospectral differentiation may run Fast Fourier Transforms in two directions and use differencing in the third. Many numerical simulation and signal processing problems bear great similarities to the problems described here. Once developed, a parallel computer properly designed to perform well on wave equation computations might therefore soon find a broader scope of application.

Also it would be relatively straightforward to map traditional seismic data processing algorithms such as stack and stack-oriented velocity analysis onto a one-dimensional architecture. For two reasons this has not been considered here. Firstly, these algorithms are not strictly computationally demanding and would be *I/O* bound on most powerful parallel systems. Secondly, they are not based on realistic physical assumptions and break down they are needed mostly, i.e. when the subsurface geology is complex. It is therefore reasonable to believe that these conventional methods will eventually be abandoned when modern depth imaging techniques based on wave propagation become available through proper parallel processing.

APPENDIX I: TIME EVOLUTION OF ELASTIC AND ACOUSTIC WAVES

By combining the equation of motion and Hooke's law for an isotropic elastic solid, we obtain a system of equations as given in (1) with

$$\mathbf{B}_{\mathrm{I}} = [u_1, u_2, u_3, \rho\dot{u}_1, \rho\dot{u}_2, \rho\dot{u}_3]^T \qquad \text{(I1a)}$$

and

$$\mathbf{A}_{\mathrm{I}}=\begin{bmatrix} 0 & 0 & 0 & \rho^{-1} & 0 & 0 \\ 0 & 0 & 0 & 0 & \rho^{-1} & 0 \\ 0 & 0 & 0 & 0 & 0 & \rho^{-1} \\ \partial_1(\lambda+2\mu)\,\partial_1+\partial_2\mu\partial_2+\partial_3\mu\partial_3, & \partial_1\lambda\partial_2+\partial_2\mu\partial_1, & \partial_1\lambda\partial_3+\partial_3\mu\partial_1, & 0 & 0 & 0 \\ \partial_2\lambda\partial_1+\partial_1\mu\partial_2, & \partial_1\mu\partial_1+\partial_2(\lambda+2\mu)\,\partial_2+\partial_3\mu\partial_3, & \partial_2\lambda\partial_3+\partial_3\mu\partial_2, & 0 & 0 & 0 \\ \partial_3\lambda\partial_1+\partial_1\mu\partial_3, & \partial_3\lambda\partial_2+\partial_2\mu\partial_3, & \partial_1\mu\partial_1+\partial_2\mu\partial_2+\partial_3(\lambda+2\mu)\,\partial_3, & 0 & 0 & 0 \end{bmatrix} \tag{I1b}$$

Where u_j, $j=1, 2, 3$ are the components of the displacement vector, $\dot{u}_j$, $j=1, 2, 3$ are the components of the displacements velocity vector, ρ is the density, and λ and μ are the Lame parameters of the particular medium. ∂_j denotes differentiation in the x_j direction. In this formulation the parenthesis on the arguments for the differentiators have been dropped for convenience and it is implicitly assumed that the operator order is from right to left. A similar formulation can be given for the stresses (Kosloff, Reshef and Loewenthal, 1984). For an acoustic (liquid) medium (I1) is replaced by

$$\mathbf{B}_{\mathrm{I}}=[P, \rho^{-1}\dot{P}]^T \tag{I2a}$$

$$\mathbf{A}_{\mathrm{I}}=\begin{bmatrix} 0 & \rho \\ c^2(\partial_1\rho^{-1}\partial_1+\partial_2\rho^{-1}\partial_2+\partial_3\rho^{-1}\partial_3) & 0 \end{bmatrix} \tag{I2b}$$

where P denotes pressure and c is the compressional wave velocity.

APPENDIX II: DEPTH EXTRAPOLATION OF ELASTIC AND ACOUSTIC WAVES

After a Fourier transform with respect to time, the equation of motion and Hooke's law for an isotropic elastic medium can be combined to give a system of equations of the form (4) with

$$\mathbf{B}_{\mathrm{II}}=[s_3, u_1, u_2, u_3, s_1, s_2]^T \tag{II1a}$$

and

$$
\mathrm{A}_{\mathrm{II}} = -\begin{bmatrix}
0 & 0 & 0 & \rho\omega^2 & \partial_1 & \partial_2 \\
0 & 0 & 0 & \partial_1 & -\mu^{-1} & 0 \\
0 & 0 & 0 & \partial_2 & 0 & -\mu^{-1} \\
-(\lambda+2\mu)^{-1} & \lambda\partial_1(\lambda+2\mu)^{-1} & \lambda\partial_2(\lambda+2\mu)^{-1} & 0 & 0 & 0 \\
\partial_1\lambda(\lambda+2\mu)^{-1}, & \rho\omega^2+\partial_1\lambda\mu(\lambda+\mu)(\lambda+2\mu)^{-1}\partial_1+\partial_2\mu\partial_2, & \partial_1 2\mu\lambda(\lambda+2\mu)^{-1}\partial_2+\partial_2\mu\partial_1, & 0 & 0 & 0 \\
\partial_2\lambda(\lambda+2\mu)^{-1}, & \partial_2 2\mu\lambda(\lambda+2\mu)^{-1}\partial_1+\partial_1\mu\partial_2, & \rho\omega^2+\partial_2\lambda\mu(\lambda+\mu)(\lambda+2\mu)^{-1}\partial_2+\partial_1\mu\partial_1, & 0 & 0 & 0
\end{bmatrix}
$$

where u_j are the components of the transformed displacement vector and s_j are the transformed components of the stress tensor δ_{3j}, $j=1, 2, 3$. ω is the temporal frequency. For acoustic waves (II1) is replaced by

$$\mathrm{B}_{\mathrm{II}} = [P, \rho^{-1}\partial P/\partial x_3]^T \tag{II2a}$$

$$\mathrm{A}_{\mathrm{II}} = \begin{bmatrix} 0 & \rho \\ \omega^2\rho^{-1}c^{-2}+\partial_1\rho^{-1}\partial_1+\partial_2\rho^{-1}\partial_2 & 0 \end{bmatrix} \tag{II2b}$$

ACKNOWLEDGMENTS

I wish to thank Lasse Amundsen, Børge Arntsen and Rune Mittet of IKU A/S, and my colleagues Erik Rosncss and Svein Sæther at A/S Informasjonskontroll for many provocative and interesting discussions on seismic inversion and parallel processing respectively. Also I acknowledge Chr. Michelsens Institute for making possible a stay to test some of the concepts outlined here on the Intel Hypercube. Finally, I would like to thank Elmer Eisner of Texaco for his insightful comments and suggestions for an improved final version of this manuscript.

REFERENCES

Berkhout, A. J., 1985, Seismic migration. Imaging of acoustic energy by wavefield extrapolation. A. Theoretical aspects, Elsevier.

Brown, D. L., 1984, A note on the numerical solution of the wave equation with piecewise smooth coefficients, Mathematics of Computation, 42, 369–391.

Claerbout, J. F., 1971, Toward a unified theory of reflector mapping, Geophysics 36, 467–481.

Holberg, O., 1987, Computational aspects of the choice of operator and sampling interval for numerical differentiation in large-scale simulation of wave phenomena, Geophysical Prospecting 35, 629–655.

Holberg, O., 1988, Towards optimum one-way wave propagation, Geophysical Prospecting, 36, 99–114.

Kogge, P. M., 1981, The architecture of pipelined computers, McGraw-Hill.

Kosloff, D. D. and Baysal, E., 1983, Migration with the full acoustic wave equation, Geophysics 48, 677–687.

Kosloff, D., Reshef, M. and Loewenthal, D., 1984, Elastic wave calculations by the Fourier method, Bulletin of the Seismological Society of America 74, 875–891.

Kung, S. Y., 1984, On supercomputing with systolic/wavefront array processors, Proceedings of the IEEE, 2, 867–884.

Levin, S. A., 1984, Principle of reverse time migration, Geophysics 49, 581–583.

Loewenthal, L., Roberson, L. R. and Sherwood, J., 1976, The wave equation applied to migration, Geophysical Prospecting 24, 380–399.

Schneider, W. A., 1978, Integral formulation for migration in two and three dimensions, Geophysics 43, 49–76.

Shubin, G. R., Baker, L. J. and Bell, J. B., 1985, Accuracy of some techniques used in the numerical solution of the wave equation, paper pesented at the 55'th SEG meeting, Washington D.C.

Tarantola, A., 1987, Inverse Problem Theory: methods for data fitting and model parameter estimation, Elsevier.

Tarantola, A., Jobert, G., Trezeguet, D. and Denelle, E., 1988, The nonlinear inversion of seismic waveforms can be performed either by time extrapolation or by depth extrapolation, Geophysical Prospecting 36, 383–416.

Ursin, B., 1983, Review of elastic and electromagnetic wave propagation in horizontally layered media, Geophysics 48, 1063–1081.

CHAPTER 4

ADVANCES IN HIGH PERFORMANCE PROCESSING OF SEISMIC DATA

by
ERNST L. LEISS
Department of Computer Science
Research Computation Laboratory
University of Houston
and
OLIN G. JOHNSON
Department of Computer Science
University of Houston and the
Houston Area Research Center

1. INTRODUCTION

Advances in geophysical processing are dependent on advances in computer hardware and software. Hence, it is important for geophysicists to be aware of research efforts and new products in computer design, *I/O* devices, algorithms, and programs.

Here we survey these areas. Section two addresses advances in hardware. Many research projects in new computer architectures are reviewed. Some of these have already been used successfully in geophysical modeling or processing. *I/O* advances are also covered. Section three addresses software advances in languages and compilers. Section four considers the problems of implementing geophysical applications in these newer systems. The realities and pitfalls of the implementation process are briefly discussed. The subject of in-core programming versus out-of-score programming is considered in some detail. Finally, implementing vector and parallel programming is discussed.

2. HARDWARE ADVANCES

The traditional von Neumann computer consists of a memory, a processor, and a bus between them. Data and instructions are stored in the memory, and the processor controls and performs the computations, that is, it generates addresses for data and instructions, fetches them and computes on data. The bus is the most frequently used component of the system. To avoid a potential bottleneck, von Neumann machines often include a small fast local storage (local memory and/or cache) which is accessed more frequently by the processor.

The von Neumann computer is a control flow computer where the flow of control causes the execution of instructions. Central to the von Neumann machine is the concept of the stored program, the principle that instructions and data are to be stored together intermixed in a single, uniform storage medium rather than separately. The ambiguity of the interpretation of an element in storage is resolved only temporarily when it is fetched and either executed as an instruction or operated on as data. A datum, created as a result of some operations in the ALU (arithmetic logic unit), might possibly be placed in storage as other datum, but then fetched and executed as an instruction either deliberately by program design or by error. Another concept central to the von Neumann machine is the program counter, a register that is used to indicate the location of the next instruction to be executed and which is automatically incremented by each instruction fetch.

2.1 *New Architectures*

The study of architectures that utilize various types of concurrency is motivated by the need to increase the performance of computers. The new machines which will supersede the von Neumann model will have greater performance and may use very large scale integration (VLSI) to implement the concurrent architectures.

The advanced computers studied here have been classified as multiprocessors, dataflow computers, array processors, pipelined computers, supercomputers, systolic arrays, very large instruction word (VLIW) machines, and uniprocessors based on the reduced instruction set computer (RISC) architecture. This classification is based on the mode of execution of the processors, the performance and size of memory, the control mechanism, and any specialized architecture like VLIW and RISC.

2.1.1 *Pipelined Computers*

Pipelining speeds up single-threaded code. Instruction execution is broken into its components (levels) such as instruction fetch, opcode decoding, operand address calculation, operand fetch, and execution, each of which can be executed independently with simultaneous computations on different sets of data. A floating add can be pipelined as follows: sign control, exponent compare, mantissa shift, mantissa add, exponent adjust, and normalization. The EXPRESSION PROCESSOR at University of Washington, PIPE at University of Wisconsin-Madison and TIP from Japan fall in this category.

2.1.2 *Array Processors*

Array processors obtain concurrency by performing identical operations on different portions of data, that is, they are SIMD (single instruction stream, multiple data stream). They act as fast coprocessors which offload many of the repetitive calculations needed in scientific applications. They are connected/controlled by a host. The host provides the mechanisms for communications and control between the array processor and the outside world. It also performs the tasks of data management, compilation, and resource allocation/control functions commonly associated with a general-purpose operating system. Although array processors are high performance machines, they are burdened with several problems. First, structured data that are vectors of irregular strides are difficult to handle because of memory conflicts. Secondly, programs do not consist only of vector instructions. The ADAPTIVE ARRAY PROCESSOR from Japan, PARALLEL IMAGE NEIGHBORHOOD PROCESSOR at University of Missouri, MULTIPLE PARALLEL PROCESSOR at Goodyear Aerospace Corporation, RICE ARRAY PROCESSOR at Rice University, VERY FAST PARALLEL PROCESSOR at Columbia University are some of the current array processor projects.

A binary array processor is a parallel matrix processor in which each processing element is constrained to bit serial operations. A parallel matrix processor is a SIMD machine that has a set of processing elements (PE's) organized as a two-dimensional matrix such that data may only be transferred between adjacent PE's. Data interconnections between PE's are one bit wide. Binary array processors process picture data, conventionally represented by a large two-dimensional array of picture elements called Pixels. BASE at Purdue University and CLIP from England are binary array processors.

The WAVEFRONT ARRAY PROCESSOR at the University of Southern

California is a specialized array processor based on the wavefront concept. The wavefront notion drastically reduces the complexity in the description of parallel algorithms. The mechanism provided for this description is a special-purpose, wavefront-oriented language. Rather than requiring a program for each processor in the array, this language allows the programmer to address an entire front of processors. The wavefront architecture can provide asynchronous waiting capability and consequently can cope with timing uncertainties such as local clocking, random delay in communications, and fluctuations of computing times. In short, the wavefront notion lends itself to a (asynchronous) dataflow computing structure that conforms well with the constraints of VLSI. The integration of the wavefront concept, the wavefront language, and the wavefront architecture leads to a programmable computer network called the wavefront array processor (WAP). The WAP is in a sense an optional trade off between the globally synchronized and dedicated systolic array and the general-purpose dataflow multiprocessor. It is mainly aimed at incorporating the vast VLSI computational capability into modern signal processing applications.

2.1.3 *Dataflow Computers*

In a dataflow computer the availability of input operands triggers the execution of the instruction which consumes the inputs. It is associated with single-assignment languages in which data flows from one statement to another, execution of statements is data-driven and identifiers obey the single-assignment rule. A node is said to be firable (enabled) if a token arrives on each of the incoming arcs representing the necessary operands for the node, and if no tokens are present on the outgoing arcs where the resulting tokens are to be emitted. To hold the database of a large scale computation, the dataflow computer has array memories. The processing elements consist of two kinds of units—cell blocks and functional units. Cell blocks hold the instructions and perform the basic function of recognizing which instructions are ready for execution. The functional units perform the execution of enabled instructions.

Dataflow machines can be static or dynamic (tagged), based on the method by which they pass tokens from node to node. A static dataflow machine allows only one token on an arc at a time. A program, as stored in the computer's memory, consists of instructions linked together. Each instruction has an operation code, spaces for holding operand values as they arrive, and destination fields that indicate what is to be done with the results of instruction execution. The routing network provides pathways needed to send result packets to instructions residing in other processing elements. If a processor has many independent activities waiting for its attention, then delay can be tolerated in the interconnection network. MULTIUSER DATAFLOW MACHINE from Canada, DENNIS DATAFLOW MACHINE at MIT, DATA DRIVEN MACHINE #1 at the University of Utah, CHICAGO DATAFLOW MACHINE at the University of Chicago, HUGHES DATAFLOW MULTIPROCESSOR at Hughes Aircraft Company, IRVINE DATAFLOW MACHINE at University of California, Irvine, PIECEWISE DATAFLOW MACHINE at Lawrence Livermore National Laboratory are some of the static dataflow projects.

In a dynamic dataflow computer, multiple tokens on an arc at a time are allowed. Tokens carry distinguishing tags which identify their individual context. This method allows for maximum parallelism in execution of programs. ARVIND DATAFLOW MACHINE at MIT, DATAFLOW COMPUTER FOR SIGNAL PROCESSING at the University of North Carolina, MANCHESTER DATAFLOW MACHINE from England, PROGRAMMABLE MODULAR SIGNAL PROCESSOR at RCA Government Systems Division are some of the dynamic dataflow projects.

2.1.4 *Multiprocessors*

Most of the present architecture research projects are multiprocessors, either shared-memory or message-passing. Multiprocessors use several processors (homogeneous or heterogeneous) concurrently to solve one or more problems. The early development of multiprocessor hardware and the operating systems necessary to make it effective in applications were largely oriented toward increased system throughput over single processor systems. They have the most flexible computer architecture in exploiting arbitrarily structured parallelism. Multiprocessor systems have multiple instruction streams over a set of interactive processors with shared resources such as memories and databases of autonomous processors with no shared resources, but with an inter-processor communication network. Multiprocessors offer another dimension of parallelism, namely multitasking (capability of a system to support two or more active tasks simultaneously) in addition to vectorization (the process of replacing sequential code by vector instructions). They are mainly two types of multiprocessors, shared-memory and message-passing.

In the shared-memory model, the data is in preallocated locations in the shared-memory where it can be accessed by each processor and operated upon without interruptions from other processors. These machines are structured with a switching network, either a crossbar connection of buses or a multistage network between processors and memory. Processor-memory communication can also be via a multiported memory. An interleaved memory is very suitable for shared-memory multiprocessors to avoid some of the memory contentions. Communication between processes running concurrently in different processors occurs through shared variables and common access to one large address space. An advantage of shared-memory multiprocessors is the memory space saving since one copy of the operating system suffices. There is a limit on the number of processors in a shared-memory multiprocessor due to the memory contentions that increase with an increasing number of processors. Some of the shared-memory multiprocessor projects are BUTTERFLY at Bolt, Beranek, and Newman, CEDAR at the University of Illinois, at Urbana-Champaign, CM* and C.MMP at Carnegie-Mellon University, CONCERT at MIT (Massachusetts Institute of Technology), HOMOGENEOUS MULTIPROCESSOR from Canada, GIGA COMPUTER at Argonne National Laboratory, MIDAS at the University of California at Berkeley, PUMPS at Purdue University and Rice University, REMPS at the University of Southern California, TAMIPS from Japan, TRAC at the University of Texas at Austin, and ULTRA at New York University. CEDAR has processor clusters where a processor can access its own local memory or the local memory of other processors in the cluster. CEDAR combines the control mechanism of dataflow architecture and the storage mechanism of von Neumann machines. DIRECT, a multiprocessor developed at the University of Wisconsin has an associative memory. An associative memory is a content addressable storage, that is, cells in memory are addressed not by location, but by content. TRAC has a special property called varistructurability which means that an n-byte operand can be processed by one or more byte-wide processors. The opcode that directs these operations must be independent of the physical structure of the machine.

The message-passing multiprocessors do not have any globally shared memory. Each processor has a local memory and an interprocessor connection network. The advantage of the message-passing model is that data is passed only once through the connection network while two passes (write and read) are needed for the shared-memory model unless the data is in the local storage. Yet another advantage is that for data-driven computation, data is passed through the network at generation time and not when it is needed. Thus longer delays through the network can be tolerated in the case when data is not used immediately after its generation.

These machines can have a very large number of processors, thus potentially having a very high performance. Message-passing multiprocessors are difficult to program since a programmer must know the code executed by each processor in order to pass the data between processors correctly. Some of the message-passing multiprocessor projects are CHIP at the University of Washington and Purdue University, CONNECTION MACHINE at MIT and Thinking Machines, Inc., COSMIC CUBE at California Institute of Technology, DADO at Columbia University, DON from Japan, MANIP at Purdue University, MU6V from England, and ZMOB at the University of Maryland. PASM is a message-passing multiprocessor at Purdue University with a partitionable SIMD/MIMD architecture. A partitionable SIMD/MIMD system is a parallel processing system which can be structured as one or more independent SIMD and/or MIMD machines of various sizes. FAIM-1 at Fairchild Laboratory for Artificial Intelligence has a number of processors where each processor is a fanatically reduced instruction set computer (FRISC). FRISC supports low-level symbol processing in ways similar to uniprocessor Lisp-Machines: tagged-memory architecture, stack caches, and a tailored instruction set.

The WAFER SCALE INTEGRATED MULTIPROCESSOR at the University of Illinois at Urbana-Champaign has the multiprocessor placed on a wafer. A wafer scale integrated multiprocessor is a macro-circuit consisting of a rectangular array of interconnected modules arranged on a large piece of silicon. Each of these modules could be as complex as the very large scale integrated (VLSI) multiprocessor. These modules are not separately manufactured, tested and then assembled as VLSI chips are. They are fabricated as a single unit, the VLSI wafer.

RP3 at IBM, T. J. Watson Research Center, CHOPP at Columbia University, HM2P at Rennsselaer Polytechnic Institute, MULTI PROCESSOR/COMPUTER at Princeton University have a organizational duality of shared-memory multiprocessors and message-passing multiprocessors. They incorporate the advantages of both models and hence serve more applications. ULTRA and RP3 have a special switch feature called combining. In this process, memory requests aimed at the same memory location are combined into one request at the switch they are passing by.

FFPM at the University of North Carolina, MULTIPROCESSOR REDUCTION MACHINE from England, SERFRE from France, REDIFLOW at the University of Utah are all Reduction multiprocessors. In a reduction computer, the requirement for a result triggers the execution of the instruction that will generate the value. It is associated with applicative (reduction of functional) languages. The reduction computer maps the functional language expressions onto hardware storage dynamically. This is a machine-wide process which involves interrupting computations in the machine, determining where resources are available or needed, and finally redistributing the available resources. The reduction languages attempt to relieve the programming problems, such as explicitly specifying flows of control and managing memory cells, normally associated with conventional computers. The style of programming is strictly functional, based on a few elementary mathematical constructs featuring a binary tree structure, from which complex expressions are built up by recursive application.

2.1.5 *Supercomputers*

Supercomputers are computers with colossal computational speeds, large memory, and high cost. Based on today's technology, a computer is considered to be a supercomputer if it can perform hundreds of millions of floating point operations per second (100 MFlops) with a word length of approximately 64 bits and a main memory capacity of millions of bytes. Supercomputers are structured in three architectural classes: pipelined computers, array processors, and multi-processors. A supercomputer is implemented using the fastest and most sophisticated circuits available and it is also architecturally balanced for the highest economy of throughput. A supercomputer's usefulness is not entirely determined by its hardware capabilities. In fact, the efficiency relies to a large extent on the availability of "super-software" that is easy to use and can obtain maximum parallelism from the hardware. The process of replacing a block of sequential code by a few vector instructions is called vectorization. The portion of the compiler that regenerates this parallelism is known as vectorizer. A vectorizing compiler regenerates the parallelism lost by using sequential languages. NON-VON at Columbia University, EMSY from the Federal Republic of Germany, NAVIER-STOKES COMPUTER at Princeton University, GF11 at IBM, T. J. Watson Research Center, PAX from Japan, S1 at Lawrence Livermore National Laboratory and Stanford University are some of the newer supercomputer projects being pursued. All these are message-passing multiprocessors. They have a very large number (up to 1,000,000) of processors communicating via an efficient communication network. NAVIER-STOKES COMPUTER is expected to have a speed of 60 GFlops and PAX, a speed of 100 GFlops. Commercial supercomputers include the Cray X-MP, Cray 2, NEC SX Series, ETA-10, Fujitsu Facom Series and Hitachi 800 Series.

2.1.6 *Systolic Arrays*

WARP at Carnegie-Mellon University and GE and SYSTOLIC PROCESSOR at ESL Incorporated (TRW subsidiaries) are systolic array projects. The systolic array is an array of processing elements (cells) of the same type, except that the boundary cells may be different. Simultaneous computations that are short and execute synchronously are said to be systolic. Every processor pumps data in and out, each time performing some short computation, so that a regular flow of data is kept up in the network. Communication is between adjacent processing elements and external communication is via the boundary processing elements. Processors are attached to a host. The systolic array processor executes computation intensive, but regular routines, and the host runs the main application programs. The cells are programmable so that the processor array can implement different algorithms. Each data item can be used a number of times once it is accessed, and thus, a high computation throughput can be achieved with only modest bandwidth. These processors are especially suited to algorithms with regular data movement patterns.

2.1.7 *Very Large Instruction Word (VLIW) Machines*

ELI-512, designed at Yale University, is a Very Large Instruction Word (VLIW) machine. VLIW machines are highly parallel architectures that offer an alternative to multiprocessors and array processors. They resemble ordinary multiprocessors but have a tightly coupled, single-flow control mechanism. Programs for VLIWs must specify fine-grained hardware control. It is impossible to hand code VLIW machines. VLIW machines have one central control unit issuing a single wide instruction per cycle. Each wide instruction consists of many independent operations. Each operand requires a small, statically predictable number of cycles to execute. Operations are pipelined. The underlying sequential architecture is invariably a reduced instruction set computer. The instructions in the underlying RISC-level are called operations, while the term instruction is reserved for the very long instruction words, which are collections of operations. The instructions are in a single flow of control. Thus a single long instruction word is fetched, and all the processors do their individual operations. The operations differ for the various processors. After an instruction is executed, the next instruction is chosen and fetched. The instruction word completely controls all communications among the processors. Data transfers and their timings are completely choreographed in the code. Compaction is the process of generating very long instructions from some sequential source. A compacting compiler is a compiler that takes some sequential high-level source and generates compacted code. A compiler (Bulldog) exists (at Yale) that can product highly parallel code from a broad range of ordinary sequential programs. This compiler uses a technique called Trace Scheduling. Trace scheduling is a complex procedure. To handle conditional jumps in a program, a trace scheduling compiler uses information about the dynamic behavior of the program to do greedy scheduling of operations. The compiler can make good guesses when jumps are weighed heavily towards one leg—because in this case it is productive to be greedy. Otherwise VLIWs are probably the wrong architecture to use.

2.1.8 *Reduced Instruction Set Computer (RISC) Uniprocessors*

RISC at University of California at Berkeley and MIPS at Stanford University are uniprocessors based on a Reduced Instruction Set Computer (RISC) architecture. RISC architecture features a simple, regular instruction set which allows a combination of instructions to be executed faster than the equivalent complex instructions. A traditional complex instruction set computer relies on hundreds of specialized instructions, dozens of addressing modes, and several high-level languages implemented in hardware. In such a computer the compiler must consider the many possibilities inherent in a complex instruction and perform a number of memory transfers to execute it. This requires identifying the ideal addressing mode and the shortest instruction format to add the operands in memory. Yet only a small number of instruction types takes up most of a computer's execution time. Load, call and branch instructions are found in compiled code more often than any other instruction type. Complex operations can actually be executed faster by breaking each one down into a series of simple instructions that move data between registers and memory. This is the principle behind the RISC approach. Some salient features of a RISC-based machine are register to register operations that allow optimization of compilers through reuse of operands with instruction formats, and addressing modes that permit instructions to be decoded in a single-machine cycle. Memory reference instructions consisting of load and store operations are also typical. A RISC machine has a high performance memory hierarchy including general purpose register and cache. One of the advantages of the RISC approach is the potential to reuse any result without computing it.

2.2 *I/O Advances*

Seismic processing is indisputably one of the most data intensive applications to be found. Western Geophysical often claimed that its tape library was second only to that of the U.S. government in size. Data collection, processing and storage is thus a matter of considerable importance. Clearly, a computer with the fastest of processors is unequal to the task of commercial seismic processing if its *I/O* components are inadequate. The seismic industry has not been just a consumer of *I/O* devices. It has, instead, been a primary motivating force in the development of new devices. It has long been standard operating procedure for *I/O* manufacturers to arrange early experiments and tests of their equipment in a seismic environment.

I/O advances have occurred in many types of hardware: channels, cartridge tapes, optical disks, hyperdisks, solid state devices, rasterizers, plotters and CRT graphic displays. It is possible only to summarize the latest status of these types of devices without large chapters of technical detail.

CHANNELS

It should be mentioned that mainframes and supercomputers use channels whereas minicomputers use busses. The essential difference in these is that a bus handles all data traffic between units of a computer system whereas channels handle only the traffic to and from specific *I/O* controllers and memory. The standard channel speed over the past several years for IBM-like systems has been three Mbytes/sec with a maximum of 32 channels. Recently, IBM, Amdahl and others have announced 4.5 Mbyte channels. CDC-like systems (CDC, Cray, ETA) have allowed only 16 channels but at essentially twice the speed.

Cray pioneered the development of 100 Mbyte channels between memory and *I/O* subsystems which in effect are computers in their own right. There are some similarities in this idea with the earlier "directly coupled system" developed by IBM for NASA. The *I/O* subsystems in turn have special channels for high performance disk units, "hyperdisks," such as the Ibis and Hydra drives. Cray also developed a 1.25 Gbyte channel for data transfers between its Solid State Device (SSD) and memory on its X-MP series. The following figure shows relative speeds in Mbytes/sec for the various data paths in a typical modern supercomputer.

TAPE	1.25	CHANNEL	1.8	MEMORY	11000	CPU
DISK	3.0	CHANNEL	3.0	MEMORY	11000	CPU
HYPERDISK	10.0	SUBSYSTEM	100	MEMORY	11000	CPU
SSD	1300	CHANNEL	1300	MEMORY	11000	CPU

Tape channels, though slower and cheaper than disk channels, are usually rated at a higher speed than the tapes themselves. Perhaps faster tapes are to be expected shortly.

Computers of different vendors can also be connected by high speed devices such as Network System's HYPERchannel and CDC's Loosely Coupled Network which operate at 50 Mbits/sec (6.25 Mbytes/sec or less).

By comparison, the speed of a DEC Unibus is essentially 1 Mbyte/sec and an Ethernet is 10 Mbits/sec (1.25 Mbytes/sec or less). Wide area networks operate at 56 Kbits/sec and user terminals at no more than 19.2 Kbits/sec. A few networks now operate at T1 speeds of 1.54 Mbit/sec.

TAPES

Tape advances have not shown the same magnitude in improvements as one finds in computations. The following table summarizes the relative performance rates at the beginning of each of the last three decades in tape technology and in computational performance.

TABLE 1.

Year	Tape Speed	Tape Density	MIPS
	(in/sec)	(bpi)	
1960	75	800	1
1970	125	1600	20
1980	200	6250	200

Thus, tapes are 20 times as fast whereas computers are 200 times as fast. The present decade has widened this difference with computers operating at one gigaflop (approximately the equivalent of 3000 mips) with no substantial improvement in tape *I/O*. Fortunately, the computing is more sophisticated now, with more arithmetic per unit *I/O*.

The recent cartridge tapes represent improvements in the handling and storage of tape archives. Not only do they load automatically and are smaller but also they can store up to 3 Gbytes of data which rivals the capacity of the optical disks.

OPTICAL DISKS

Optical storage technology is gradually becoming more important. Its characteristics make it an interesting alternative to conventional magnetic storage technology, especially magnetic tape.

Optical storage was first used commercially for video and audio compact disks. Whereas in magnetic medium, information is recorded and read by changing magnetic properties, optical storage technology uses tiny solid-state lasers to create (write) and sense (read) microscopic pits in the disk's surface. Typically, the disk is coated with a reflective material; writing then consists of burning a pit into that surface material using the laser at a higher power setting, while reading is done by measuring the reflectivity of a particular position. Thus, high reflectivity (no pit) might represent a 0 and low reflectivity (pit) at 1. This set-up is the basis for all of the currenly (1987) commercially available laser disks; it follows from this that information can be recorded only once, but read many times, given rise to the acronym WORM ("write once, read many"). This indicates the major disadvantage of current optical storage technology: it is generally not possible to change information stored on such a laser disk.

(Strictly speaking, this is not quite true; if one uses certain non-standard codes to record information, a certain number of changes of information recorded on a WORM laser disk is possible. For a discussion of this issue and how to guarantee that such changes can be prevented, see [LEISS84]. However, since this would require changes in the recording software and firmware, this possibility is ignored here.)

The ability to rewrite information seems crucial, mainly because one is accustomed to it. However, upon examining the requirements of seismic data storage (as well as those of many other types of information), it should be obvious that the WORM medium laser disk is quite acceptable, especially since it has several interesting features that are quite attractive for storage of seismic data:

1. *Permanence and Robustness:* Compared with magnetic media, information

stored on laser disks is far less affected by environmental factors. A laser disk can be removed and stored much like a magnetic tape but unlike a magnetic disk. Magnetic fields, heat, humidity, and within limits dust do not affect a laser disk that is stored for long periods of time in an office or a warehouse. Magnetic tape on the other hand must be stored in a very controlled environment if it is to survive reliably for even only five years.

2. *Information Density:* Because information is optically recorded, the information density is significantly higher than that of magnetic media. For example, a single one of the ubiquitous audio compact disks holds 540 Megabytes or 4.32 Gigabits of information (about 300,000 pages of double-space copy). Kodak recently introduced a system that stores one trillion bytes (8 Terabits) on four 14-inch disks [HECH87].

3. *Elimination of Head Crashes:* The technical set-up allows a distance on the order of millimeters between head and disk; thus the dreaded head crashes of magnetic storage media, where distance is one order of magnitude smaller, is eliminated. (Head crashes occur when dust particles are caught between the head and the disk surface; they destroy the disk and the head, but even more damaging, they irretrievably erase the data. They can be avoided by keeping the environment dust free).

4. *Fast Access:* Compared with magnetic tape, which is perhaps the most comparable storage medium, laser disks provide much faster access to individual portions of the data. This is due to the fact that laser disks allow direct access to tracks simply by moving the read/write head. In this, they behave just like magnetic disks. Magnetic tape on the other hand provides only sequential access.

5. *Removability:* Laser disks containing sensitive data can be removed from the disk drives; they are moreover small enough to fit into safes.

There are other advantages that are not directly relevant to seismic data storage, in particular the fact that prerecorded compact disks are cheap to mass produce. It can cost between $3000 and $5000 to create a master disk of a conventional audio compact disk, but copies from it can be manufactured for less than $5 per copy [MATT87]. Encyclopedias are already being distributed in this way.

Among the current main players in laser disks (for information storage for use with computers) are Laser Magnetic Storage Technology (LMS) (a joint venture between N.V. Philips (Netherlands) and Control Data (Colorado), Kodak, and Toshiba). A significant number of companies are also manufacturing laser disk drives for personal computers and workstations, with prices for the drives starting around $2500 and the 5 1/4 inch disks costing on the order of $100 [HECH87].

Erasable optical disks have been announced every year since at least 1984, always for the next year. They are expected to use a magneto-optic technology whereby a laser is used to change the configuration of a magnetic field on the recording surface [MATT87]. The major problem so far seems that the number of phase changes (changes of the structure of the alloy on the recording surface) that the materials permit is not high enough to yield truly erasable laser disks. Another problem is related to the information density that can be achieved in this way. At present (1987), no erasable optical disks are commercially available [HECH87].

For these reasons, we expect laser disks of WORM type to be phased in gradually and in some cases to replace magnetic tapes for the storage of seismic data. While technologically laser disks are superior to magnetic tape, the large investment in both magnetic tape drives and even more so in magnetic tapes (all of which would have to be copied to laser disks, were one to change over completely to optical storage), will slow this development.

HYPERDISKS

The standard high performance disks for the CDC and Cray systems have been manufactured by CDC. The DD-29 series transfers data at 4 Mbytes/sec and has a capacity of .6 Gbytes. The newer DD-49 series has a speed of 10 Mbytes/sec and a capacity of 1.2 Gbytes.

Since 1982, Ibis Systems of Westlake, California has produced a parallel-transfer disk drive made with a proprietary 14-inch thin film medium. Its first product, the Model 1400, has a 12 Mbyte/sec data transfer rate and a 1.4 Gbyte storage capacity. In order to make these disks useful to industry in general, Ibis has developed two industry standard interfaces, Ibis-I and Ibis-II. Both of these interfaces satisfy the requirements of the Intelligent Standard Interface (ISI). Ibis has shipped over 1000 of these units to Cray, its single largest customer.

In order to use these disks even more effectively than simply relying on their inherent speed, the concept of disk striping has arisen. In this technique, sequential elements of a file are divided into small groups so that one group occupies one track of a disk. Sequential groups are stored across the disk units so that several groups can be read in parallel. Using a multidimensional variation of this technique along with other programming techniques Lhemann [LHEM85] was able to convert an *I/O* bound three dimensional migration algorithm into a compute bound program.

RASTERIZERS AND PLOTTERS

Rasterizers, such as the Houston Scientific HSR series, are hardware devices which convert pictures stored in the form of vector move draw files into display files called rasters. In these rasters, each pixel is represented by as little as one bit of data up to several bytes. Often there is one byte for black and white rasters and up to three for color. Seismic software vendors are split as to whether it is better to rasterize with the software of a supercomputer or to use the rasterizer boxes and be tied to one vendor. It is now common practice to provide both alternatives and let the user select.

3. ADVANCES IN SOFTWARE

3.1 *Languages and Extensions*

Fortran remains the most commonly used programming language for scientific computing. While other languages are being used (Pascal, C, Ada), they should not present major challenges to Fortran's domination (strangle-hold?) on this field for the near future. Of importance however, is the fact that Cray seems intent to phase in UNIX as main operating system; this should give C a certain advantage. The emphasis placed by the US Department of Defense (DoD) on Ada does not seem to be shared by the manufacturers of high-performance computing equipment nor their software suppliers, mainly because DoD has not (yet) materialized as a major buyer. On the other hand, the proposed Fortran Standard, hopefully called Fortran 8X (the X to be replaced by either 8 or 9—this is where the hope come in: if final adoption does not take place in this decade, it will be Fortran 9X!), will incorporate certain language features that will aid in utilizing vector, and to a lesser extent, parallel, computers. Fortran is highly suitable for vector processing because its main program structure is DO-loop, and this is precisely the construct that vectorizes best automatically. The proposed SEG seismic subroutines (Seismic Subroutine Standard) are basically a library of subroutines which facilitates seismic processing; they are formulated language-independently but are clearly aimed at Fortran. Fortran however, although excellent for vectorization, is a poor vehicle for parallel computations. For this reason, various languages have been designed with the aim of facilitating the use of parallelism that is available in the hardware; they enable the programmer to control parallelism explicitly. None of them however has reached a level of acceptance that promises significant prospects for becoming a standard (or even only dominating).

3.2 *Compilers*

There are two kinds of compilers of interest, compilers that automatically produce *vectorized* code (V-compilers) and compilers that automatically produce *parallelized* code (P-compilers). In both cases, the source program is written in some standard language, usually Fortran. V-compilers have been in use for a number of years; they are the major reason for the roaring success of vector computers. Their main advantage is that they automatically transform standard language into vectorized code, with relatively little programmer interaction. Initially (six to eight years ago), V-compilers were rather simple-minded and primitive; now, there are fairly sophisticated V-compilers available for all major machines which approach reasonably well hand vectorization and are therefore highly cost-effective. Vectorization is the alpha and the omega of seismic processing and will remain so for quite some time.

P-compilers (compilers that automatically detect parallelism and generate code to take advantage of this) are an entirely different story. To date, most of the parallelization must be done by hand; in other words the programmer must explicitly code for parallelism. Automatic parallelization to date is limited to individual loops [FERR85]; parallelism at a higher language construct level must still be specified by the programmer [KARP87]. Several projects, in academia and in industry, are under way, but the problem of detecting inherent parallelism in a program is substantially more difficult than vectorization. Even a rather primitive P-compiler is still relatively far away. On the other hand, it is questionable whether parallel computer systems will ever by variable without a reasonably smart P-compiler; the cost of recoding existing application programs for parallelism by hand is simply too high.

4. IMPLEMENTATION: REALITIES AND PITFALLS

Problems in seismic data processing are characterized by huge data sets, occurring both as input and as output. For example, a 3D migration program may have as input a data set consisting of 240 traces on 240 lines, with each trace containing 3000 samples (SALNOR7; see Nelson, 1982). Consequently, the input file contains 172.8 million numbers; if each number (word) has 32 bits, the input file is of size 5.5 Gigabits, with the output file being of the same order of magnitude. Therefore, processing realistic seismic data sets is very likely to at least severely strain, if not exceed the capacity of most current computer systems.

Three issues are of major importance in this context:

- The amount of primary or main memory available for processing
- The availability of vector processing
- The possibility of utilizing parallelism, especially macro parallelism.

In the following sections, we discuss each of these issues and outline their implications for the present and the future of seismic data processing.

4.1 *In-Core and Out-of-Core Programming*

A program whose data in their entirely can be read into main memory from secondary storage devices (disks, tapes) is called in-core. In contrast, an out-of-core program requires that the operations performed by the program be grouped together into program parts in such a way that the data set can be partitioned into subsets with the following properties:

- Each subset fits into the available main memory
- The operations in one program part require only the data in the corresponding data subset.

Therefore, at different times during the execution of the program, different data subsets will reside in main memory.

With the exception of the Cray 2, currently available computer systems are unable to accommodate in main memory data sets of size in excess of 5 Gigabits; therefore in-core programs are not feasible. This leaves two alternatives, namely out-of-core programming and virtual memory management.

A virtual memory environment provides automatic paging; this means that the data set is uniformly subdivided into relatively small portions (in the VAX, 512 words), called pages. These pages initially reside on disk. Whenever a data item is needed during execution, the operating system determines automatically in which page the item resides and reads that page from disk into main memory. While this is done, the program waits. The retrieval of a page from disk may require two orders of magnitude (or more) more time than the operation that is eventually performed on the requested item. Since the number of pages that fit into main memory is limited, the request for another page may necessitate the removal of a page currently in main memory. Also, the same page may have to be retrieved again, even if a different data item is requested, because many different items reside in the same page. If the page has been removed in the meantime, it will have to be read from disk again in this case. As an illustration consider the following two functionally identical Fortran loops:

DO 10 $I = 1{,}512$	DO 10 $J = 1{,}512$
DO 20 $J = 1{,}512$	DO 20 $I = 1{,}512$
$A(I, J) = B(I, J) + C(I, J)$	$A(I, J) = B(I, J) + C(I, J)$
20 CONTINUE	20 CONTINUE
10 CONTINUE	10 CONTINUE
Loops (L1)	Loops (L2)

If we assume that 512 array elements fit into one page, then (L1) performs over a quarter of a million page retrievals, whereas in (L2) only 512 page retrievals are necessary because arrays in Fortran are stored in columns. Running the two programs on a VAX-11/780 yields the following timings: (L1) requires 293 sec, (L2) requires 9 sec.

Virtual memory is not at all the same as out-of-core programming: in an out-of-core version, the emphasis is at least as much on partitioning the operations of the program as it is on partitioning the data; in fact the two have to be very well coordinated. In a virtual memory environment, no attention is paid at all to the partitioning of the operations, and as the example above shows, vastly different data transfer requirements and consequently vastly different timings may result.

In a virtual memory environment the programmer is less able to control precisely the flow of input and output; this may result in inefficient use of the computer resources. For this reason, virtual memory has not been preferred for high-performance data processing. Indeed, supercomputers such as the Cray systems at present do not support virtual memory management; instead the programmer is required to partition data and operations explicitly. This results in a tradeoff between savings in computer resources (at the cost of additional programmer effort) and savings in people resources (at the cost of computer time).

At present, out-of-core programming is still necessary in realistic seismic processing. To give a concrete example of the amount of computer time that can be saved by intelligently restructuring data and instructions coordinately, consider an implementation of the 3D Phase Shift migration of the SALNOR7 model on the Cray X-MP [LHEM85]. A perfectly competent initial implementation has an estimated CPU time of 130 sec for lines of 256 traces, each trace with 2048 samples; however, closer inspection indicated that the *I/O* waiting time (the time the program spends in waiting unproductively for requested data to be transferred) was approximately 2800 sec! This was due to the fact the initial implementation required the transfer of approximately 4 million disk sectors (similar to a page). Restructuring the algorithm resulted in the same CPU time, but the number of disk sectors that had to be transferred was now reduced to 250,000, resulting in an *I/O* waiting time of only 175 sec.

In general, a careful analysis of the data transfers should be made, with special emphasis on the fact that items occur in blocks (sectors, pages) and that it is the block which contains an item that is transferred, not the individual item. As a rule of thumb, any program requiring that items (i.e., the blocks that contain them) be transferred more than once from secondary storage to main memory or more than once from main memory to secondary storage must be considered a candidate for restructuring.

Several supercomputers have superfast large secondary storage (e.g., the Cray X-MP has the SSD—Solid-State Storage Device; the NEC SX has the XMU—Extended Memory Unit). This storage is typically significantly larger than the main memory and access time to it is much shorter than that to disk. The intent is to store all data required for the program in that storage (from disk or tape) and then use it, instead of the disk or tape, as secondary storage medium. While the access time to this superfast secondary storage is less than that to disk, a data transfer analysis is still advisable since access timings and type of access are still closer to those of disk than of main memory. (Clearly, the transfer from disk or tape to this device should occur only once; similarly for the transfer to disk or tape).

4.2 *Vector Processing*

Vector processing is currently the mainstay of all serious seismic data processing. This is due to the following observation:

Any Fortran program that:

- uses large amounts of memory,
- has large input and output data sets, and
- performs at least 10^{12} operations

can be vectorized with a rather modest amount of effort, to such an extent that a speed-up of at least one order of magnitude is achieved.

Speed-up is defined as the CPU-time of the scalar version divided by the CPU-time of the vectorized version (everything else unchanged). Modest amount of effort means 5 % or less of the time required to develop the (scalar version of the) program. Indeed with today's vectorizers it is possible to submit a scalar version of a (Fortran 77) program and obtain a program that is substantially vectorized; for certain vectorizers (Convex Fortran Vectorizing Compiler), it is claimed that the resulting code approaches 90 % efficiently of hand-coded vector code. Moreover, those parts that can not be vectorized by the software tool can be flagged so that the programmer may attempt to restructure the code according to well understood rules. There are "catalogues" of these rules which can be applied without great difficulty.

To give a concrete example, a 2D PSPI algorithm was run based on that described in [MAJO86] where the velocity varies only in the x-direction, from 4000 ft/sec to 5800 ft/sec at the midpoint and then back to 4000 ft/sec (linearly). The synthetic time section consists of a row of 1's at the 10th row; the size is 512×512. This program was run in two versions on a VAX-11/780, one version using the VAX alone, with the FFTs in scalar mode, the other version using one FPS 100 as vector processor. The vector processor was only used for the FFTs involved in the vectorized PSPI version, the remainder of that program was unchanged, i.e., not vectorized. The *I/O* waiting times are identical for the two versions, but the CPU timings are not: the scalar version took approximately 42,670 sec (11:51:09.15), whereas the vectorized version took about 2670 sec (0:44:27.38). Consequently, the speed-up obtained by using a library routine that uses the FPS 100 for the FFTs only is 16! This clearly constitutes a significant performance increase at a rather modest increase in cost.

4.3 *Parallelism*

At the hardware level, parallelism denotes the presence of several processors, each with its own instruction stream and under its own control. Each processor may use a shared memory (common memory) and/or have its own private memory. Since there are several independent agents, provisions must exist for the communication between processors. This may be achieved through common memory or by message passing. In the former case, the system is called tightly-coupled (an example is the Cray X-MP/4 where up to four processors use the same large main memory), in the latter case the system is called loosely-coupled (an example is provided by the Intel Hypercube). The underlying idea is to provide N processors and thereby to achieve a speed-up of N; this is clearly also the theoretical upper bound on any speed-up.

In contrast to vector processing where one vector instruction acts on many data items, in parallel systems each processor executes independently. Therefore, in contrast to vector processing, where most of the vectorization is done automatically, in order to exploit parallelism efficiently one must specify explicitly which portion of the program is to be executed on which processor using which portion of the data. The software tools (called vectorizers) that allow the user to submit scalar code and perform the rewriting necessary to utilize the vector capabilities of the target machine do not exist yet for automatically parallelizing code. In addition, some questions have been raised as to whether the currently available loosely coupled systems are suitable for processing seismic data because of their limitations on interprocessor communication and *I/O* [KAOL87].

Implementations on the Cray X-MP/4 of migration algorithms such as PSPI [AMES87] and finite difference methods [TERK87] indicate that a speed-up of 3.5 is quite attainable; this is close to the theoretical upper bound of 4. However, four processors are still manageable for the programmer so that the code for these applications can be carefully hand-coded. For more processors, we would expect the actual speed-up to be significantly less than 80 % of the theoretical upper bound. Also unclear is how one might achieve similar results automatically, i.e., with a software tool akin to a vectorizer.

At the present time, loosely-coupled systems do not appear competitive for production processing of seismic data. No software that would automatically parallelize uniprocessor code is commercially available. The lack of parallelizers is particularly damaging because debugging parallel code is significantly harder than debugging uniprocessor code. The existing processing software, almost exclusively written in Fortran (unless a lower-level language is used), is written for uniprocessors and will not be allowed to become obsolete with the arrival of new processing hardware. Fortran is a poor vehicle for parallel programming (in contrast to vectorizing, for which it is very well suited since the only data structure it supports is the array).

Proposals have been advanced of systems that are specifically designed for seismic processing but do not serve any other purpose. For example, it is technologically feasible to design and manufacture a chip for migration. It is safe to expect that a chip can be designed that will beat any software implementation of migration. There are however two major problems with this approach. One is obviously cost—since the market for such a system is quite restricted, the development cost per sold unit might be prohibitive. Also, such a system would severely stifle work on new processing methods, since a chip containing a migration algorithm will render unattractive work on improved migration methods. The field is not mature (stagnant?) enough that anyone company could make a decision to use one processing method, and one only, for the next decade or so.

5. CONCLUSION

High-performance processing of seismic data must clearly start with an efficient algorithm. There is a host of efficient methods that can be tailored to a given situation. Most applications use vector processing, and with very good reason: at present, this is the single most important factor in the performance of a competently written application program. However, in realistic implementations, questions such as the *I/O* behavior and the inherent parallelism of a program become of concern since they can very seriously affect the performance of the program if they are not properly considered. At present, *I/O* analysis and detection of parallelism must be carried out manually. We expect that in the next few years, software tools will become available that assist in these tasks. However, the actual restructuring of the code will require knowledge of the application and therefore it is highly unlikely that restructuring can be fully automated, in the near of in the long-term future. Therefore, programming the new machines will place a significant burden on the programmers. The reason why vectorization is such a success is that it can be done syntactically, i.e., without any understanding of the underlying application. This is not the case for the restructuring of a program in order to improve its *I/O* behavior or to exploit inherent parallelism.

In particular, there are two major problems associated with parallelism at the hardware level, one related to hardware, the other related to software. The hardware problem is one exclusively associated with loosely-coupled systems, while the software problem is common to both loosely- and tightly-coupled systems. The hardware problem is that of interprocessor communication; at present the bandwidth is simply too small for realistic seismic processing. While the remedy is obvious, it is also costly and may seriously affect the price/performance ratio of the resulting systems. Nevertheless, improvements here are expected as soon as the manufacturers realize that interprocessor communication bandwidth is a major bottleneck. This should be in the near future; indeed there are indications that the

Connection Machine has addressed this problem. The software problem is one that cannot be solved that fast. The objective are software tools that parallelize uniprocessor code automatically; this implies that it must be based on purely syntactic considerations. While this appears feasible, the first reasonably efficient parallelizer is probably several years away. Until then, parallelization will have to be done by hand, which is time consuming, not least of all because debugging parallel code is at least one order of magnitude harder than debugging uniprocessor code. Also, the larger the number of processors, the more difficult will it be to design efficient parallel code; this is again more in favor of the tightly-coupled systems which typically have fewer processors (four for the Cray X-MP/4; eight for the ETA-10 for the time being) than of the loosely-coupled systems which may have up to 65000 processors.

REFERENCES

Amestoy, P., Larsonneur, J. L., Leiss, E. L., and Gardner, G. H. F., 1987, Prestack Migration with Phase Shift Methods on the Cray X-MP: Research Computation Laboratory, Annual Progress Review, 3, 80–129.

Ashton-Tate, 1984, The dBase III Reference Guide, Ashton-Tate.

Basart, E., 1985, RISC design streamlines high power CPU's: Computer Design, July Issue.

Date, C. J., An Introduction to Database Systems: 1981, Addison-Wesley Publication, 1981.

Dettmer, R., 1985, Chip architecture for Parallel Processing: Electronics and Power, March Issue.

Fathi, E. T. and Krieger, M., 1983, Multiple Microprocessor Systems: What, Why, and When: IEEE Computer, March Issue.

Ferrante, M. W., 1985, Taking Parallel Processors to the scientific community: Computer Design, December Issue.

Fisher, J. A., Donnel, J. O., 1984, VLIW machines: multiprocessors we can actually program: Spring Compcon.

Folger, D., 1985, RISC architecture as an alternative to parallel processing: Computer Design, August Issue.

Gajski, D. D., Parallel Processing: Problems and solutions: University of Illinois at Urbana-Champaign, Technical Report.

Hecht, J., 1987, Optical Memories Vie for Data Storages, High Technology, August Issue, pp. 43–47.

Hennessey, J., 1985, VLSI RISC processors: VLSI Systems Design, October Issue.

Hwang, K., 1985, Multiprocessor Supercomputers for scientific/engineering applications: IEEE Computer, June Issue.

Kao, S. T. and Leiss, E. L., 1987, An Experimental Implementation of Migration Algorithms on the Intel Hypercube: Research Computation Laboratory, Annual Progress Review, 3; The International Journal of Supercomputer Applications Vol. 1, No. 2, 1987, pp. 75–99.

Karp, A. H., 1987, Programming for Parallelism, IEEE Computer, May Issue, pp. 43–57.

Kuck, D. J., Supercomputers: Encyclopedia of Computer Science, Second edition. Van Nostrandt Reinhold, Inc.

Leiss, E. L., 1984, Data Integrity in Digital Optical Disks, IEEE Transactions on Computers, Sept. Issue, Vol. C-33, No. 9, pp. 818–827.

Lhemann, O., 1985, A 3D PSPI Migration, Research Computation Laboratory, Annual Progress Review, 1, 86–108.

Ma, H. H. and Johnson, O. G., 1986, Implementation of PSPI and Prestack Migration on the CYBER 205: Research Computation Laboratory, Annual Progress Review, 2, 148–170.

Matthews, M., 1987, A Permanent Record, Logic Vol. 2, No. 2, Summer Issue, pp. 8–13.

Nelson, H. R., Jr., 1982, SALNOR North Sea Model: Building, Data Acquisition and Interpretation: Seismic Acoustics Laboratory, Semiannual Progress Review, 9, 321–360.

Patton, P. C., 1985, Multiprocessors: Architectures and Applications: IEEE Computer, June Issue.

Polavarapu, U. R. and Johnson, O. G., 1986, A Database on Advanced Computer Research Projects, Research Computation Laboratory, pp. 289–307.

Raguskus, A. G., 1985, *I/O* computer supercharges minisystems: Computer Design, July issue.

Sashti, J., Johnson, O. G., and Leiss, 1986, From Superminis to Supercomputers—A Survey, Research Computation Laboratory, pp. 213–238.

Schwartz, J., 1983, A taxonomic table of parallel computers based on 55 designs: New York University note #69, November issue.

Siewiorek, D. P., Anzelmo, T., and Moore, R., 1985, Multiprocessor computers expand user vistas: Computer Design, August Issue.

Terki-Hassaine, O. and Leiss, E. L., 1987, A Multitasking Implementation of 3D Forward Modeling using High-Order Finite Difference Methods on the Cray X-MP/416: Research Computation Laboratory Annual Progress Review, 3; 190–216, The International Journal of Supercomputer Applications (to appear).

Treleaven, P. C., 1984, Control-driven, data-driven, and demand-driven computer architecture: IEEE Computer, March Issue.

Wallich, P., 1985, Toward simpler faster computers: IEEE Spectrum, August Issue.

Wilson, A., 1985, Array Processors–Increasing speed by MIPS, MOPS, FLOPS, and GOPS: Computer Design, August Issue.

CHAPTER 5

PARALLEL AND OPTICAL ARCHITECTURES FOR MODELING AND INVERSION

by
DR. ALASTAIR D. MCAULAY
NCR Distinguished Professor and Chairman
Department of Computer Science and Engineering
Wright State University
Dayton, OH 45435

1. INTRODUCTION

Research into accurate modeling and inversion is required for significant advances in geophysics but progress is handicapped by insufficient computational capability. Computers that achieve high performance by means of massive parallelism look attractive because the natural parallelism of the computations of interest may be exploited by matching architecture and computations. Highly parallel architectures designed for application to geophysical modeling and inversion do not exist today. Progress in optical technologies also provides new possibilities for achieving fast cost-effective massive parallel architectures.

This chapter considers algorithms for modeling and inversion, alternative computer architectures, and optical computing. It then provides examples of systems in which specific algorithms are matched to specific architectures to assess performance. The aim is to permit development of architectures most suitable to meet the needs of a specific organization's requirements for seismic modeling and inversion research and/or production applications. Speed is achieved by parallelism

---

[1] Presented in part by Dr. Alastair D. McAulay at Workshop on Geophysical Modeling Computers, SEG 56th Annual Int. Mtg. Nov. 1986.

and in some cases by the use of optics. The cost of customizing machines is minimized by the use of software and hardware modularity.

Section 2 describes some of the widely used algorithms in geophysical modeling and inversion. In particular these include: spectral methods, section 2.1; 1-D modeling and inversion, section 2.2; 2-D and 3-D finite approximation methods, section 2.3; and linear equation solution, section 2.4.

Architecture issues are discussed in section 3. These include the desirable features for supercomputers, section 3.1; the rationale for parallelism, section 3.2; the importance of modularization for cost effective matching of user needs, section 3.3; the difficulties of parallelism, section 3.4; and alternative architectures, section 3.5.

Section 4 discusses issues in optical computing including: the advantages of optics, Section 4.1; the use of spatial light modulators (SLMs), section 4.2; and examples of specific SLMs, section 4.3.

Section 5 provides two examples illustrating the performance of geophysical algorithms selected from section 2 operating on specific architectures selected from section 3.5. In section 5.1, a parallel processor with 32 commercially available processing elements connected by a fast commercially available bus is proposed for solving 2-D and 3-D finite element computations [34]. Analysis shows that this configuration could be effective for grids up to 285 by 285 for 2-D problems and for grids up to 40 by 40 for 3-D problems. However, further improvement in performance by adding more processors or making the processors faster is limited by the bus.

Section 5.2 describes the use of a systolic array for performing 1-D modeling and inversion. Three parts of the computation dominate. The first part, involving forward modeling and Jacobian determination, has thousands of independent tasks permitting efficient use of a systolic array with hundreds of processors. The second part requires less computation and involves 2-D transforms which may be implemented reasonably efficiently on a systolic array. An implementation of the conjugate gradient algorithm on the systolic array suggests that the third computationally demanding part may also be computed reasonably efficiently on a systolic array.

Section 6 describes three optical architectures aimed at Gigaflop performance for seismic modeling and inversion computations. Optical components are under development that should make such designs feasible in the next ten years. The first computer, described in section 6.1 [28], consists of fast elementary semiconductor processors interconnected by a reconfigurable optical interconnection network.

Programmed dataflow is used to convert code to maximally parallel graphs and then map these graphs to the multiprocessor. Implementations of fast Fourier transforms, nonlinear spectral estimation, and matrix-vector multiplication are shown. A conjugate gradient algorithm is modified to permit 76% efficiency. Reconfiguring the switch suggests that larger and smaller problems may be handled with almost the same efficiency.

The second optical architecture, described in section 6.2 [29], is aimed at solving seismic finite approximation modeling problems in which the field may be represented on a rectangular grid or distortion of such a grid. A nearest neighbor design using residue number arithmetic and two 1000 by 1000 deformable mirror arrays would solve finite difference computations for rectangular grids of size 750 by 6000 at a rate of 15 million operations per second with 32 bit accuracy. Duplication of equipment by 64 times enables one billion operations per second because perfect parallelism is achievable with the residue number and nearest neighbor concepts employed.

Symbolic substitution may be also used for numerical computation [5]. The same approach is being investigated for use in Artificial Intelligence (AI) with the Prolog language [14], and with associative memories in the form of neural networks [17]. This approach is not considered further in this chapter.

2. ALGORITHMS FOR MODELING AND INVERSION

Algorithms are classified by 1-D, 2-D, and 3-D as shown in figure 1. The dimension refers to the earth model parameters in 3-D space. For example, a 1-D earth model corresponds to plane-layers. The earth parameters are assumed known for modeling and are sought in the case of inversion. Many of the modeling algorithms are used during inversion, in particular because Gauss-Newton inversion approaches involve repeated forward modeling [25], [32]. The field is assumed to be modeled in 3-D in all cases.

Spectral methods, section 2.1, are widely used throughout geophysical processing because, for example, wave propagation energy clusters in the frequency domain. 1-D modeling and inversion algorithms are described in section 2.2. Section 2.3 discusses finite approximation methods for 2-D and 3-D modeling and inversion. Section 2.4 describes the solution of large sparse ill-conditioned equations arising in geophysical modeling and inversion.

A few important algorithms are used for illustrating the implementation on specific architectures subsequently, section 5 and 6.

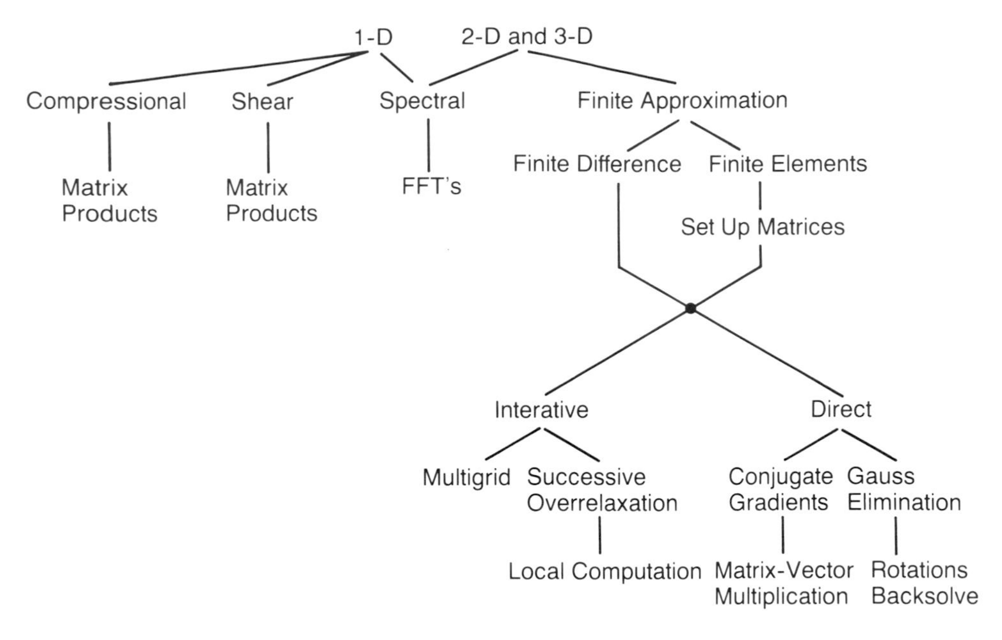

Fig. 1 Algorithms for modeling and inversion.

2.1 *Spectral methods*

Spectral methods are widely used in all areas of geophysical processing. For example, velocity filtering is applied in the frequency-wavenumber domain. Velocity filtering is used during modeling and inversion [25], [32].

Linear spectral methods. The fast Fourier transform (FFT) is used widely in signal processing and numerical computation to convert a time series to the frequency domain. Often the information sought is clustered only in the frequency domain, for example, a plane wave with a specific frequency and wave number.

The FFT of a sequence **x** is

$$X_k = \sum_{n=0}^{N-1} x_n w^{kn} \quad k = 0 \text{ to } N-1 \tag{1}$$

where $w = e^{-j(2\pi/N)}$. A specific algorithm for computing the FFT which matches the architecture selected is described in section 6.1.

Nonlinear spectral estimation methods. There are many situations for which a linear spectral estimator such as the Fourier transform is inadequate and nonlinear spectral estimation approaches such as autoregressive modeling, linear prediction, or maximum entropy give better results. In the case of one dimensional data any of these approaches may be used as they all lead to solution of the same Yule-Walker equations.

An example is predictive deconvolution in which a linear prediction is used to remove source oscillations from the data [31]. Further examples arise when a high resolution estimate of frequency is required or the sequence is short having insufficient cycles of the frequencies of interest. For example, consider a linear array of uniformly spaced sensors. Figure 2(a) shows the time trace collected at each sensor in a towed array for 2 seconds after an airgun impulsive source was activated. Figure 2(b) shows the 2-D FFT for this data, assuming independence of the space and time domain. The sharp edge of the spatial aperture resulting from too few sensors causes ringing horizontally, in the space direction. Spatial resolution may be improved by using nonlinear spectral analysis in the space direction. Figure 2(c) shows the 2-D spectral plot for which time is transformed with linear spectral analysis and distance with nonlinear spectral analysis. The latter involved computing an autocorrelation function, equation (3), solving Levinson-Durbin's

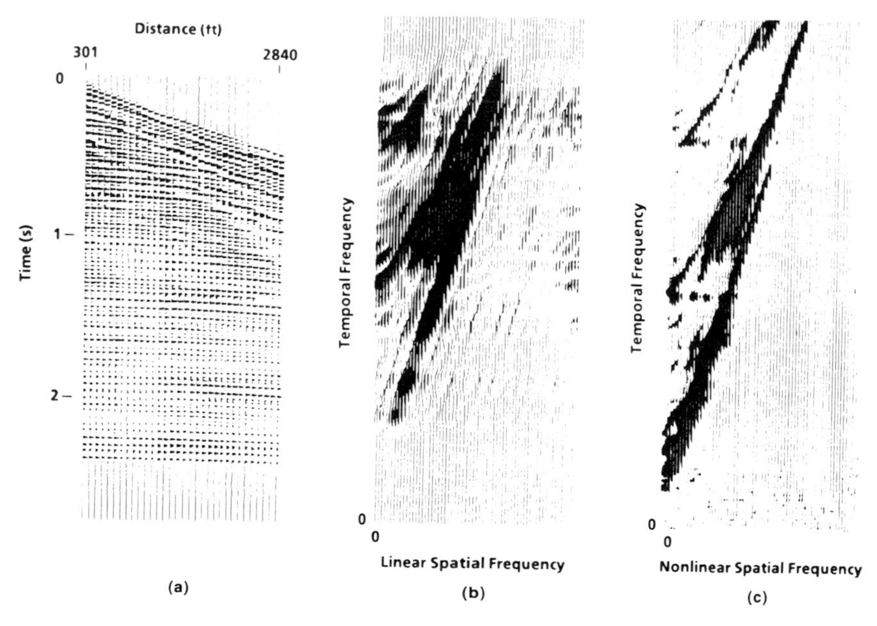

Fig. 2 Spectral processing of array data. (a) data, (b) linear spectrum, (c) nonlinear spatial spectrum.

algorithm, equations (5), (6), (7), and computing the power spectrum, equation (4). The ringing is removed and the resolution improved relative to Figure 2(b).

Nonlinear spectral estimation equations and Levinson-Durbin algorithm. The solution of the Yule-Walker equations is generally performed using the Levinson-Durbin algorithm [18], [26], [27]. This algorithm is much harder to implement on parallel machines than the FFT because it is iterative, increases dimension at each iteration, and has to compute termination criteria. For this reason other algorithms have been proposed that are less efficient on a uniprocessor but more adaptable to a multiprocessor, e.g. Schur's algorithm [22].

An autoregressive (AR) model converts white noise to a least square approximation for the data x_t. AR parameters (or prediction error filter coefficients) a are minimum phase and are computed by solving the Yule-Walker equations

$$\begin{bmatrix} r_0 & r_1 & \cdots & r_{m-1} & r_m \\ r_1 & r_0 & \cdots & r_{m-2} & r_{m-1} \\ \vdots & & & & \vdots \\ r_{m-1} & r_{m-2} & \cdots & r_2 & r_1 \\ r_m & r_{m-1} & \cdots & r_1 & r_0 \end{bmatrix} \begin{bmatrix} 1 \\ a_1 \\ \vdots \\ a_{m-1} \\ a_m \end{bmatrix} = \begin{bmatrix} v_m \\ 0 \\ \vdots \\ 0 \\ 0 \end{bmatrix}, \tag{2}$$

where V_m is the white noise power, r_τ is an estimate of the autocorrelation function at lag τ and the mean is μ,

$$r_\tau = \frac{1}{N} \sum_{j=1}^{N-\tau} (x_j - \mu)(x_{j+\tau} - \mu) \qquad \tau = 0 \text{ to } m. \tag{3}$$

The power spectrum $S(\omega)$ at angular frequency ω is computed from the AR coefficients $\mathbf{a}$ when required by using:

$$S_\omega = \frac{2v_m}{|\sum_{k=0}^{m} a_k e^{jwk}|^2} \tag{4}$$

The AR coefficients $\mathbf{a}$ are computed using the Levinson-Durbin algorithm. At the nth iteration a "reflection coefficient" is computed as the inner product

$$c(n) = \sum_{i=0}^{n-1} \frac{a_i r_{n-i}}{v(n-1)}. \tag{5}$$

The power of the white noise associated with the AR process is computed from

$$v(n) = v(n-1)(1 - |c(n)|^2). \tag{6}$$

Minimum delay is maintained by updating the AR parameters from

$$a_k(n) = a_k(n-1) - c(n)\, a_{n-k}(n-1) \qquad k = 0 \text{ to } n. \tag{7}$$

The implementation of the algorithm on a specific architecture is considered in section 6.1.

2.2 1-D *modeling and inversion*

1-D modeling. A synthetic seismogram is generated by computing the non-normal incidence reflection coefficient from a set of plane layers in the frequency-wavenumber domain, i.e., the spectral domain in time and space [25], [32], [35]. The equations for fluid layer waves follow.

The impedance of a fluid medium is

$$I_j = \rho_j \bigg/ \sqrt{\frac{1}{\tilde{v}_j^2} - \frac{p_0'^2}{\omega^2}}, \tag{8}$$

where ω is angular frequency, ρ_j is density in the jth layer and p_0' is the horizontal component of the wavenumber vector. p_0' is independent of layer because of Snell's law. $\tilde{v}_j$ is the complex velocity for the jth layer, complex to include absorption.

The reflection coefficient for a nonnormal incidence wave striking an interface between two media and approaching from the medium having impedance I_j and connecting with the medium of impedance I_{j+1} is

$$c_j = \frac{I_{j+1} - I_j}{I_{j+1} + I_j}. \tag{9}$$

The phase delay across a layer is

$$z_j = \exp\{2i\delta \sqrt{\omega^2 - p_0'^2 \tilde{v}_j^2}\}, \tag{10}$$

where δ is the time taken for a normal incidence wave to travel across a layer. The layer thicknesses are selected to make δ the same in all layers and approximately equal to half the shortest wavelength in the measurement data. The assumption is that narrower layers would not be resolvable without higher frequencies. The results shown use $\delta = 4ms$ and $J = 243$ layers.

The propagator matrix $\mathbf{M_j}$ enables computation of the up and down wave components just above the jth layer, U_j and D_j respectively, from those just above the $j+1$th layer.

$$\begin{bmatrix} U_j \\ D_j \end{bmatrix} = \mathbf{M_j} \begin{bmatrix} U_{j+1} \\ D_{j+1} \end{bmatrix}, \qquad \mathbf{M_j} = \frac{1}{t_j \sqrt{z_{j+1}}} \begin{bmatrix} z_{j+1} & c_j \\ c_j z_{j+1} & 1 \end{bmatrix}, \tag{11}$$

where t_j is the normal incidence travel time across the jth layer.

The upcoming and downgoing wave components just above the top layer U_0 and D_0 respectively, may be related to those a distance $v_{j+1}\,\delta$ below the deepest or Jth interface by the matrix

$$\mathbf{W} = \prod_{j=0}^{j=J} \mathbf{M_j}. \tag{12}$$

The reflection coefficient for the stack of layers is the ratio of the upgoing to the downgoing wave components at the top, which may be written

$$R = U_0/D_0. \tag{13}$$

Assuming that no waves enter from below $U_{j+1} = 0$ and the source strength providing the down wave D_0 is known, the reflection from the stack of layers for a specific frequency and wavelength (or plane wave angle of incidence) is

$$R = \frac{W_{12} D_{j+1}}{W_{22} D_{j+1}} = \frac{W_{12}}{W_{22}}. \tag{14}$$

The computation of R requires a string of matrix-matrix multiplications, equation 11, and is known as the propagator matrix or Thompson-Haskel method [1]. Speed is improved by computing only the elements needed at each matrix-matrix computation. For shear wave computations [33] the 2 by 2 matrices are replaced by 6 by 6 matrices.

In order to model the cylindrical symmetry about the vertical axis for a point source over plane layers, Hankel transforms are used rather then FFT's for transforming from wavenumber to distance. FFTs are used for transforming from frequency to time.

1-D inversion. *Approach and equations.* In the case of inversion a Gauss-Newton or Generalized Linear Inverse method is used which involves linearization using a Taylor series expansion and neglecting higher order terms, figure 3. An initial estimate **m** for the unknown earth layer parameters is made. A forward modeling computation involving the previously described propagator matrix method computes the measurements that would result using these parameters. The synthetic data is compared with measured field data **x**, (the 2-D data array in time-distance is written as a vector **x**), and the difference $\delta\mathbf{x}$ used in conjunction with the Jacobian matrix **J** to make a correction $\delta\mathbf{m}$ to the initial estimate. The synthetic data is computed in the frequency-wavenumber domain for convenience and must be converted to the time-distance domain in which real data is collected in order that the surface and array effects [25], [32] (finite aperture in distance and time) are correctly included for matching with the data. The same transformation must be applied to the Jacobian matrix for consistency. The Jacobian matrix has as dimensions the number of measurement values by the number of parameters to be estimated.

The Jacobian matrix has *mj*th element $\partial R_m/\partial v_j$, with R_m representing the *m*th measurement value of the reflection coefficient for a specific frequency and

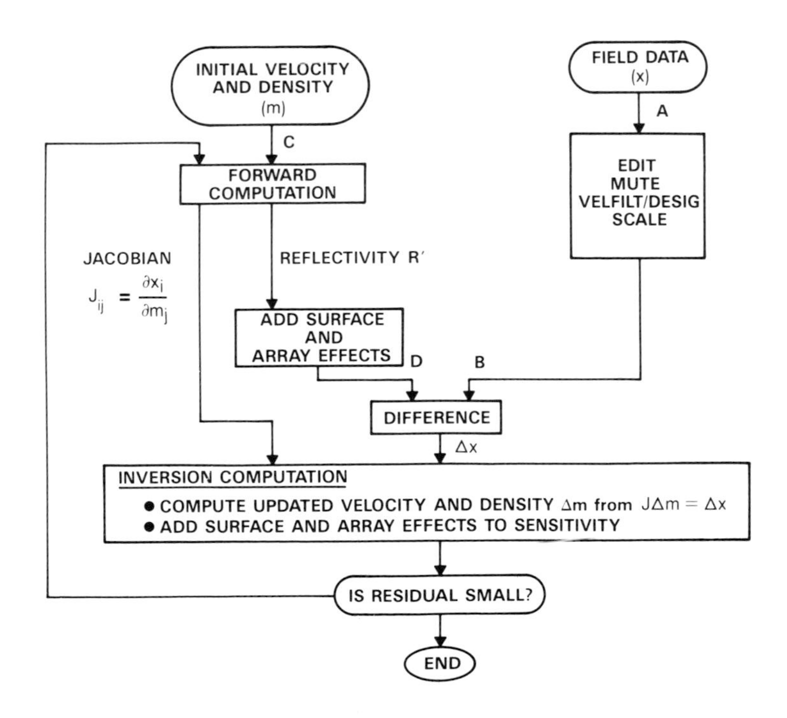

Fig. 3 Gauss-Newton method.

wavenumber of plane wave striking the stack of layers and v_j the velocity in the jth layer [32]. An element of the Jacobian matrix is

$$\frac{\partial R_m}{\partial v_j} = \left\{ W_{22m} \left(\frac{\partial W_{12m}}{\partial v_j} \right) - W_{12m} \left(\frac{\partial W_{22m}}{\partial v_j} \right) \right\} \Big/ W_{22m}^2. \tag{15}$$

The terms $\partial W_{12m}/\partial v_j$ and $\partial W_{22m}/\partial v_j$ are obtained from the appropriate elements of the following matrix equation, which has been modified to include the absorption term $\tilde{q}$. The subscript m has been dropped as all terms depend on m. The equation is

$$\begin{aligned} \frac{\partial \mathbf{W}}{\partial v_j} = & \left[\prod_{k=0}^{j-1} \mathbf{M_k} \right] \left\{ \left[\begin{matrix} 0 & 1 \\ z_{j+1} & 0 \end{matrix} \right] \left(\frac{-2I_{j+1} I_j \omega^2}{Q_{+1}^2 p_j^2 v_j^3 \tilde{q}^2} \right) \right\} \left[\prod_{k=j+1}^{J} \mathbf{M_k} \right] \\ & + \left[\prod_{k=0}^{j-2} \mathbf{M_k} \right] \left\{ \left[\begin{matrix} 1 & 0 \\ c_{j-1} & 0 \end{matrix} \right] \left(\frac{-2i\, \delta_j z_j p_0'^2 \tilde{q}}{p_j} \right) \right. \\ & \left. + \left[\begin{matrix} 0 & 1 \\ z_j & 0 \end{matrix} \right] \left(\frac{2I_{j-1} I_j \omega^2}{Q_{-1}^2 p_j^2 v_j^3 \tilde{q}^2} \right) \right\} \left[\prod_{k=j}^{J} \mathbf{M_k} \right], \end{aligned} \tag{16}$$

where I_j and $Q_{\pm 1}$ depend on layer impedances, c_j is layer interface reflection coefficient, z_j is phase delay across a layer, and p_j is the vertical component of the wavenumber. $Q_{+1} = I_{j+1} + I_j$ and $Q_{-1} = I_{j-1} + I_j$. The matrix $\mathbf{M}$ is the propagator matrix for the up and down going waves across a layer. p_0', the horizontal component of the wavenumber, is independent of layer because of Snell's law. The derivation of this expression and the meaning of the terms are given in reference [25].

Algorithmic approach. Each element of the Jacobian matrix $\mathbf{J}_{p,j}$ is computed in a similar manner to the synthetic data because the computation involves the derivative of a synthetic data point with respect to a specific layer parameter. A layer parameter affects only the matrix corresponding to that layer and an immediately preceding layer because "reflection coefficient" at a layer interface depends on media parameters on either side of the interface. Consequently, the jth element in a row of the Jacobian matrix is computed by adding the result of performing the matrix product computation twice, first with the jth matrix replaced by a specially computed derivative matrix and second with the $j-1$th matrix replaced by a specially computed derivative matrix. Other rows correspond to different frequency-wavenumber combinations represented by the index p.

The large set of linear equations is solved using least square conjugate gradients, sections 2.4 and 6.1, equation 23. Column normalization of J is used to assist parameters which have a weak effect on the observed measurements. The number of conjugate gradient iterations required is much less than the number of layer parameters (columns), consequently, two successive matrix-vector multiplications are used at each iteration in preference to one matrix-matrix multiplication to account for the least square procedure. It is also more economic to apply the linear space and temporal transformations to the vector after the first matrix-vector multiplication during the conjugate-gradient method than to each of the columns of the Jacobian in advance. A significant advantage of the conjugate gradient method for large ill conditioned matrices is that computation is terminated by convergence criteria before small eigenvalues or singular values have an opportunity to interfere with accuracy.

It has been shown that the compressional velocity of a layered earth medium may be estimated from realistic synthetic data [25], [32]. Figure 4(a) shows the velocity depth profile used to create the synthetic data, figure 4(b).

Figure 5 shows that the velocity profile may be recovered from the synthetic data, figure 4(b). The velocity is shown as a function of iteration during the Gauss-Newton method.

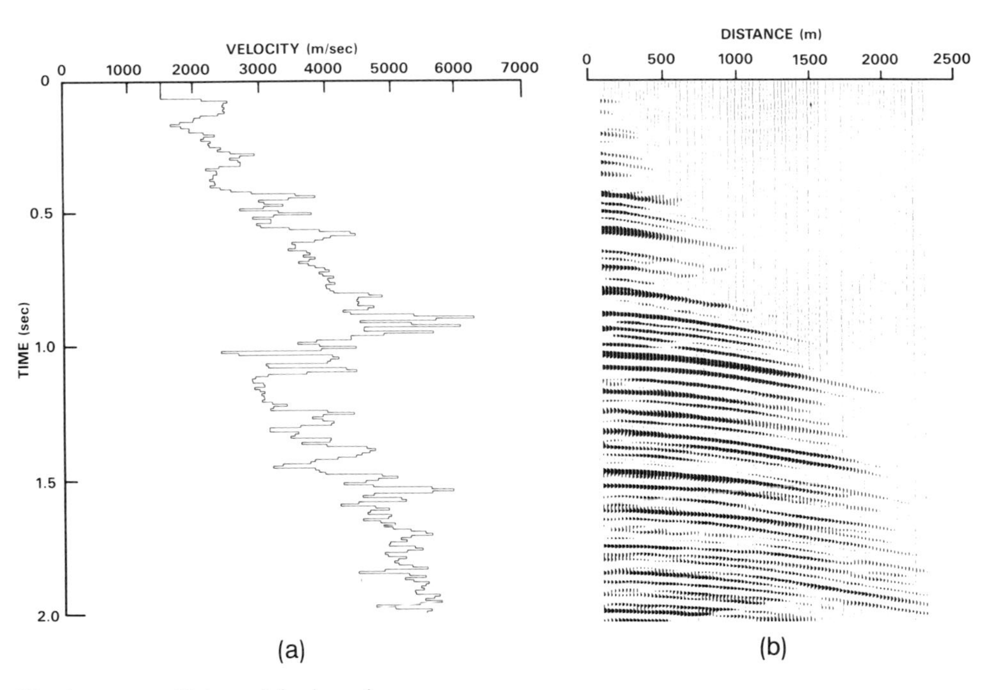

Fig. 4 Data used for inversion.

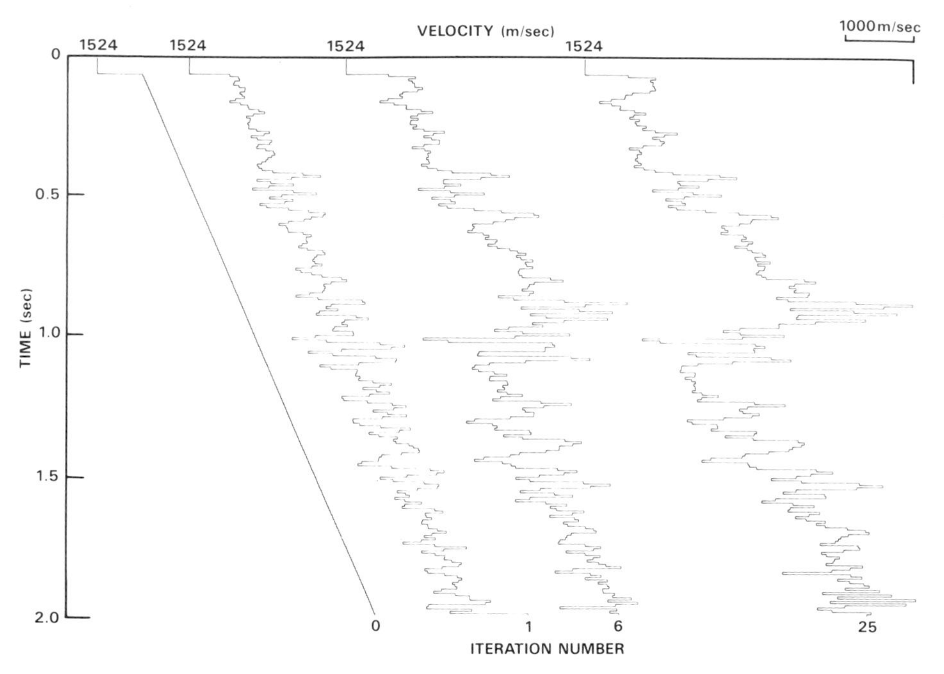

Fig. 5 Estimated velocity profiles during inversion.

Natural parallelism and timing for computation. Table 1, p. 115, shows the high level of natural parallelism, column 3, with typical values for a minimum size useful problem in brackets. The computation times involved are shown in column 7. The forward modeling using propagator matrices and the Jacobian computation must be performed for all frequencies and wavenumbers (some 2500) so that a high level of parallelism is available. The 2-D filtering has a high level of parallelism because the space transforms have to be performed for each temporal component and vice versa. The only computation which does not show a high level of parallelism is the conjugate gradient computation. Task parallelism or dimension size for the independent tasks is shown in column 4 and represents potential low level parallelism.

2.3 *Finite approximation methods*

2-D and 3-D modeling may be divided into spectral methods, as discussed in section 2.1, and finite approximation methods. Spectral methods use fewer samples than finite approximation methods. For example, spectral methods use slightly over two samples per wavelength while finite approximation methods require approximately eight samples per wavelength. However, smoother media are assumed in the case of spectral methods for the same accuracy. At greater depths, the velocity of sound tends to increase so that the wavelength is greater and fewer samples per unit distance are required.

Finite elements [34], [36], [37] is superior to finite differences because the field is approximated by summation of fields described by polynomial functions over the small element regions. For example, a first order linear polynomial represents a piecewise linear function approximation. In contrast, finite differences aim at matching fields at points on a grid. However, finite elements require computations for setting up the matrices involved which is not required in finite difference techniques.

The matrix equations to be solved in finite approximation models may be solved using iteration or direct techniques. Iterative techniques update the estimate of the field at each iteration. Equations for solving finite difference and finite element problems with regular, or distorted from regular, grids can be formulated by replacing the derivatives by difference operators. Assume a grid imposed on the field with k a north-south index, increasing in the north direction, and m an east-west one, increasing to the east. The overrelaxation algorithm for updating the k, mth value at the ith iteration in solving an elliptic partial differential equation is

$$\begin{aligned} u_{k,m}^{i+1} &= \omega[a_{1,(k,m)}u_{k+1,m}^{i} + a_{2,(k,m)}u_{k-1,m}^{i} + a_{3,(k,m)}u_{k,m+1}^{i} a_{4,(k,m)}u_{k,m-1}^{i} \\ &= + a_{5,(k,m)}u_{k+1,m+1}^{i} + a_{6,(k,m)}u_{k-1,m-1}^{i} + a_{0,(k,m)}u_{k,m}^{i}] \\ &\quad - (\omega - 1)\, u_{k,m}^{i} + f_{k,m} \end{aligned} \tag{17}$$

The coefficients **a** are scaled by the overrelaxation coefficient ω. $f_{k,m}$ represents a source applied across the grid.

Such computations appear highly parallel. However, speed ups are achieved by using grids of coarseness varying by powers of two in multigrid techniques [4], [39]. This is faster on a uniprocessor because the finest most computationally demanding grid is used only a small part of the time. This reduces the level of parallelism and introduces considerable book keeping and data transfer operations.

2.4 *Linear equation solution for geophysics*

Gauss elimination is often used to solve the linear equations arising in finite element computations because of its ability to handle negative definite matrices.

Gauss elimination may be divided into rotational and backsolve computations. These have been investigated on optical machines in reference [22]. However, Gauss elimination is too slow for most real geophysical problems which involve large sparse, ill-conditioned matrices. Conjugate gradient algorithms often provide much more rapid solution [8].

Conjugate gradient equations. Preconditioning is normally used because large systems tend to be ill-conditioned. The techniques for preconditioning [8] are not considered here but implementations of some of these, such as Choleski decomposition, have been considered for use on an optical processor [22] proposed in section 6.1.

Conjugate Gradients (CG) is considered a direct method for solving linear equations because in theory the solution is obtained in a number of operations equal to the dimension for the matrix involved. In practice, iterations are performed until negligible change occurs with further iterations. The iterative approach is appropriate for rapid approximate solution to linearized equations arising in nonlinear computations, e.g., in Gaus-Newton methods. Accurate solutions are not meaningful because the solution to the linearized equations is only an approximation to the nonlinear solution. Examples arise in geophysical inversion [25] [32] and in the finite difference computations in reservoir modeling.

A CG algorithm is presented for solving

$$\mathbf{y} = \mathbf{A}\mathbf{x}, \tag{18}$$

where $\mathbf{A}$ is an n by n matrix, $\mathbf{y}$ is a vector of length n, and $\mathbf{x}$ is the unknown vector of length n.

Initialization:

$$\begin{aligned}
&\text{Set:} && \mathbf{x}_1, \\
&\text{Residual:} && \mathbf{r}_1 = \mathbf{A}\mathbf{x}_1 - \mathbf{y}, \\
& && \textit{or } r_1 = -\mathbf{y} \textit{ if } \mathbf{x}_1 = \mathbf{0}, \\
&\text{Direction:} && \mathbf{d}_1 = -\mathbf{r}_1, \\
&\text{Compute:} && \mathbf{r}_1^T\mathbf{r}_1
\end{aligned} \tag{19}$$

do $k = 1$ *to* n:

$$
\begin{aligned}
&\mathbf{q}_k = \mathbf{A}\mathbf{d}_k, \\
&\alpha_k = \frac{\mathbf{r}_k^T \mathbf{r}_k}{\mathbf{d}_k^T \mathbf{q}_k}, \\
&\mathbf{x}_{k+1} = \mathbf{x}_k + \alpha_k \mathbf{d}_k, \\
&\mathbf{r}_{k+1} = \mathbf{r}_k + \alpha_k \mathbf{q}_k, \\
&\text{Compute: } \mathbf{r}_{k+1}^T \mathbf{r}_{k+1}.
\end{aligned}
\tag{20}
$$

If $\mathbf{r}_{k+1}^T \mathbf{r}_{k+1} < \varepsilon \mathbf{r}_1^T \mathbf{r}_1$ then exit, else,

$$
\begin{aligned}
&\beta_k = \frac{\mathbf{r}_{k+1}^T \mathbf{r}_{k+1}}{\mathbf{r}_k^T \mathbf{r}_k}, \\
&\mathbf{d}_{k+1} = -\mathbf{r}_{k+1} + \beta_k \mathbf{d}_k,
\end{aligned}
\tag{21}
$$

end do

The algorithm improves an estimate of **x** by adding an amount determined by a step α in a direction **d** in equation (20). α_n is chosen such that $\mathbf{r}_{n+1} = \mathbf{r}_n + \alpha_n \mathbf{q}_n$ is perpendicular to $\mathbf{d}_n$. A new direction is selected in equation (21) such that it is conjugate with respect to **A** to all previous directions **d** used.

A least square version of this algorithm was successfully used for 1-D geophysical inversion simulations [25], [32]. The least square solution to equation (18) is the solution to the normal equation

$$
\mathbf{A}^*\mathbf{y} = \mathbf{A}^*\mathbf{A}\mathbf{x}. \tag{22}
$$

The algorithm updates the vector **x** in such a way as to reduce toward zero the residual

$$
\mathbf{r} = \mathbf{A}^*\mathbf{A}\mathbf{x} - \mathbf{A}^*\mathbf{y}. \tag{23}
$$

It is only necessary to reduce the residual to some fixed percentage ε of its initial value in the application to prestack inversion.

A form of the conjugate algorithm is presented for the least square solution of equation (18) or equivalently the solution for equation (22).

Initialization:

$$\begin{aligned}
&\text{Set:} && \mathbf{x}_1, \\
&\text{Residual:} && \mathbf{r}_1 = \mathrm{Re}(\mathbf{A}^*\mathbf{A}\mathbf{x}_1 - \mathbf{A}^*\mathbf{y}), \\
& && \textit{or } \mathbf{r}_1 = -\mathrm{Re}(\mathbf{A}^*\mathbf{y}) \textit{ if } \mathbf{x}_1 = \mathbf{0}, \\
&\text{Direction:} && \mathbf{d}_1 = -\mathbf{r}_1, \\
&\text{Compute:} && |\mathbf{r}_1|
\end{aligned} \tag{24}$$

do $n = 1$ *to* N:

$$\begin{aligned}
\mathbf{q}_n &= \mathrm{Re}(\mathbf{A}^*\mathbf{A}\mathbf{d}_n), \\
\alpha_n &= \frac{|\mathbf{r}_n|^2}{\mathbf{d}_n^T \mathbf{q}_n}, \\
\mathbf{x}_{n+1} &= \mathbf{x}_n + \alpha_n \mathbf{d}_n, \\
\mathbf{r}_{n+1} &= \mathbf{r}_n + \alpha_n \mathbf{q}_n.
\end{aligned} \tag{25}$$

If $|\mathbf{r}_{n+1}| < \varepsilon |\mathbf{r}_1|$ **then exit, else,**

$$\begin{aligned}
\beta_n &= \frac{|\mathbf{r}_{n+1}|^2}{|\mathbf{r}_n|^2}, \\
\mathbf{d}_{n+1} &= -\mathbf{r}_{n+1} + \beta_n \mathbf{d}_n,
\end{aligned} \tag{26}$$

end do

The algorithm presented differs from that in reference [8], by collecting together the matrix related terms into one line in equation (20), by involving complex terms and by using $\mathbf{r}$ for the residual to equation (22) rather than equation (18).

3. COMPUTER ARCHITECTURE ISSUES

3.1 *Desired features for supercomputer*

The desirable features of a supercomputer are: high speed, extendability, scalability, flexibility, user friendliness, automatic partioning, fault tolerance, and reliability [23]. Most of these features require parallelism. User friendliness and flexibility suggest symbolic capability because interfaces involve symbol processing for editing, debugging, and natural language.

High performance involves considerations of throughput and latency. Throughput is the number of computations per unit time and latency is the time taken for one computation to be completed. Throughput can be many times higher than latency when temporal overlap is used in the form of pipelining. High throughput is suitable for many large problems because repetitive computations are often required. However, minimum latency is also required because of those situations where results are needed before subsequent computations can be performed.

Extendability implies that more processors may be added to the multiprocessor together with corresponding interconnections without requiring new software and with performance approaching linear improvements with increasing number of processors. Scalability suggests that machines of different computational capability may be manufactured. Together with extendability this will enable the same architecture and associated software to apply to a wide product range and provide longer life for customers and products.

Flexibility implies that a wide range of algorithms must run efficiently. Also, new algorithms and as yet undiscovered algorithms must be easily entered into the machine and run efficiently.

User friendliness and the ability to automatically handle some of the parallelism is required because software costs are a major part of the cost of computation. For example, as much as 80% of the cost of a new computer and the maintenance is for software. Further, research is hampered if code has to be written at a low language level or in detail specifically for an unusual architecture.

Fault tolerance is increasingly sought in systems and implies that the system will be able to continue operation, normally in a degraded manner, after a significant equipment failure. High reliability is aimed at minimizing failures.

3.2 *Rationale for parallelism*

Many of the desirable features for supercomputers discussed in the last section cannot be achieved without the use of parallel computation. For example, extendability, scalability, and fault tolerance. The following shows that higher speed also requires the use of parallelism and that modeling and inversion in geophysics are also amenable to the use of parallelism.

Geophysical modeling and in particular inversion tend to be compute bound with present day computers. Further, the computations appear to have considerable parallelism 2.2. Therefore, parallel computation is appropriate. In contrast, geophysical data processing such as wavelet deconvolution, is frequently input/output bound and faster computations may have little benefit. Increasing input/output bandwidth with more or faster channels is required in such cases before computation parallelism can be used effectively.

Previously, computers have increased performance by approximately an order of magnitude every five years for the same cost. This was accomplished by shrinking active elements on semiconductor chips to provide more faster elements for the same cost. The ability to shrink elements further has reached a natural limitation. The time to communicate between active elements is determined by an R-C (resistance-capacitance) time constant that does not decrease with size and is now comparable with the device switching speed. As the connection decreases in length and area the capacitance decreases but the resistance increases, keeping R-C con-

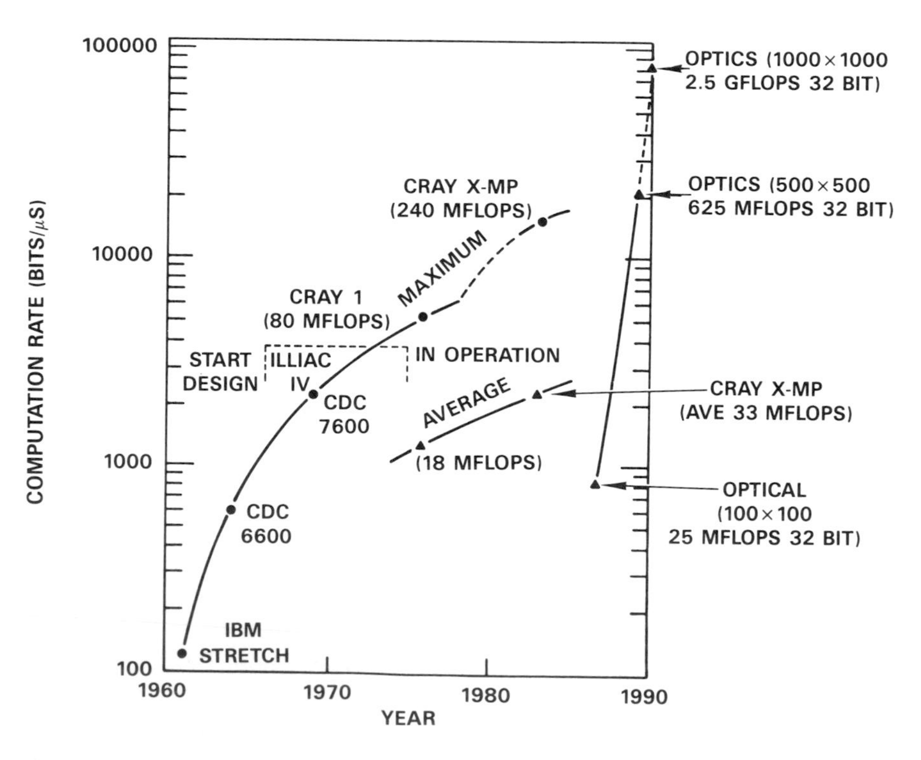

Fig. 6 Supercomputer performance with time.

stant. Another problem with more active elements on a VLSI device is due to the limited number of pins and input/output bandwidth.

Further gains in performance must be achieved by using new materials or technologies, such as GaAs, or optics, and/or by the use of many processors in parallel. Figure 6 shows how supercomputer performance was flattening off with time until Cray switched to the use of two processors for the Cray XMP. The computation speeds marked are for LINPAC, a widely used linear equation solver. The lower curve shows the performance, 33 Mflops, before hand vectorization. After careful vectorization 240 Mflop performance was achieved.

Current computers are based on the Von Neuman architecture that uses a single processor and memory unit and a program in the form of a serially arranged set of instructions. When an operand is to be called from memory an address is sent to the memory unit to save interconnection paths. An address of length $\log_2 N$ bits can select one of N memory locations. A single connection returns the result to the processor for the next processor operation. Current computers use pipelining and overlapping at every opportunity such as in arithmetic unit and in instruction processing. Hence no more opportunities arise for gaining increased performance using these techniques.

3.3 *Modularization of computer architecture*

Computers have developed with multiple levels of abstraction in hardware and software that are typical of complex systems. For example, a programmer using a high level language, such as Fortran, does not need to know the details of operation of the inside of the computer. These levels of abstraction were developed for the benefit of the user and developer. In particular, hardware and software modules may be replaced with new technology components without impacting the remainder of the system. The idea of modules is extended here to permit the user to tailor or alter his computer to meet specific computation needs in order to achieve higher performance for cost. Modularity permits the user to acquire a machine to match his needs while still having the benefit of large scale production of hardware and software modules.

Typical software modules include: operating systems and utilities, such as editors, debuggers, compilers, as well as math packages and geophysical application programs. Hardware modules include: processors, memory units, interconnection networks, and input/output processors. The software and hardware modules may be limited to optimize the performance-cost ratio for the user.

Figure 7 shows a generic parallel architecture from which a user can design a specific computer. Many of the decisions involved are dictated by currently available technologies, for example, the cache speed must match the processor speed. However, many factors may still be selected such as the number of processors units. These should be fairly powerful for good performance because of the communication overhead associated with parallelism. The size of cache is determined by the locality of addresses during operation of typical algorithms. Statistics may be collected using a simulator to determine such parameters.

The speed, throughput, and type of interconnection network must be specified and choices are discussed in the following section. The interconnection network must have sufficient bandwidth to support traffic required. However, the cost of interconnections is high and it should not be overdesigned.

The number of memory modules is determined by the need to decrease average access time available to the processors. The size is determined by the typical requirements of large computations and must minimize access to slower disk through the input/output for a virtual memory system.

The number of input/output processors and devices is selected to match the machine usage. For example, if many users are running small programs more input/output processors are useful. On the other hand inversion often requires multiple applications of a modeling algorithm and is therefore more likely to be compute bound. Therefore, fewer input/output modules relative to computation modules are required if the system is to be used mostly for inversion as distinct from modeling.

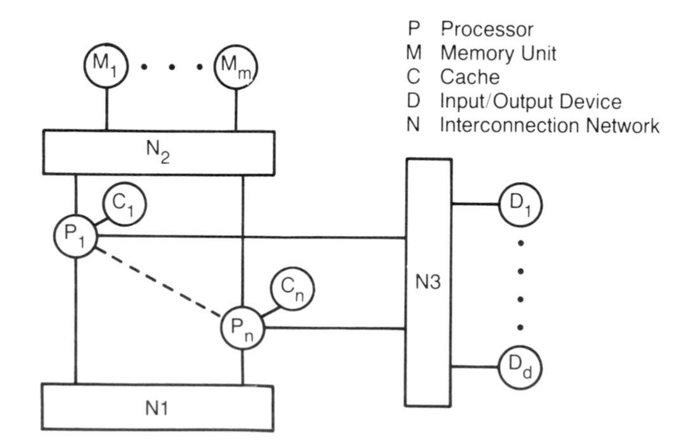

Fig. 7 Generic multiprocessor architecture.

It is advisable to model the machine at a coarse level and emulate execution of typical programs in order to select a balanced architecture. The above procedure is useful in aiding with the selection of a commercial machine because the vendors often have little data regarding the performance of the machine with a specific proposed usage.

3.4 *Difficulties of parallelism*

The distribution of effort amongst a number of processors does not remove the need for some minimum level of central control, although, for fault tolerance purposes this may not always be the same physical part of the system. The idea of a single program which alone determines the complete operation of the machine is replaced by numerous such programs running concurrently in different processors, (MIMD). The communication channel to the central control must be sufficient to prevent it from becoming a bottleneck.

Common memory is frequently used in the process of communicating information from one processor to another. A potential difficulty, memory contention, arises when two or more processors request the same piece of information from a common memory at the same time. Some arbitration is now required and one processor will have to remain idle or make the memory request again later. This increases complexity, cost and inefficiency. A simple example arises in matrix-matrix multiplication where a single row of the first matrix is required in all processors for simultaneous multiplication with each column of the second matrix. Memory contention for such well defined operations should be taken care of in the design. Optical memory may reduce the problem of memory contention because in theory several users can read the same section of memory simultaneously.

Great skill is required to partition problems such that the various processors complete their tasks at the appropriate time to provide information for the next stage. Synchronization forces everything to wait for the slowest link with resulting inefficiency. A parallel algorithm may involve more steps than a commonly used serial algorithm even though it is more efficient on a specific parallel machine. The overhead reduces the efficiency of the algorithm. Efficiency is measured as the speed with the fastest algorithm on a single processor divided by the product of the speed on the multi-processor and the number of processors. The stability and accuracy of the parallel algorithm relative to the serial algorithm must also be considered in comparisons.

Designs using the fastest economic processors are encouraged because of lack of experience in extensive parallelism and because of the need for fast scalar processing to cover non parallelizable operations. The latter operations could be performed in a separate scalar processor or host, but this could be less desirable because of data transfer constraints. Parallelism is now used to further increase performance.

A major difficulty involves balancing computation power of a single processor, inter processor communication and input/output communication. At present, the cost of fast processors is generally greater than that of the communication links, so cost effectiveness is reduced if processors must remain idle because of intercommunication delays. Conceivably, decreasing processor costs, resulting from advancing technology or the use of less powerful processors based on better understanding of parallelism, could alter the relation between desired utilization of processors and communication channels. Balancing the communication and computation speeds is accomplished by suitable selection for the processor, communication bandwidths, and amounts of local memory at each processor. Other factors will arise during the discussion of specific architectures following.

3.5 *Architecture alternatives*

The large problems of interest involve repetitive operations on different data sets so that overhead is reduced by decoding instructions once and applying the instruction to many processing elements in a single instruction stream multiple data stream machine (SIMD). A similar feature is used in vector machines except that pipeline length is generally limited by the number of stages into which an operation may be subdivided. In a SIMD machine the number of processors that may be usefully employed is limited only by the average parallelism of the application.

A fixed word length is appropriate for these problems because most computations require at least 64 bit floating point. In a serial machine or uniprocessor, longer words are required for faster machines because more computations tend to be performed on the same data. Roundoff error is more likely to occur, e.g., there is more chance that numbers of similar size will be subtracted.

More flexibility is desirable for research applications. In this case, multiple instruction stream (MIMD) machines are suitable. The granularity is likely to be less as more complex control and communication protocols are applied. The advantages and disadvantages of alternative interconnection networks shown in figure 8 are discussed [9]. The complexity of the interconnections affect development time, speed, latency, cost, and performance.

Bus Architectures. Bus architectures, figure 8(a), section 5.1, are expected to become dominant as a means of permitting expandability up to a relatively small number of processors, (around 20), each processor having some local memory. Existing processors and peripheral devices may be attached to a common bus. Multiple busses may be used to extend capability. Examples of machines available today are: Sequent, Flex, Encore, and ELXSI. Inherently parallel problems, e.g. multi-task, run efficiently on the largest number of processors. The ability to efficiently run a single problem on many processors requires sophisticated software environments including parallel operating system, parallel language extensions and math libraries. Analysis in section 5.1 shows the limitations for performing geophysical finite approximation computation on a bus connected system.

Simply Connected Arrays. Simply connected arrays, figure 8(b), section 6.2, are also readily manufactured in a cost effective manner. The minimization of connections and cross overs reduces power and cost and increases reliability. The class may be categorized by the number of ways a processor is connected and

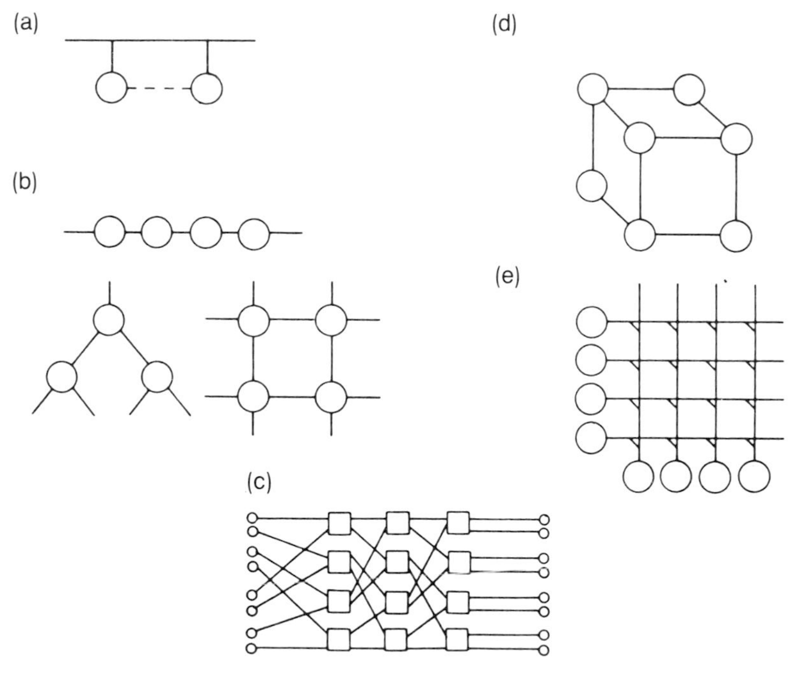

Fig. 8 Interconnection network alternatives.

includes 2-way, 3-way and 4-way or mesh. The Non-Von machine at Columbia has a tree connected structure. An upper level bottleneck tends to arise with trees.

Meshes are appropriate for image processing (Goodyear-NASA Massive Parallel Processor) and physical spatial problems which may be mapped to a grid, such as many finite approximation problems. An optical mesh connected computer is considered for relaxation algorithms with finite approximation computations in section 6.2. Symbolic substitution is another nearest neighbor approach used in optical computing [5], [14].

Nearest neighbor arrays have severe limitations for general computation. The processing of information must exceed the time to load and unload the results if the array is to achieve significant gains in performance. Unfortunately, many algorithms require the amalgamation of information at some stage. For example, an inner product, convolution or correlation, requires many multiplications which may be conducted in parallel, followed by summation of the results. If n results are summed in odd-even pairs, also referred to as recursive doubling, only $\log_2 n$ steps are required but it is necessary to move the results an increasing distance across the array after each step. Communication time dominates after a certain size array is reached. A similar problem may exist with many algorithms which achieve speed in this way, e.g. FFT's.

Systolic Arrays. Systolic arrays [10], figure 8(b), section 5.2, use information many times in sequence once it has been fetched from memory. This permits the efficient use of a large number of processors, perhaps 100, for well structured problems, as arise in real time signal processing or numerical subroutines, e.g. solving linear equations. 2-D arrays may be constructed on a similar principle. The structure permits data and addresses to enter the perimeters of the array and flow through the array while generating results at the boundaries. Systolic arrays are operating at the Naval Ocean Systems Center, Carnegie Mellon University (CMU), ESL, Lincoln Labs., and SAXPY. The CMU array is being manufactured and marketed by GE. Section 5.2 describes the use of a linear systolic array for 1-D modeling and inversion.

Interconnection Network Systems. Popular interconnection networks are discussed. The hypercube, figure 8(d), was developed at CalTech as the Cosmic cube and has been adopted in the Intel iPSC, N-Cube, and Floating Point architectures. The increase in connections to nearest neighbor in higher dimensions increase interconnection bandwidth. The multistage shuffle, figure 8(c), also provides more paths than a nearest neighbor and is representative of multistage interconnection networks used in the Bolt Beranek and Newman (BBN) Butterfly machine. The crossbar switch, figure 8(e), provides the highest interconnection capability and also has the highest cost. CMU has performed experiments with such a switch in the past. An optical crossbar switch architecture is considered in section 6.1. Optics could provide a performance/cost ratio required to make crossbar switches competitive.

Dataflow and Other Architectures. A computation is performed as soon as a processor has all the necessary information in a data flow oriented system. This avoids overhead in many applications. Data flow is used as an upper level control mechanism in some signal processors. Data flow machines, such as that under investigation at MIT, pass a computation to an operation or decision box as soon as all the data is available. The optical crossbar switch system described in section 6.1 is used in a static dataflow manner.

4. OPTICAL COMPUTING ISSUES

4.1 *Advantages of optics*

Optics has several critical advantages over electronics for interconnections. The principle advantage is immunity to interference. Unlike electrons, photons in linear media tend not to interact with one another. Optical signals may be transmitted through space or fibers. Large numbers of optical fibers carrying high frequency modulated signals may be mingled without crosstalk and electromagnetic fields passing through loops of fibers do not cause interference.

A second advantage of optics is virtual freedom from capacitative loading due to the optical interconnection line. This permits higher bandwidths than with electrical interconnections. Higher bandwidths permit a few fibers to replace a larger number of coax cables. Faster signal propagation between units decreases sensitivity to different interconnection lengths.

The optical computer designs discussed in this chapter use free space to perform optical interconnections between 2-D and 1-D arrays of devices. A large number of high bandwidth optical interconnections may be made in parallel when arrays of devices with a large number of elements are used. One of the goals for spatial light modulator (SLM) technology, section 4.2, is to produce devices with over a million elements. Connecting more than a million elements in a device is difficult using electronics. The free space coupling through spatial light modulators permits dynamic reconfigurability of the interconnection network, a feature not easily accomplished with electronics. This is powerful for computing and forms the basis for the system in section 6.1 in which the interconnection network is reconfigured to match the algorithm graph interconnections.

Other advantages of optical computers follow [38], [42]. First, optical computers interface more easily with increasingly used optical communication networks. Second, GaAs is optically active and much research is directed toward GaAs semiconductors for high speed electronics. This will permit multiple-channel high-bandwidth optical connections into chips. Third, the parallelism needed to meet continuing demand for higher computer performance has led to a desire for fast interconnection networks, for which optical technologies look promising. Optical disks are expected to replace magnetic disks and optical disks lend themselves to parallel addressing which decreases access time. Fourth, there is an increasing interest in symbolic computing for software development and artificial intelligence. Optics is suited to the 2-D pattern matching required as well as to graphical input/output [11], [14], [15], [20], [21].

4.2 *Using Spatial light modulators (SLM)*

A feature of optical processors is the desire to process information in 2-D planes using spatial light modulators (SLMs). These are like film with the additional capability to modify the reflectivity or transmissitivity electronically or optically. Typical examples arise in digital watches and miniature televisions. This matches seismic processing in which processing is often performed on 2-D arrays. A generic all-optical computer processes the 2-D data in an SLM by repeated operations until the result is obtained.

SLMs may be classified according to whether they are transmissive or reflective and whether they are settable optically or electronically. Reflecting SLMs require beam splitters to separate the light incident on the SLM from that reflected. Discussion relating to transmitting SLMs is applicable to reflecting ones. The transmittance of each pixel in a 2-D grid of pixels on an SLM may be varied electronically or optically. Electronically set devices perform the interface from electronics to optics, for example in input/output devices such as displays.

Figure 9 shows how an SLM may be used for storage, multiplication, or interconnections. Binary numbers are stored in the SLM in the optical computer

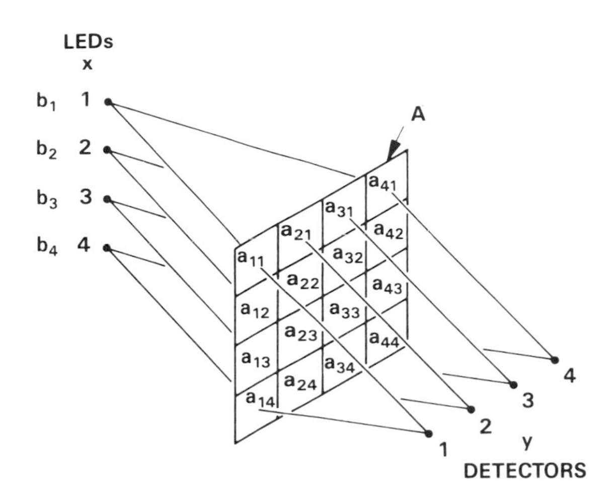

Fig. 9 Spatial light modulator functionality.

described in section 6.1. A one is associated with full transmittance and a zero with no transmittance. Hence the SLM acts as a switch at each pixel to permit or to block incoming light at each pixel position. These switches are used to form a crossbar switch.

The optical computer described in section 6.2 uses SLMs as pixel arrays for storing analog numbers and for multiplication as light passes through the SLM during interconnection [18], [20]. An image consisting of an array of pixels will be modified on passing through an SLM, the effect being that of analog multiplication of the intensity of each incoming pixel by the transmittance of the SLM pixel through which the light passes. Analog multiplication is performed naturally when light passes through transparent media, for example, transparencies laid on top of each other provide the product of the two intensities at every point. Summation is accomplished in optics by means of lenses. Focussing of light by a cylindrical lens represents the summation of pixels displaced in the direction at right angles to the axis of the lens.

4.3 *Types of spatial light modulator*

Numerous different phenomena are being investigated for use as spatial light modulators [3]. Features of interest in selecting an SLM are the throughput, speed of setting, and whether optically or electronically addressable. In general throughput is much higher than for electronic interconnections but the time required to reconfigure the switches is longer. Liquid crystal and magneto-optic devices are currently available. The liquid crystal display in miniature commercial television sets have tens of thousands of elements and have been used in experiments.

The deformable mirror device (DMD), figure 10, is constructed from a 2-D array of silicon MOS transistors with a reflecting layer deposited over the surface. The layer is etched to produce cantilever beams hanging over the transistors. Activation of a transistor causes the cantilever beams to deflect down due to the electrostatic attraction. A collimated beam of light striking the surface will have its phase delayed where the surface is indented due to the additional distance travelled by the wave. Spatial phase variation $\theta(x, y)$ is not visible in an intensity plane as $|e^{j\theta(x, y)}|^2 = 1$. Phase is converted to intensity in the applications of interest by means of a spatial filter [6], [18]. The spatial frequency filter also removes low spatial frequency components arising from undeflected pixels and from regions between pixels.

Arrays of GaAs devices that are in the early stages of development include the Self Electro-Optic Effect Device (SEED) [40] and the Digital Opto-Electronic Switch Device (DOES) [44]. The SEED device is a three optical port transitor

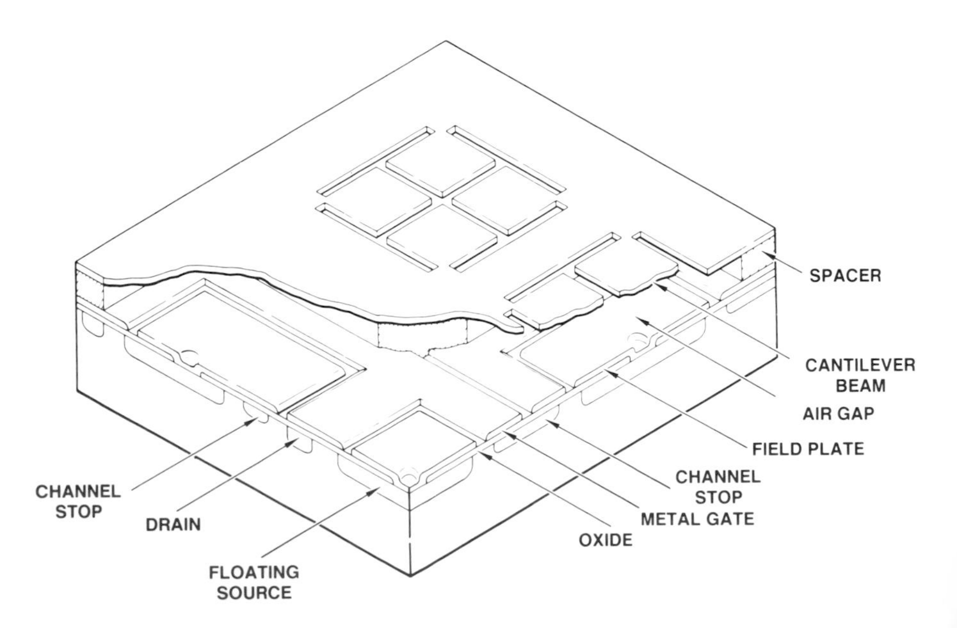

Fig. 10 Deformable mirror device.

consisting of layers of GaAs-GaAlAs. A low level optical signal illuminating a photodetector on the edge of the top surface will prevent light from passing through the device. Removal of this signal light will permit light to pass through the device. The device has hysteresis, permitting light to pass or not pass through the device according to the past history of the input intensities. This bistable operation permits the device to act as an optically addressable memory.

The DOES device is a transistor with an opening by the gate input that permits light to enter the device. Hence, the device may be switched *on* with light or with an electronic signal. A device that is switched *on* will have a low electronic signal out and will not emit light. A device that is switched *off* will have a high electrical output and will also emit light. Devices in this family are characterized by whether the emitted light is incoherent (LEDISTOR) or coherent (LASISTOR).

There are other optical devices under development that can perform some of the same functions as SLMs. Examples are photonic rebroadcasting devices that have the effect of storing photons in a 2-D pattern under exposure by one frequency and of releasing photons under exposure by another frequency. Such devices have high resolution and speed and are expected to impact read/write optical disk technologies. They can also be used in place of SLMs in the systems described in this paper. Experiments involving these devices are underway at Wright State University [12], [13].

5. PERFORMANCE OF ARCHITECTURES WITH ALGORITHMS

Two algorithm-architecture combinations are considered in more detail in order to assess performance. The first, section 5.1, involves mapping finite approximation techniques to a bus connected multiprocessor system. The bus limits the number of processors, preventing massive parallelism. The second case, section 5.2, considers mapping of 1-D modeling and inversion to a systolic array. This is limited by the fixed interconnection of the array. More flexible computers are considered in the following section 6 where optics is used.

5.1 *Finite elements with a bus architecture*

Architecture. A parallel architecture was hypothesized for solving finite approximation computations arising in 2-D and 3-D seismic modeling and inversion, section 2.3, [34]. The system has eight boards containing four processors and a control unit. The processors on the boards are connected by nearest neighbors. The boards are interconnected via a standard 37.5 Mbyte/second bus. Each processor has 32,000 words of 64 bit memory. A range of different processor speeds are assumed between 0.1 Mflops and 10 Mflops for 64 bit multiplications in order to reflect devices currently available and under development.

Operation. Successive overrelaxation, section 2.3, equation 17, is used to solve for field points on a 2-D or 3-D mesh. The mesh is subdivided evenly among the processors and the matrix coefficients, **a**, and overrelaxation factors, ω, $(\omega - 1)$, are loaded into the appropriate processor memories. Colors are assigned to the nodes in the sequence Red, Black and Green from left to right (see section 6.2). Successive rows are shifted one place to the right and wrapped around [2], [34]. All processors compute their assigned color first and then cycle synchronously through the other colors in the order Red, Black and Green. After a predetermined number of iterations, checks are made to determine whether adequate convergence has occurred. This involves computing the sum of the squares of the node values at a processor and the sum of the squares of the node value changes since the last iteration. This information is assembled by the host in order to compute whether convergence is adequate and may be computed while the array boards continue with the next iteration. A 3-D SOR algorithm may be developed similarly for an element shape selected.

Performance analysis. The amount of data that must be transferred from one board to another, for one of the 3 colors, during computation of one iteration of equation 17, is the perimeter of the data stored on the board for that color.

For a 3-D mesh, the number of colors required also depends on the element shape. 6 colors are required for tetrahedral elements because 6 lines pass through every point and 13 colors are required for cubic elements. We will use 6 colors as this is more efficient for our machine. However, we assume the mesh of elements is fitted inside a cube for simplicity of calculations.

For a 2-D triangular mesh and a 2-D configuration of processors on a board, the only data transfer required on a board is between neighbors and occurs concurrently for all processors.

For a 3-D mesh and a 2-D configuration of processors we consider a simple, though not necessarily efficient arrangement, in which the board handles a part of one horizontal plane of the 3-D mesh. In this case, information involving the upper and lower faces of a cube is transferred to the board edge from every processor via others during one iteration. A third cube face is included for the nearest neighbor horizontal path. I proposed a new algorithm in reference [34] which improves performance by permitting overlap between data transfer and computation. This can be used to reduce the effect of bus and nearest neighbor communications.

We observe that the above approach is equally applicable to a finite difference algorithm. In this case the red-black algorithm would be extended to a four color algorithm in order to overlap data transfer and computation.

Performance results. Figure 11 shows the performance for solving the equations for a 285 by 285 2-D mesh problem as a function of log processor speed. The slowest processor, 0.1 Mflops, is available today. The fastest is slightly faster than single chip processors anticipated shortly. The efficiency is shown in figure 11(a) for the conventional 3 color algorithm and a new 6 color algorithm. The dashed curve shows the efficiency if the bus is assumed perfect, requiring zero delay to communicate data. For the 3 color algorithm, the efficiency for a 0.1 Mflop processor, is 98% and falls off for hypothesized future processors because the bus communication time increases relative to the computation time. The new algorithm permits overlap of bus communication and computation. Processors of speed less than 5 Mflops now no longer have to wait for the bus. We observe that for a 0.63 Mflop processor, the efficiency is improved from 92% to 99% by the new algorithm. Figure 11(b) shows the estimated time to solve the equations for the 2-D mesh for the conventional algorithm and the new algorithm. The time

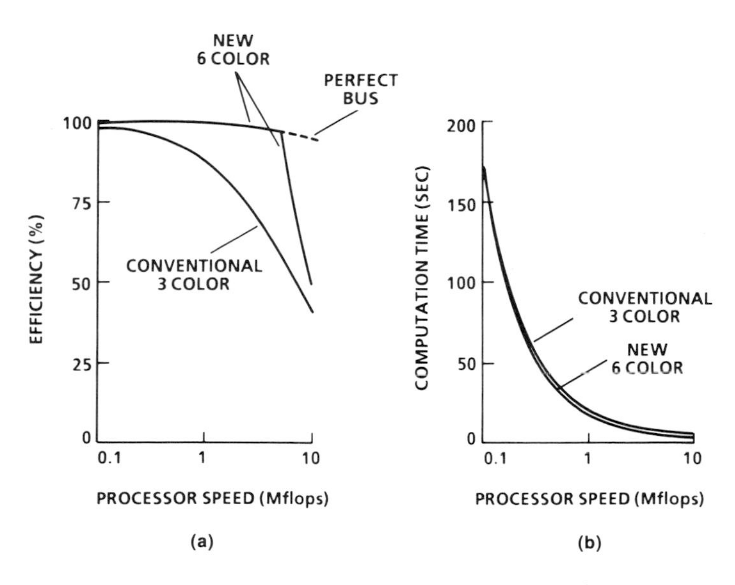

Fig. 11 Performance estimates for 285 by 285 mesh on a bus architecture.

decreases even when the efficiency is falling off. For a 5 Mflop processor the new algorithm is approximately 40% faster.

Figure 12(a) shows the estimated performance for solving the equations for a 40 by 40 by 40 3-D mesh as a function of processor speed for a conventional 6 color algorithm and a new 12 color algorithm. The efficiency of 89% for the conventional algorithm and a 0.1 Mflop processor is lower than for the 2-D mesh because more data has to be transferred on the bus relative to computation. The efficiency drops off rapidly. The use of the new algorithm provides an impressive improvement up to 0.63 Mflops. Above this, the bus is essentially saturated for the algorithm chosen.

The dashed curve shows the situation with no bus delays. This shows that with the small number of processors per board, the communication delays on the board, although worse than for the 2-D case, are not significant until closer to 10 Mflops. This suggests that the balance between bus and board communication delay could be improved by having more processors on a board and fewer boards. This is entirely feasible with future VLSI. The impact on board communication may be further reduced by applying cyclic reduction in a similar manner as was used to permit computation and bus transfer. The interfaces are now considered to be between processors on a board rather than between boards.

Figure 12(b) shows the computation time for the conventional 3-D 6 color algorithm and the new 12 color algorithm. There is a 40% reduction in computation time for a 0.63 Mflop processor. At this point bus transfer equals com-

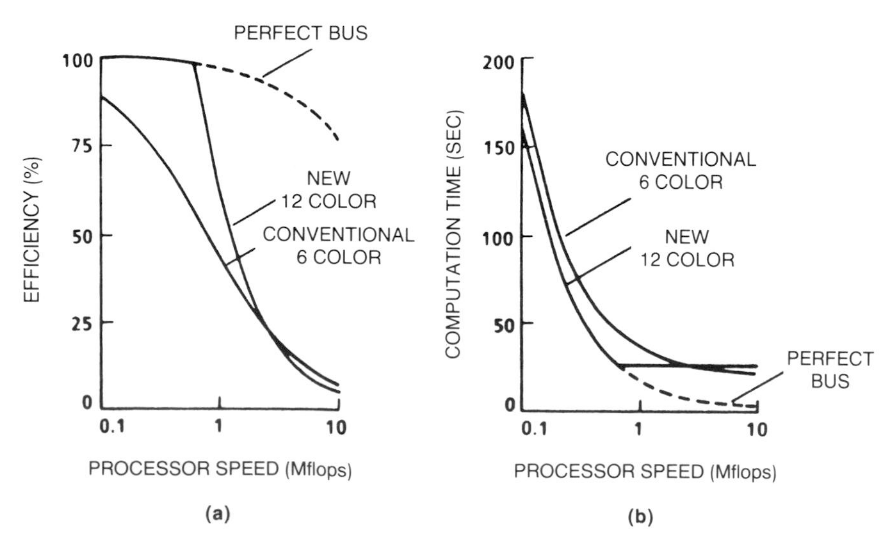

Fig. 12 Performance estimates for 40 by 40 by 40 mesh on a bus architecture.

putation time. Consequently, the use of faster processors will not improve speed with the algorithm selected.

Significantly faster performance may also be achieved by using the more complex multigrid method, which on a serial machine can reduce the number of operations from $O(N_n N_l)$ to $O(N_n)$, [4], [39]. Multigrid loses some of its efficiency on a parallel machine [7] but for the machine discussed, there are many nodes relative to the number of processors and multigrid techniques should still be efficient.

5.2 1-D *modeling and inversion with a systolic architecture*

Three parts of the 1-D inverse estimation computation are dominant, section 2.2, [25], [30], [32]. An implementation is selected that attempts to configure these parts on the systolic array in such a manner that the pipelining feature of the systolic array is fully utilized and at the same time other data transfers are minimal. Preliminary performance estimates suggests that it is possible to use high levels of parallelism of this simple form for this type of problem.

Computation of propagator matrix with systolic array. Figure 13 shows the computation of the synthetic data by implementing the propagator matrix method on a

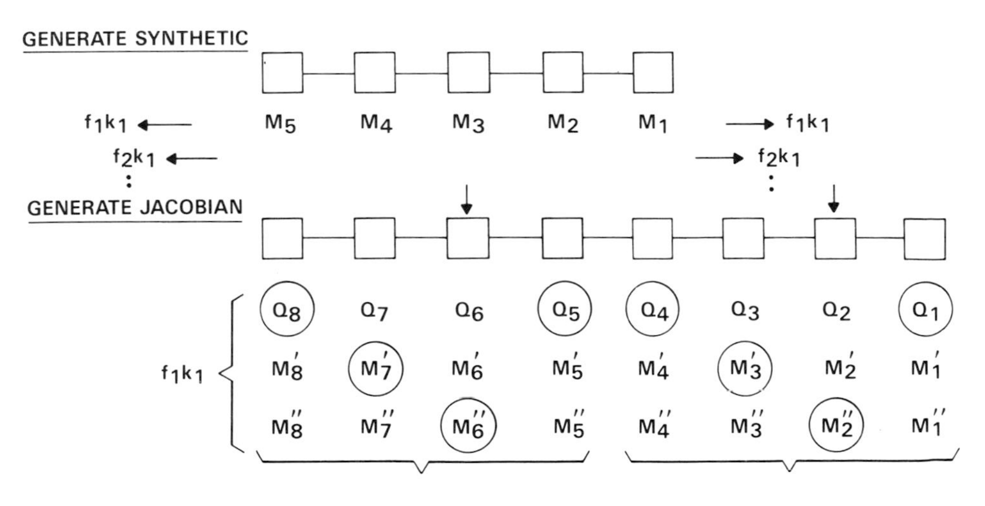

Fig. 13 Propagation and Jacobian matrix computations with a systolic array.

systolic array. The layers are mapped to processors, the jth processor handling the jth layer. Several layers may be assigned to one processor if necessary. The number of processors is assumed equal to the number of layer parameters to be estimated and only one parameter (sound velocity) is assumed per layer for the purpose of simplification. Each processor is supplied with density and compressional velocity for its layer. It then computes the 2 by 2 matrix elements for its layer. The product of matrices is accomplished by taking results from the right, multiplying by the 2 by 2 matrix and passing the result to the left. The 2500 independent computations indicated in table 1 for frequency and wavenumber follow in a pipelined manner. The synthetic data results flow out of one end of the array and are stored for future use.

The result of computation up to a processor is stored in the processor for use in computing the Jacobian matrix, equation 15. The computation is performed in the reverse manner to obtain partial matrix products for the other direction. The special derivative matrices are computed at every processor. The elements of the Jacobian matrix are now computed. For example, at the box marked with an arrow above $Q6$, the Jacobian element is obtained by adding 2 terms. The first is obtained by multiplying the left side partial product $Q7$ by the derivative matrix $M6'$ and then by the partial product for the right $Q5$. The second is obtained by multiplying the left partial product $Q8$ by the special derivative matrix $M7$ and then by the right partial product $Q6$. A row of the Jacobian, the elements of which correspond to derivatives with respect to different layer parameters, may be computed in parallel with the array.

Computation of conjugate gradients with systolic array. A row of the sensitivity matrix is spread across the array as a result of the computation in the previous section. One row corresponds to one frequency and wavenumber combination and the jth element of the row corresponds to the derivative with respect to the jth layer parameter and is located in the jth processor. For example, matrix elements $A_{p,1}$, $A_{p,2}$, $A_{p,3}$, and $A_{p,4}$ would be stored in processors 1 to 4. Similar results are obtained for other frequency-wavenumber combinations p, therefore each processor stores one column of the sensitivity matrix. The following procedure permits the Jacobian matrix elements to remain in the processors in which they were computed throughout the iterations of the conjugate gradient method. Column normalization is performed by each processor independently.

The other computationally demanding steps in the least square conjugate gradient algorithm are the two matrix-vector multiplications, equation 24. The first matrix-vector multiplication involves a vector of length the number of layers. This

TABLE 1

Natural parallelism for prestack layer inversion.

f ~ FREQUENCY (250), k ~ WAVENUMBER (20), r ~ RECEIVERS (90), l ~ LAYERS (250)

TASK	COMPUTATION	INDEP. TASKS TYPICAL	TASK DEP.	TASK PARALLELISM	TYP. VAX TIME (sec)	
					TASK	TOTAL
FORWARD MODELING						
GENERATE SYNTHETIC DATA	PROPAGATOR MATRIX	fk/2 (2500)	100l	l	0.1	250
INSERT APERTURE:						
SPATIAL	HANKEL T	f (250)	2rk	r,k	0.02	5
TEMPORAL	FOURIER T	r (90)	$2f \log_2 f$	f	0.02	2
INVERSE COMPUTATION						
GENERATE JACOBIAN	MODIFIED PROPAGATOR MATRIX	fk/2 (2500)	300l	4	0.3	750
INSERT APERTURE:						
SPATIAL	HANKEL T	f (250)	2nrk	r,k	0.2	50
TEMPORAL	FOURIER T	r (90)	$2nf \log_2 f$	f	0.2	20
SOLVE LINEAR EQUATIONS (n iterations)	CONJUGATE GRADIENTS	1	20nfkl	fk,l	1000	1000

vector is passed into the processor array so that the jth element is at the jth processor. Each processor multiplies its Jacobian matrix column by the corresponding vector element. The first entry of the resulting vector is obtained by starting at one end of the array and moving across the array while accumulating the values obtained using the 1st row of the Jacobian matrix. The second entry of the resulting vector uses values computed from the 2nd row of the Jacobian and lags the accumulation of the 1st entry by one processor. Further pipelining provides the remaining elements of the resulting vector.

The resulting vector has length corresponding to all combinations of frequency and wavenumber. The vector may be considered to be a 2-D temporal and spatial array. A 2-D transform to time and space is now applied to add aperture effects and an inverse transform is used to return the data to the frequency-wavenumber domain. These computations may be implemented reasonably efficiently on the array. This is adequate because they are not as computationally demanding as the propagator matrix and conjugate gradient computations, table 1.

The second matrix-vector multiplication involves the transpose of the Jacobian matrix and the vector obtained from the last step. In this case, rows of the transpose Jacobian are already stored in each processor. The vector is entered into the end of the array one element at a time. Each processor multiplies the next element of the vector by the conjugate of the next element of the column of the Jacobian matrix and accumulates the result at that processor. The resulting vector is spread across the array ready to add the incremental addition for updating required in the conjugate gradient method.

The remaining parts of the conjugate gradient algorithm 25, 26, are much less computationally demanding. However, they may become significant if a gain of several hundred is achieved with several hundred processors. Prior work has shown that even the inner product computation may perform better than possible with recursive doubling by expanding in terms of the inner product from previous iterations [46], section 6.1.

6. OPTICAL ARCHITECTURES

The above architectures have limits to their ability to use massive parallelism, mainly because of interconnection delays between processing elements. This section considers optical architectures that use optics to provide higher bandwidths between processing elements. However, the use of optics in computing is not yet well developed and is dependent on progress in the development of suitable devices for interfacing with electronic computing.

Two architectures are considered. The first, section 6.1, has an optical crossbar switch for interconnecting electronic processors. This permits efficient mapping of a wide variety of algorithms. It requires efficient devices for conversion to and from electronics and optics. Such devices are being developed. The second, section 6.2, uses optics to perform computation as well as interconnection. However, it has a nearest neighbor connection that limits its application to finite approximation computations that can be distorted to form a regular grid.

6.1 *Optical crossbar interconnected computer*

Implementations are considered of geophysical spectral (section 2.1) and conjugate gradient (section 2.4) algorithms on an optical interconnected computer for the purpose of illustration and performance estimates.

System description. A novel computer architecture was proposed previously that uses a deformable mirror SLM as an optical crossbar switch [28], figure 14.

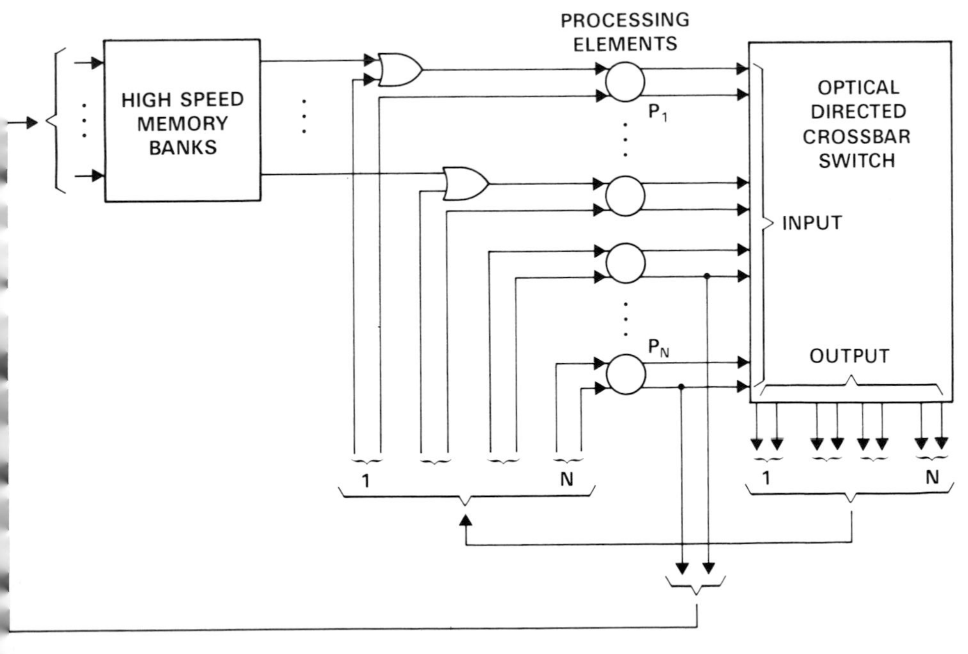

Fig. 14 Optical crossbar interconnected processor.

Hundreds of processing elements, P_i, $i=1$ to N, are connected to the optical switch of size $2N$ by $2N$ by means of commercially available fiber optic links of bandwidth 160 MHz or more. This bandwidth is adequate to keep up with the output converted to serial from a state of the art 32 bit floating point multiplier. Higher bandwidth links are feasible at greater cost, the limitation being the parallel/serial conversion.

The processors perform elementary operations such as multiply or add and therefore have two input connections for the two operands. This fine granularity permits the maximum amount of parallelism to be extracted from algorithms. The processing element output is converted from parallel to serial in a shift register and used to drive a laser diode. A second output is provided for systolic array implementations. The laser diodes are connected via optical fiber links to the optical crossbar switch. Fibers returning from the switch connect to light sensors at the processor inputs. A second fiber optic loop between processors and main memory banks provides input/output. The logic permitting input/output may be mounted alongside the processor. The flow of data is prearranged so as to minimize run time overhead.

Implementation of a number of different algorithms on the proposed system was considered in order to evaluate performance: fast Fourier transform (FFT), systolic filter, matrix-vector multiplier [28], nonlinear spectral estimation, conjugate-gradients [19], and symbolic processing. The crossbar switch and extension of the system are described next. FFT and nonlinear spectral estimation algorithms are used to illustrate algorithm mapping and programming methodology.

SLM crossbar switch. Estimates indicate that software often accounts for 80% of computer development cost and also dominates maintenance. Therefore, more expensive hardware may be justified if it reduces software costs significantly. Crossbar switches are more expensive than incomplete switches but should make software development significantly easier because all paths are equal, any combination of connections is possible and switch conflicts are reduced. The latter reduces queuing and protocol complexity. Automatic and efficient mapping of algorithms is more feasible because searching for optimal paths, configurations and sequences is minimized.

Limitations in increasing the performance of uniprocessors and requirements for fault tolerance are driving computer technology to multiprocessing systems that require consideration of interconnection networks. Electronic crossbar switches are appearing in some applications, e.g., in the Texas Instruments VHSIC array processor and in the Carnegie Mellon systolic array processor. The advantage of using optics relative to electronics for a crossbar switch is that is should be possible to construct larger higher throughput switches [43]. Optical fibers may be used to lower cost by reducing the number of connections to an electronic switch by using serial transmission and/or multiplexing. In this case it makes sense to use an optical crossbar rather than an electronic one. The optical switch avoids the use of N^2 wires for an N by N switch by using connections across free space. Lenses can provide large fan-in and fan-out.

Large optical switches require gain to recover energy lost in fan-out and fan-in. Devices with gain such as the SEED and DOES device are needed. Alternatively, image intensifiers may be used to restore gain.

Figure 15 shows how a transparent spatial light modulator acts as a crossbar switch. Each intersection in a crossbar switch, figure 15(a), has a switch permitting a horizontal input line to be coupled with a vertical output one. One output receives information from one input but one input may broadcast to several outputs in a *generalized* crossbar switch. Figure 15(b) shows a diagrammatic crossbar switch implemented with a spatial light modulator (SLM) and dots indicate transparent regions consistent with the closed switch settings marked by dots in figure 15(a). An optical lens system is used to spread the light from the input sources horizontally without spreading the light vertically. Light passing through the spatial light modulator is collapsed onto receiving diodes by means of a lens system which focusses vertically without spreading horizontally. Arbitration is avoided at

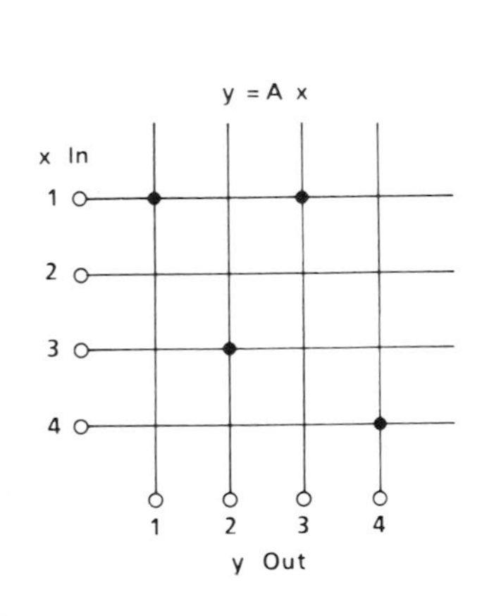

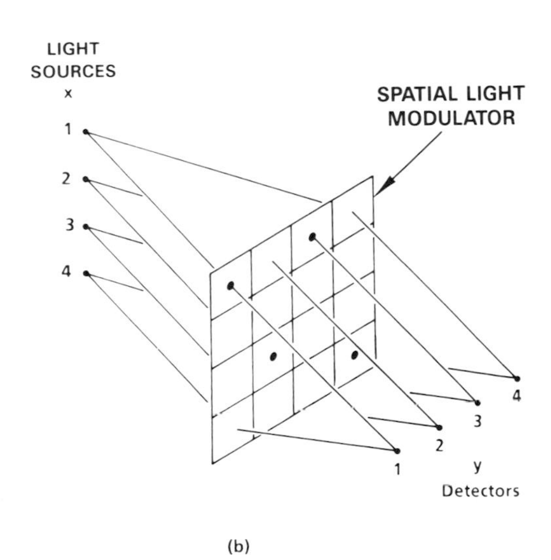

Fig. 15 Optical crossbar switch.

execution time by the use of static data flow. The size of switch may be analyzed or measured in experiments and is limited by the crosstalk or leakage between cells, for example, arising in the spatial filtering system discussed earlier. The next section describes methods of producing larger switches.

Extensions to larger systems. Larger crossbar systems may be constructed from groups of individual crossbar switches [23]. This permits expandability, scalability, and fault tolerance, section 3.1. One approach is to use four crossbar switches of size $N/2$ by $N/2$ to construct a double size N by N crossbar switch system. $N/2$ inputs are connected in parallel into crossbar switches one and two and the other $N/2$ inputs are connected in parallel into switches three and four. The outputs of switches one and three are connected in parallel to produce $N/2$ outputs and the outputs of switches two and four are connected in parallel to produce the other $N/2$ outputs.

An alternative is to use only two $N/2$ by $N/2$ switches together with two sets of $N/2$ switches, figure 16. This uses less switches than the four crossbar switch system. The loss of some broadcasting capability is not significant. The fixed interconnections between switches and crossbars can be seen to be shuffle and inverse shuffle networks. These can be readily implemented in optics. Larger crossbars are obtained by further doubling the system and adding more shuffle networks and switches [23].

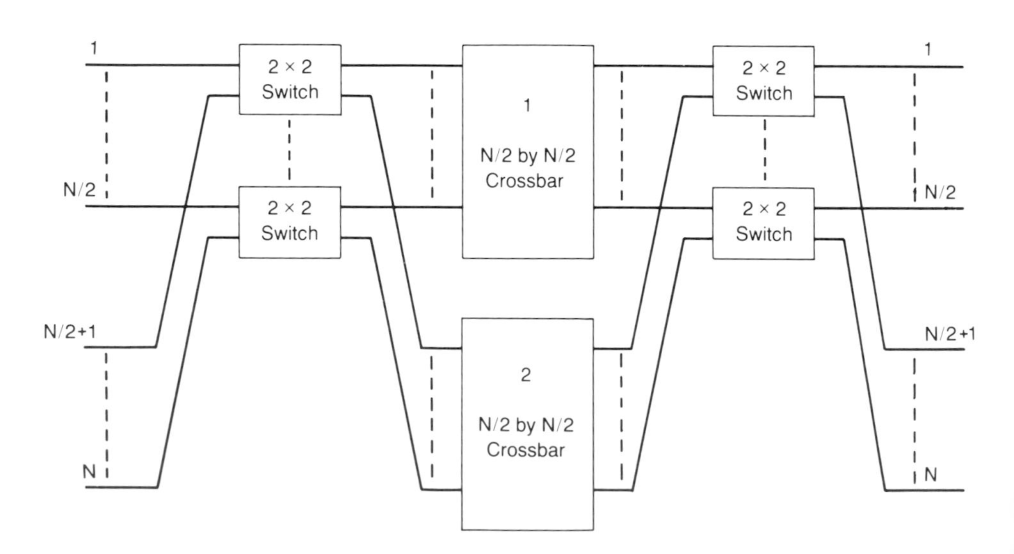

Fig. 16 Extended multiple crossbar system.

Fast Fourier transform illustrates mapping. Figure 17 shows a graph for a decimation in time FFT, section 2.1. Figure 17a shows the bit reversal for the start of this configuration. Figure 17b shows the fixed configuration stage which is used at each iteration by feeding the output at the right back to the input $\log_2 N$ times for an FFT of length N. The weights $\mathbf{w}$, which correspond to the appropriate exponential terms in equation (1), must be altered on each iteration. The output for the particular interconnection strategy shown in figure 17 as a function of the input $\mathbf{x}$ may be summarized by

$$\mathbf{y} = \left[\prod_{i=1}^{\log_2 N} \mathbf{s}_{2N,N} \gamma'_{N,2N} \mathbf{w}_{iN} \right] \rho'_N \mathbf{x} \tag{27}$$

where ρ'_N represents a bit reversal interconnection network, ($'$ is used to represent a network), $\gamma'_{N,2N}(x)$ is a shift by N and straight across network, $\mathbf{w}_{iN}$ is an N long vector that is dependent on loop number and used for parallel multiplication, and $\mathbf{s}_{2N,N}$ represents summing in pairs to reduce vector dimension from $2N$ to N.

Figure 18 shows the FFT implementation on a 24 by 24 crossbar switch. The FFT input is fed into processors 1 through 8, which pass the data through to the

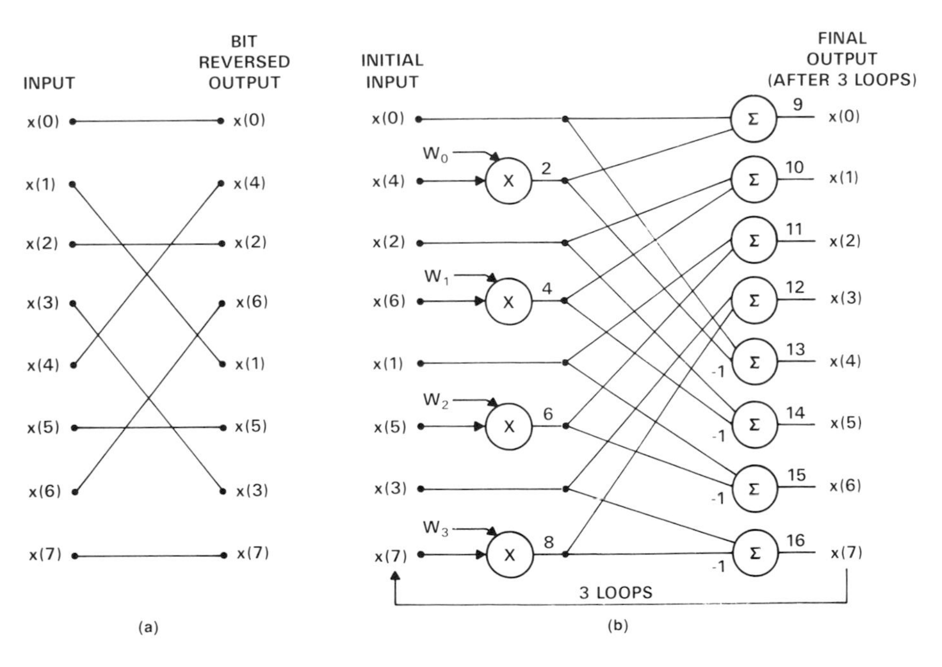

Fig. 17 Fast Fourier transform flow graph (a) Bit reversal (b) Fixed configuration stage.

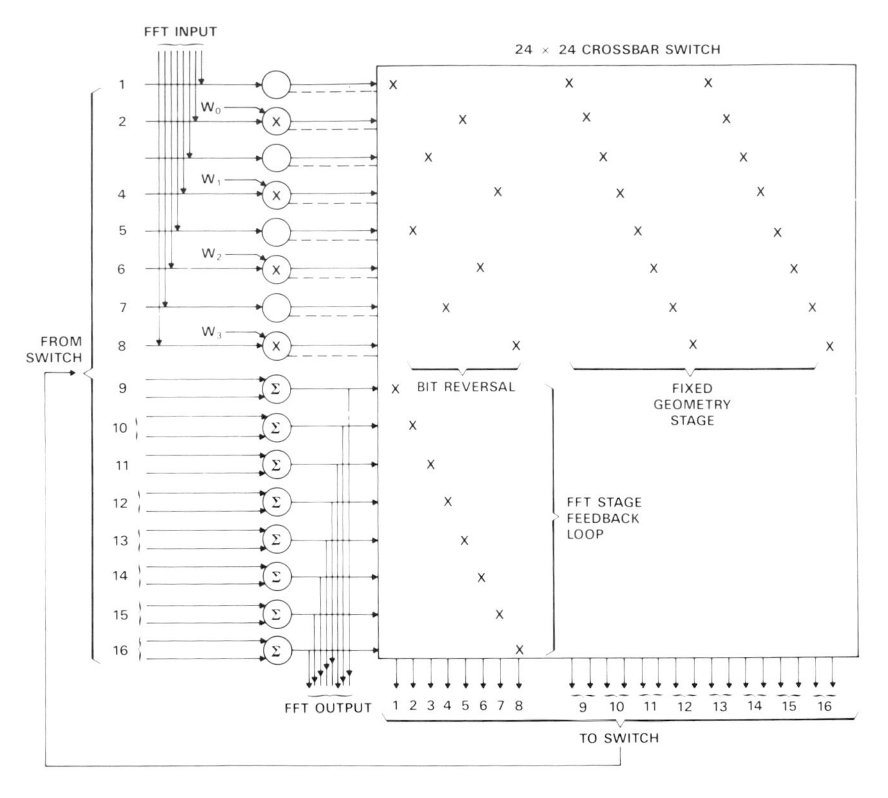

Fig. 18 8 point FFT implementation on 24 by 24 crossbar switch.

switch (i.e., multiply by one). The left upper quadrant of the switch converts the data to the bit reversed sequence as required in figure 17a and returns it to processors 1 to 8. The first set of weights **w** are used for the first loop of the FFT stage, figure 17b. The data is passed via the top right quadrant of the switch which implements the fixed configuration FFT graph, to the adders, processors 9 through 16. The lower left quadrant of the switch is used to return the data to processors 1 through 8 for the next loop of the FFT. After $\log_2 N$ loops the output is taken from the adder processors 9 through 16. Memory management is simplified because a crossbar can provide the correctly ordered output.

Nonlinear spectral estimation illustrates programming methodology. The Levinson-Durbin algorithm, equations (5), (6), and (7), is difficult to parallelize because at each iteration the dimension increases and cross interconnections are required in equation (7). Specifically, computation of a_k at the nth iteration requires a_{n-k} from the previous iteration. As a large number of processors are considered in the parallel system it is possible to unroll the iterations into a long section of code. Unrolling permits the overhead, such as that associated with loop counters, to be significantly reduced. In order to simplify the figures and explanation we unroll four iterations. The resulting code is shown in figure 19. A second subscript was added to the autoregressive coefficients to indicate the loop number.

Figure 20 shows the flow graph for implementing the code in figure 19. Autocorrelation function values r_1, r_2, and r_3 for lags, 1, 2 and 3 are input at the top of the graph. The four autoregressive coefficients, $[1, a_{13}, a_{23}, a_{33}]$, and the power v_3 are output from the bottom of the graph. Nodes marked with subtraction imply the subtraction of the right hand input from the left hand input. The triangular arrows indicate negation and this is accomplished at the input to the appropriate node rather than with an extra node. Identity instructions permit fanout. Results are forwarded at each clock cycle. Consequently, delays are inserted on those edges of the graph that pass through stages without being used. The amount of delay is indicated at the node input. Parallelism is evident in the flow graph by the number of nodes that occur side by side. For the example selected, every processor performs an operation at every cycle once the pipeline is filled, representing an efficiency of 100%.

LOOP 1

$$v_1 = 1 - r_1^2$$
$$a_{11} = -r_1 \qquad (1)$$

AR COEFFICIENTS $[1, a_{11}]$

LOOP 2

$$e_1 = -r_1^2 + r_2$$
$$c_2 = e_1/v_1$$
$$v_2 = v_1 - e_1 c_2 \qquad (2)$$
$$a_{22} = -c_2$$
$$a_{12} = a_{11} - c_2 a_{11}$$

AR COEFFICIENTS $[1, a_{12}, a_{22}]$

LOOP 3

$$e_2 = r_1 a_{22} + r_2 a_{12} + r_3$$
$$c_3 = e_2/v_2$$
$$v_3 = v_2 - e_2 c_3 \qquad (3)$$
$$a_{33} = -c_3$$
$$\begin{vmatrix} a_{13} \\ a_{23} \end{vmatrix} = \begin{vmatrix} a_{12} \\ a_{22} \end{vmatrix} - c_3 \begin{vmatrix} a_{22} \\ a_{12} \end{vmatrix}$$

AR COEFFICIENTS $1, a_{13}, a_{23}, a_{33}$

Fig. 19 Code for Levinson-Durbin algorithm.

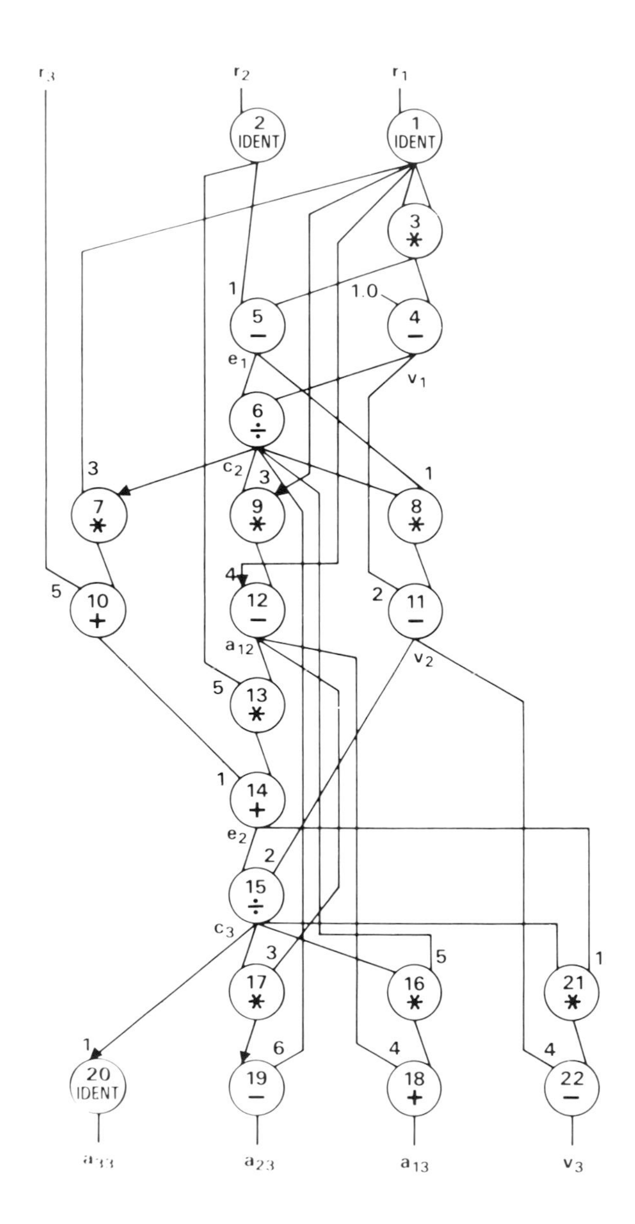

Fig. 20 Flow graph for Levinson-Durbin algorithm.

An ideal is to be able to automatically map the flow graph of figure 20 directly to the machine without writing computer code in a traditional language, i.e., in a sequential manner. This preserves parallelism. The level of parallelism may be observed from the flow graph and algorithm changes made to increase the parallelism [19]. Computation and control nodes in the flow graph are assigned to processing elements in the system and links in the flow graph to settings of the crossbar switch. Data flows into the switch during operation and is routed to the appropriate processor. A processor will perform the operation for which it is programmed on the next clock cycle after receiving its operands. The output is routed via the switch to the next processor. Overhead associated with instruction decodes, address computation, and data fetch and store, are significantly reduced relative to a conventional machine.

Matrix-vector multiplication. Figure 21 shows flow graphs for three alternative implementations of matrix-vector multiplication, $\mathbf{y} = \mathbf{Ab}$, where $\mathbf{A}$ is an n by m matrix. The proposed optical crossbar connected processor has the advantage that it may be configured to represent any of the three flow graphs shown. Selection between these depends on whether serial or parallel inputs and outputs are preferable and on the relative importance of storage versus latency. The flow graph implementations in figure 21b and figure 21c are used in the conjugate gradient implementations following.

Figure 21a shows a systolic matrix-vector multiplier [10]. In this case the input vector is supplied serially and the output vector is produced serially. Most

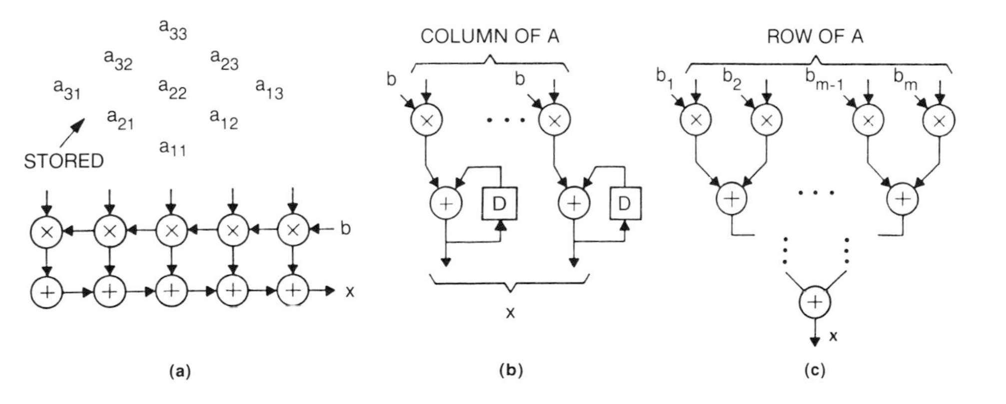

Fig. 21 Matrix-vector multiplication. (a) Systolic implementation (b) Accumulator implementation (c) Doubling implementation.

linear systolic arrays have fast nearest neighbor links and slow links for loading the processors during initialization. Consequently, the whole matrix must be stored in the multiprocessor. Latency is also larger for this configuration.

Figure 21b shows the approach taken in the Floating Point Systems 164. The vector **b** is broadcast to each multiplier element. A row of the n by m matrix **A** is entered serially into each multiplier element. The resulting inner product is accumulated in each parallel pipeline. After m cycles the output vector is available in parallel at the output. The storage required is n^2.

Figure 21c shows a tree or doubling implementation. The elements of vector **b** are spread across the multipliers and a row of matrix **A** is entered across the multipliers at each clock cycle. While one vector is being multiplied in the top row of processors the result of the preceding multiplication is being summed in the row of summers directly below the multipliers. Consequently, a new element of the output vector for the matrix-vector multiplication is obtained at each compute-move clock cycle once the summation pipeline is filled. Only m storage locations are used in this configuration. Pipelining of several correlations against the same reference is readily accomplished with this configuration. The implementation of the flow graph, figure 21c on the processor of figure 14 is given in reference [28].

Conjugate gradient algorithm implementation. *Modification of conjugate gradient equations.* Inspection of the algorithm, equation 18 through 21, section 2.4, [19], shows that the matrix-vector product $\mathbf{q}_k = \mathbf{A}\mathbf{d}_k$ must be computed before the inner product $\mathbf{d}_k^T\mathbf{q}_k$. Also, this last inner product must be completed before computing the second inner product $\mathbf{r}_{k+1}^T\mathbf{r}_{k+1}$. Note that the inner product $\mathbf{r}_k^T\mathbf{r}_k$ is available from the previous iteration. Computing the two inner products in series is undesirable with a parallel machine. Consequently, we expand the second inner product using the update equation for the residual, $\mathbf{r}_{k+1} = \mathbf{r}_k + \alpha_k\mathbf{q}_k$, to provide the following new version for each iteration,

do $k = 1$ *to* n:

$$
\begin{aligned}
&\mathbf{q}_k\mathbf{A}\mathbf{d}_k,\\
&\alpha_k = \frac{\mathbf{r}_k^T\mathbf{r}_k}{\mathbf{d}_k^T\mathbf{q}_k},\\
&\mathbf{x}_{k+1} = \mathbf{x}_k + \alpha_k\mathbf{d}_k,\\
&\mathbf{r}_{k+1} = \mathbf{r}_k + \alpha_k\mathbf{q}_k,\\
&\mathbf{r}_{k+1}^T\mathbf{r}_{k+1} = \mathbf{r}_k^T\mathbf{r}_k + 2\alpha_k\mathbf{r}_k^T\mathbf{q}_k + \alpha_k^2\mathbf{q}_k^T\mathbf{q}_k.
\end{aligned}
\tag{28}
$$

If $\mathbf{r}_{k+1}^T\mathbf{r}_{k+1} < \varepsilon\mathbf{r}_1^T\mathbf{r}_1$ then exit, else,

$$\beta_k = \frac{\mathbf{r}_{k+1}^T\mathbf{r}_{k+1}}{\mathbf{r}_k^T\mathbf{r}_k},$$
$$\mathbf{d}_{k+1} = -\mathbf{r}_{k+1} + \beta_k\mathbf{d}_k, \tag{29}$$

end do

This version replaces the second inner product $\mathbf{r}_{k+1}^T\mathbf{r}_{k+1}$ with a summation involving two inner products $\mathbf{r}_k^T\mathbf{q}_k$ and $\mathbf{q}_k^T\mathbf{q}_k$ that depend only on values from the previous iteration. Consequently, these inner products may be computed at the same time as the first inner product, $\mathbf{d}_k^T\mathbf{q}_k$.

Flow graph for matching problem and processor dimension. Figure 22 shows a flow graph for implementing the modified conjugate gradient algorithm on the

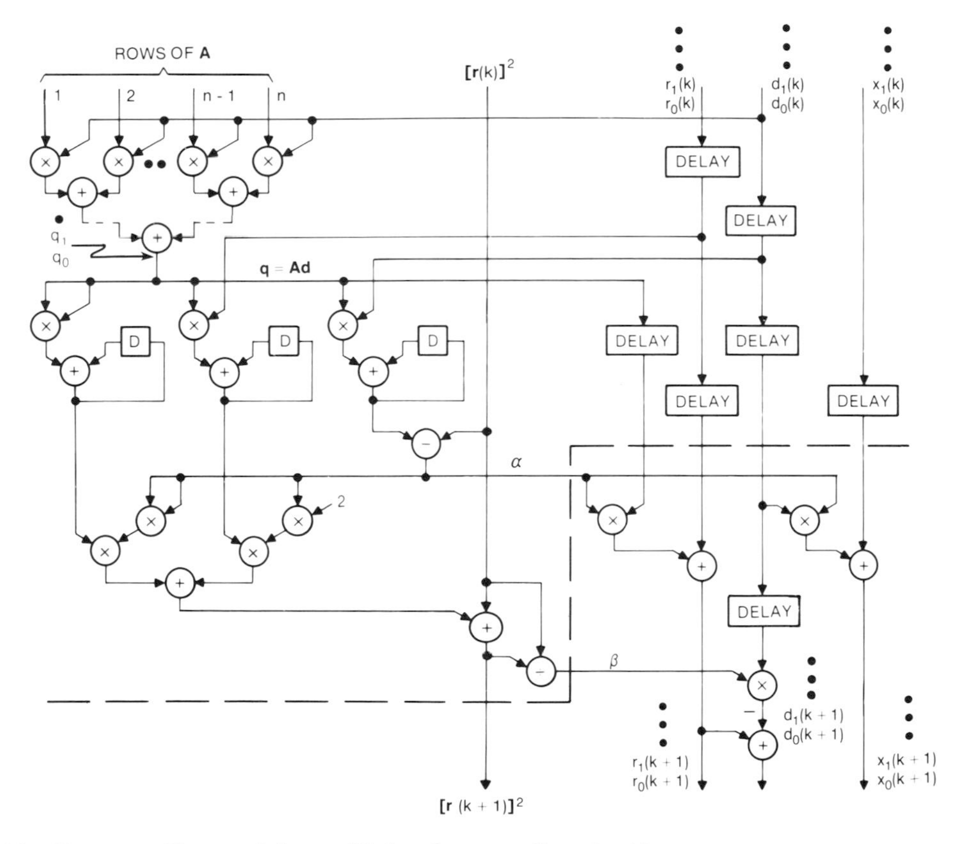

Fig. 22 Flow graph for modified conjugate gradient algorithm.

system figure 14, having hundreds of processors, when the problem dimension is approximately half the number of processors. In an actual implementation additional processors are required to provide the control [22]. These have been omitted here for simplicity. The matrix-vector product is computed at the top left using the implementation of figure 21c. At each cycle, once the $\log_2 n$ pipeline is full, a new element of the vector **q** is fed to the three inner product accumulators at the left. These use the configuration of figure 21b. Consequently, the matrix-vector multiplication and inner products take $\log_2 n + 2 + n$ cycles. A further 6 cycles are needed to compute β. Updating the unknown vector **x**, the direction vector **d** and the residual **r** are shown at the bottom right. These operations may be performed in parallel in 6 cycles using the row of processors at the top left rather than by assigning other processing elements.

The total operations for one iteration of the conjugate gradient algorithm is therefore approximately $n + \log_2 n + 14$. The speed up relative to a uniprocessor is approximately 228 and the efficiency of processor utilization is over 76%. For example, a system using 300 processors consisting of Weitek multipliers and adders with 240 nsec. cycle time and 32 bit floating point capability would compute an iteration for $n = 128$ in approximately 36 microsec., assuming a switch of size 600×600 and that the processors unused for computation perform the control functions. This corresponds to 4.6 msec. to compute 128 iterations, though we would normally expect convergence to occur more rapidly [8]. The processor would be running at close to a Gigaflop for this computation or thousands of times faster than a Vax 11/780.

Flow graph for differing problem and processor dimension. Most parallel processors change performance dramatically when the problem dimension deviates from the number of processing elements. The ability to reconfigure the interconnections permits high efficiency to be maintained for any relation between problem size and number of processing elements.

Figure 23a shows the case for which the problem is larger than the number of processing elements. The switch is configured to perform inner products as in figure 21b for the matrix-vector multiplication stage. The number of rows of the matrix **A** to be used at one time is selected so as to fully utilize the processing elements. The elements of the resulting vector **q** are stored. Doubling summation networks 21c are used for the inner product stage as the inner products are divided to make more efficient use of the processing elements.

When the problem dimension is much larger than the number of processing elements, say 20 times larger, it is possible to reconfigure the crossbar during each

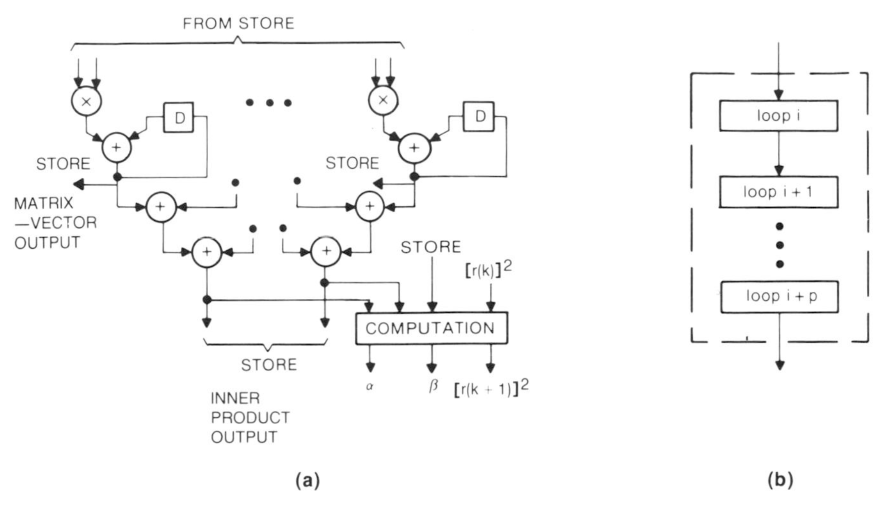

Fig. 23 Conjugate gradient flow graph. (a) Large problems (b) Small problems.

iteration because the parts of each iteration have sufficient length to make the switching time negligible. Figure 21b shows the configuration that appears most desirable for the matrix-vector multiplication stage. The switch is changed to perform the inner products as soon as the matrix-vector computation is complete.

Problems that have a dimension smaller than that of the processor will be completed rapidly. Efficiency is a concern where there are large numbers of these smaller problems to be computed. In this case the conjugate gradient loops are unfolded so that many iterations are pipelined through the processor 22b. The number of iterations unrolled is selected to provide the optimum efficiency of utilization of the processing elements.

The methods presented make use of reconfigurability to maintain high efficiency for problems having dimension close to a multiple or divisor of the number of processing elements. Between these problem sizes efficiency will fall off somewhat. It would be difficult to achieve this high efficiency over a wide range of problem dimension with a multiprocessor not capable of reconfigurability.

6.2 *Finite elements with a nearest neighbor optical architecture*

The following optical architecture is efficient for finite approximation computations for which the grid may be distorted to regular. It uses spatial light modulators currently under development.

System Overview. A host computer with a terminal is shown at the top of figure 24. This machine is suited to computationally demanding problems in which the initial setting up of the matrices and the output operations are not a significant part of the computation time and may therefore be accomplished in the host computer.

The box at the lower left in figure 24 uses parallel 2-D element by element optical multiplication to perform the main calculation. Each element of one matrix, representing say north coefficients, is multiplied by the corresponding element of the second, representing say the field. Multiplication for south, east and west are performed in sequence. The results are accumulated in a charge-coupled-device (CCD) to provide the solution to equation 17 at each point in the grid in parallel. If numbers are represented by 8 length residue codes occupying 8 adjacent deformable-mirror-device (DMD) elements in a row, then the nearest neighbor value in the east-west direction is actually 8 DMD elements separated, while in the north-south direction the nearest neighbor is only 1 DMD element away. The processor and its use for performing nearest neighbor computations are described later.

The output intensity measured by the CCD is converted back to an intensity within the range of the residue number system in the lower center box by division with the appropriate moduli. This is explained in the sections on the residue numbers and the modulo operation.

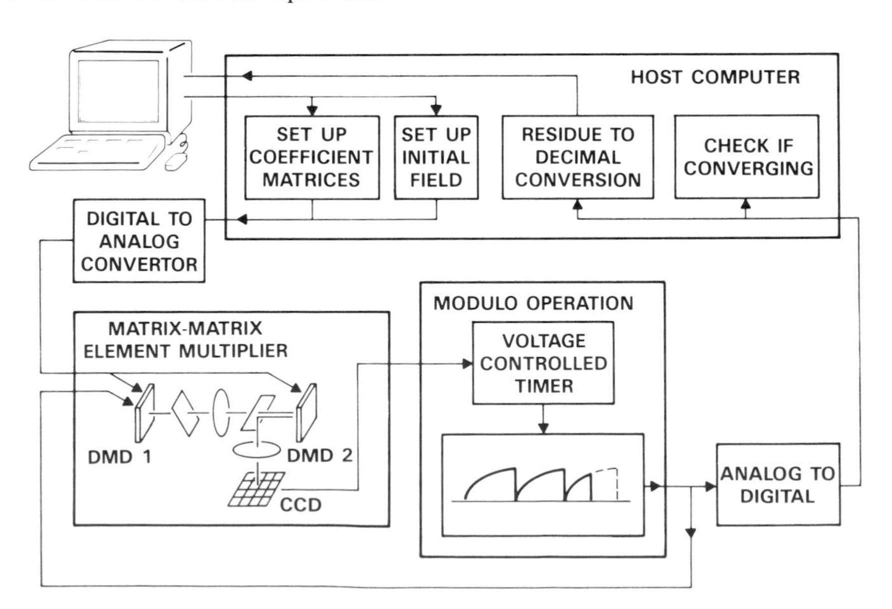

Fig. 24 Deformable mirror nearest neighbor optical computer.

The new intensity is also fed via an analog to digital converter [16] to the host for convergence evaluation and display purposes. The convergence evaluation and intermediate displays may be performed at the same time as the optical part of the processor is continuing to compute further iterations. The conversion time is small relative to the computation time because large problems requiring the speed of this processor will perform thousands of optical iterations before the results are of interest. Consequently, the conversion, to decimal for display may be performed with special electronics or in the host if it is sufficiently fast.

Residue numbers. Residue number systems have been built in digital electronics at Lockheed, Martin-Marietta and Mitre [45] and optical systems have been proposed. The principle advantage is that once numbers have been converted to a code, coded numbers can be added and multiplied by separate operations on each element of the code and no carries between elements of the code are necessary. Let us perform operations $7+3$ and 7×3 for the purpose of illustration. Select a set of relatively prime numbers such as (3, 4, 5) for a moduli set. The residue number code for a number is determined by dividing it by each element of the moduli set and keeping the remainders. The code for 7 is (1, 3, 2) because 7 divided by 3 has a remainder of 1, 7 divided by 4 has a remainder of 3 and 7 divided by 5 has a remainder of 2. Similarly 3 becomes (0, 3, 3). The sum $7+3$ has the code $(1+0, 3+3, 2+3)$, while the product 7×3 has code$(1 \times 0, 3 \times 3, 2 \times 3)$. The results may exceed the moduli values. Each term that exceeds its modulo value must be divided by the modulo value to bring it back within the correct range. Thus the sum (1, 6, 5) becomes (1, 2, 0) and the product (0, 9, 6) becomes (0, 1, 1). I refer to this as a modulo operation. Note that no carries were required between elements of the code, unlike fixed radix arithmetic. This enables parallel operation by avoiding sequential operation required for carries. It is necessary to convert back to decimal after completing the computation.

The first reason for using RNS rather than digital multiplication by analog convolution or some other technique to extend dynamic range is that I wish to use the dynamic range already available in the DMD to reduce the number of DMD elements required to represent a number. 5 bits, available on a DMD, permits numbers upt to $2^5 = 32$ for each of the code numbers in the residue number code. Consequently, numbers may be represented in a residue number system up to $7 \cdot 11 \cdot 13 \cdot 17 \cdot 19 \cdot 23 \cdot 29 \cdot 31 = 6 \cdot 7 \cdot 10^9$ which is greater than 2^{32}. These 8 relative prime numbers require 8 elements of the DMD as distinct from 32 that would be required for a 32 bit binary convolver.

Secondly, the residue number approach permits numbers to be multiplied in a straight forward manner by imaging light from a DMD element, set by the first operand on to a second DMD element, set by the second operand. This is described in the next section. The numbers out of the matrix element multiplier are converted back to the remainders by division with the appropriate relative prime numbers for the moduli set selected. No other operations are required with residue numbers because they may be processed independently of each other. In contrast, an alternative approach using acousto-optic cells referred to as the systolic acousto-optic binary convolver involves more complexity. It requires high frequencies to drive the acousto-optic cell and a sequential shift and add after converting the numbers back to binary, following a computation.

Outer product multipliers require very large dynamic range for the receiving time integration device and this is a major limitation. In addition, the idea of using binary numbers in a 32 bit outer product multiplier is also complicated because it requires the addition of the diagonal elements for each 32 by 32 submatrix. This involves extracting and adding many sets of 32 numbers. This complication is totally avoided by using residue numbers.

Matrix-matrix element multiplier. A lens is used to image DMD 1 on to DMD 2 to perform matrix-matrix element multiplication in figure 24. It is necessary to use Schlieren type imaging to remove the reflection from inactivated mirrors and from the flat mirror surface between elements as well as convert phase to amplitude. This is accomplished by placing a stop at the lens focal point (FT plane) to filter out low spatial frequencies. The stop is circular for a deformable membrane device and a cross for a cantilever beam deformable mirror. The same lens performs the imaging of the array of dots remaining, the dots having intensities dependent on the mirror deflections.

There is a nonlinear relation between the voltage applied to a drain line of the DMD and the intensity of the spot imaged on the CCD. A look up table is used to convert data values to voltage so as to provide linear intensity of spots with data values for multiplication. Only positive values may be multiplied with an incoherent light source. However, operations in which only one of the operands is permitted to be negative may be handled by separation into two parts, that involving negative numbers and that involving positive numbers. The answers are known to be negative and positive respectively. This applies to image processing, where the image is always positive, and only the filter may have negative values. The addition of a large constant to the initial field, when solving partial differential equations, ensures that the field is always positive.

A matrix with elements $a_{i,j}$ is entered on DMD 1 and a matrix with elements

$b_{i,j}$ is entered on DMD 2. DMD 1 is then imaged on to DMD 2 and the result imaged on to a CCD. An element of the resulting matrix $c_{i,j}$ is the product of the two elements in the same matrix location in the two matrices,

$$c_{i,j} = a_{i,j} b_{i,j} \tag{30}$$

Element products of other matrices may be added to the result because of the time integrating properties of the CCD. However, the number of matrix-matrix element products that may be added is limited by the dynamic range of the CCD. Only five products are accumulated on the CCD when implementing equation (17) even though the matrices have dimensions up to 1000. Note that matrix-matrix multiplication is not being performed as this would involve accumulating 1000 products and the result would exceed the dynamic range of the CCD.

The matrix-matrix element multiplier requires 1000 analog entries (with dynamic range 32 or equivalent to 5 bits) to be made every 8 msec on a 1000 channels. This is equivalent to supplying $5 \cdot 10^6$ bits/sec in 8 msec or a total bandwidth over 1000 channels of 78 Mbytes/sec. This would require 2 VME or NuBus busses.

Modulo operation. Multiplying and adding numbers in a residue number system will generate some numbers that exceed the moduli for the number system. The maximum number for a relative prime number in the moduli set is approximately 5 times the product of the relative prime number in the case of a 4 point finite difference scheme. Consequently, it is necessary to reduce these numbers back into the appropriate ranges for the number system. Assume a set of moduli m_i, $i=1, M$ and that the result of a computation for the ith member of the set is x_i, then the modulo operation corresponds to computing residuals,

$$R_i = x_i - n_i m_i \tag{31}$$

where n_i is the number of times that x_i is divisible by m_i.

An analog system is used for the modulo function, figure 24. A voltage representing intensity from the CCD in the matrix-matrix element multiplier is used to control the length of time for which a saw tooth generator runs. The height of the waveform corresponds to the relative prime number in the set of moduli. The generator signal is shaped to compensate for the nonlinearity of the DMD, as described earlier. The height of the generator voltage signal when the timer switches off represents the residue for the specific relative prime number. The largest number which may arise in a four point finite element computation, equation 17, is the square of the largest number, 31, in the modulo set times the number of directions, 5, (N, E, S, W, self). Two additional directions are required for finite elements with triangular elements to produce $7 \cdot 31^2$. This number requires a 14 bit analog to digital converter. The look up table provides the corresponding residue number adjusted to compensate for the DMD nonlinearity.

Nearest neighbor finite element computations. Equation 17 is implemented efficiently by means of a nearest neighbor computer such as the one described. For a four point finite difference computation the $(k-1, m-1)$ and $(k+1, m+1)$ terms are omitted. The field u is maintained positive and the whole computation must be performed for positive number coefficients and negative number coefficients separately. The field $u^i(k+1, m)$ is placed on DMD 2, displaced in the south direction by one row as suggested by the $k+1$ index, and the coefficients $a_1(k, m)$ for the north direction are set on DMD 1. The result of element by element multiplication is accumulated on the CCD. The field for $u^i(k-1, m)$ is now set on DMD 2, displace by one row north, and the coefficients $a_2(k, m)$ for the south direction are set on DMD 1. This process is repeated for east, west, load and previous iteration field, while the CCD accumulates the results. Note that the nearest neighbor is displaced by 8 elements in the east-west direction if 8 long residue numbers codes are arranged in rows.

In a serial machine, I assume that the grid is transversed with a four point finite difference operator from west to east and from north to south so that when a point is updated, points above and to the left have just been updated during this sweep. If all points are updated in parallel, old values would have to be used and convergence would be slower. The serial convergence rate is recovered by dividing the grid into red and black nodes to look like a checker board, section 5.1. The red nodes are updated in one 'half' iteration and the black on the next 'half' iteration. As red (black) nodes have no red (black) neighbors the most recently updated nodes are used everywhere on each iteration. The CCD is now read out on each iteration. The DMD on which the field is entered holds only red nodes on one half iteration and only black nodes on the next, so that all the DMD elements are used in each half iteration for a problem having twice as many nodes as there are DMD elements. Three colors in place of two are required to apply the same principles for finite elements using triangular elements.

An initial estimate of the field is also required for iterative schemes [24]. In cases such as that arising in geophysical prestack inversion, where a region to be estimated overlaps that estimated earlier, a good estimate is known and the iterative technique is advantageous compared with a direct method.

Performance. DMD's having 1000 by 1000 elements are anticipated in the next few years. The performance of the proposed system with such a device is considered. The time to set the device in parallel is assumed to be 8 msec. Assuming a moduli set of size 8 in order to produce a 32 bit machine, the DMD's can process an array of size 125 by 1000. 125×1000 numbers are multiplied with the matrix-matrix element multiplier in 8 mseconds, corresponding to a rate of 15.6 million operations per second. Different matrix form factors are accommodated by rearranging the allocation of the residue number code to the DMD array.

The overall speed need not be much slower than the matrix computation time if sufficient parallel electronics is used in the iteration loop. Perfect parallel efficiency is possible because the modulo operation is completely independent for every value. It is possible to apply a further cycle of cyclic reduction to the red-black algorithm, (or three color algorithm for finite elements), that is, take the red points and impose a further checkerboard on them and repeat this with the black points. This permits the modulo operation on one set of field values to occur in parallel with the calculation of another set. This would be efficient for problem meshes that are twice as large again. The problem mesh matching the DMD is now 500 by 4000 for finite differences or 750 by 6000 for finite elements with triangular elements. 100 electronic circuits each has to perform 10,000 modulo operations in 8 mseconds, corresponding to a rate of 1.25 MHz, if there is an electronic circuit for every 10 rows of the CCD. Floating point capability is desirable and this has not yet been explored for this processor.

It is possible to increase the machine speed to 1 billion operations per second by simply duplicating the equipment for each residue number modulus ($\times 8$) and for each of the eight sums performed in a finite element problem with triangular elements ($\times 8$). The multiplications for the nearest neighbor computation for different directions may be performed in parallel with perfect efficiency as suggested in reference [29]. The advantage of the residue number system is that each of the code elements may be treated completely independently providing perfect parallel efficiency. The algorithm is expected to take $0.72N \log_{10} N$ iterations on the average for an N by N grid. For the 1 billion operation per second machine, with cyclic reduction applied twice to a triangular element finite element problem, this involves approximately 6,000 iterations. Consequently, it takes approximately 288 seconds to solve the finite element equations, assuming six colors (overlap matrix-matrix and modulo operation, $(6 \cdot 8 \cdot 10^{-3} \cdot 6000 = 288)$ for a problem of mesh size 1000 by 6000. The conversion time to convert from residue numbers back to decimal is much less.

7. CONCLUSION

A diverse selection of algorithms used in geophysical modeling and inversion were presented. Massively parallel computer architectures are required to produce a significant improvement in performance. Alternative architectures were discussed. Modularity is required to permit tailoring in a cost effective manner.

Two electronic architectures were considered, a bus architecture and a systolic architecture. The performance of the first was limited by the bus as illustrated by mapping a finite approximation algorithm on to the machine. The performance of the second was limited by the fixed interconnection structure.

Two optical architectures were considered. The first used an optical crossbar interconnection switch to provide high bandwith interconnections to many electronic processors. This permits efficient mapping of a wide range of algorithms as illustrated. However, while efficient in terms of hardware utilization and speed it would not be efficient in terms of energy until devices are perfected that permit low loss conversion between optics and electronics.

The second optical architecture minimizes the conversion difficulty by computing in optics. However, it is restricted to nearest neighbor connections and still requires fast, dense, spatial light modulators. Thus it is only efficient for finite approximation computations for which the grid may be distorted to regular.

Progress in optical spatial light modulators and conversion devices suggest that it will soon be possible to build machines that utilise optics to overcome limitation in electronics, thus permitting massive parallelism and significant improvements in computing performance.

8. REFERENCES

1. Aki, K., and Richards, P. G., 1980, *Quantitative seismology: Theory and methods*, W. H. Freeman and Company.
2. Adams, L. M. and Ortega, J. M., A Multicolor SOR Method for Parallel Computation. NASA. ICASE 82–9, April 1982.
3. Bell, T., "Optical Computing: a Field in Flux", IEEE Spectrum, Vol. **23**, Aug. 1986, pp. 34–57.
4. Brandt, A., Multi-level Adaptive Solution to Boundary Value Problems. Math of Comp. 31, 1977.

5. Brenner, K. H., Huang, A., and Streibel, N., "Digital Optical Computing with Symbolic Substitution", Applied Optics, Vol. **25**, Sep. 1986, pp. 3054–3064.
6. Goodman, J. W., *Introduction to Fourier Optics*, McGraw-Hill, 1986.
7. Grosch, C. E., Poisson Solvers on a Large Array Computer. Dept. Math and Computer Science, Old Dominion University, Norfolk, Va. TR 78–4 Oct. 1978.
8. Hestenes, M. R., *Conjugate direction methods in optimization*, Springer Verlag, New York, 1980.
9. Hwang, K., and Briggs, F. A., *Computer Architecture and Parallel Processors*, McGraw-Hill, New York, 1984.
10. Kung, H. T., Why systolic architectures, *Computer*, **15**, 1, (1982), 37–46.
11. McAulay, A. D., "Optical Interconnections for Real Time Symbolic and Numeric Processing", Invited Chapter in Book *Optical Computing: Digital and Symbolic*, Marcel Dekker. NY. 1988.
12. McAulay, A. D., "Application of luminescent devices to electronically controlled optical image processors", SPIE, Hybrid Image and Signal Processing Conference, April 1988.
13. McAulay, A. D., "Logic and arithmetic with luminescent rebroadcasting devices", SPIE, Advances in Optical Processing Conference, April 1988.
14. McAulay, A. D., "Optical Prolog computer using symbolic substitution", Digital Optical Computing Symposium, SPIE O-E LASE, Jan. 1988.
15. McAulay, A. D., "Application of optical spatial light modulators to expert systems", Wright State University Fifth International Conference on Systems Engineering, Sep. 1987, pp. 507–510.
16. McAulay, A. D., "Optical A/D converters and application to digital multiplication by analog convolution", SPIE Real Time Signal Processing, Aug. 1987.
17. McAulay, A. D., "Engineering Design Neural Networks using Split Inversion Learning", IEEE First International Conference on Neural Networks, June 1987.
18. McAulay, A. D., "Spatial light modulator interconnected computers", IEEE Computer, Vol. **20**, No. 9, Oct. 1987.
19. McAulay, A. D., "Conjugate Gradients on Optical Crossbar Interconnected Multiprocessor", to appear in Journal of Parallel and Distributed Processing.
20. McAulay, A. D., "Real-time optical expert systems", Invited paper, Applied Optics, Vol. **26**, May 1987, pp. 1927–1934.
21. McAulay, A. D., "Optical diagnostic processor for flight control", NAECON May 1987, pp. 610–615.

22. McAulay, A. D., "Performance of Schur's algorithm on an optically connected multiprocessor", (with Eric Parsons) IEEE Internat. Conf. on Acoustics, Speech and Signal Processing, April, 1987, pp. 1019–1022.
23. McAulay, A. D., "An extendable optically interconnected parallel computer", IEEE-ACM Fall Joint Computer Conference, Proceedings, Vol. **1**, (1986), pp. 441–447.
24. McAulay, A. D., "Plane-layer point-source prestack inversion of marine data with unknown initial profile", 56th Annual Intern. Society of Exploration Geophysicists Meeting, Vol. **1**, (1986), pp. 537–539.
25. McAulay, A. D., "Plane-Layer Prestack Inversion in the Presence of Surface Reverberation", Geophysics, Vol. **51**, No. 9, (1986), pp. 1789–1800.
26. McAulay, A. D., "Parallel AR Computation with a Reconfigurable Signal Processor", IEEE Internat. Conf. on Acoustics, Speech and Signal Processing, 86CH2243-4, Vol. 2, (April 1986), pp. 1365–1368.
27. McAulay, A. D., "Optimal Least Square Filtering with a Digital Optically Interconnected Processor", SPIE Technical Symposium SE, Advances in Optical Information Processing II, Vol. 639, (March 1986), pp. 118–124.
28. McAulay, A. D., "Optical Crossbar Interconnected Signal Processor with Basic Algorithms", Optical Engineering, Vol. **25**, No. 1, (Jan. 1986), pp. 82–90.
29. McAulay, A. D., "Deformable Mirror Nearest Neighbor Optical Computer", Optical Engineering, Invited, Vol. **25**, No. 1, (Jan. 1986), pp. 76–81.
30. McAulay, A. D., "Influence of Surface and Noise on Plane-Layer Prestack Inversion", 55th Annual Intern. Society of Exploration Geophysicists Meeting, Washington, (Oct. 1985), pp. 372–375.
31. McAulay, A. D., "Predictive Deconvolution of Seismic Array Data for Inversion", IEEE International Conference on Acoustics, Speech and Signal Processing, Publn. 85CH2118–8, Vol. **1**, 180–183, (March 1985), pp. 180–183.
32. McAulay, A. D., "Prestack Inversion with Plane-Layer Point Source Modeling", Geophysics, Vol. **50**, No. 1, pp. 77–89, (Jan. 1985).
33. McAulay, A. D., "Importance of Shear in Plane-Layer Point Source Modeling", 54th Annual Intern. Society of Exploration Geophysicists Meeting, Atlanta, pp. 541–544, (Dec. 1984), (Abstract in Geophysics, (Feb. 1985)).
34. McAulay, A. D., "Finite Element Computation on Nearest Neighbor Connected Machines", NASA Symposium on Advances and Trends in Structures and Dynamics, NASA Langley Research Center, Research in Structures and Dynamics-1984 Proc., NASA Publn. **2335**, 15–29, (Oct. 22, 1984).
35. McAulay, A. D., "Generation of Accurate Synthetic Seismograms", IEEE Inter-

national Conference on Acoustics, Speech and Signal Processing, IEEE Publn. 80CH1559–4, Vol. **1**, 111–115, (April 1980). (Joint Author)

36. McAulay, A. D., "Variational Finite Element Solution of Dissipative Waveguides and Transportation Application", IEEE Trans. Microwave Theory and Techniques Vol. **MMT–25**, No. 5, 382–392 (May 1977).
37. McAulay, A. D., "The Finite Element Solution of Dissipative Electromagnetic Surface Waveguides", International Journal for Numerical Methods in Engineering Vol. **11**, No. 1, 11–25, 382–392, (January, 1977).
38. Neff, J. A., "Major initiatives for optical computing", Optical Engineering, **26**, 1, 1987.
39. Nicolaides, R. A., On the 1 Squared Convergence of an Algorithm for Solving Finite Element Equations. Math of Comp., Vol. 31, no. 140, Oct 1977.
40. Miller, D. A. B., et al., "The Quantum Well Self-electrooptic Effect Device: Optoelectronic Bistability and Oscillation, and Self-linearized Modulation", IEEE J. of Quantum Electronics, Vol. **QE-21**, Sept. 1985, pp. 1462–1476.
41. Pape, D. R., and Hornbeck, L. J., Characteristics of the deformable mirror device for optical information processing, *Opt. Eng.*, **22**, 6, (1983), pp. 675–681.
42. Sawchuck A. A., and Strand, T. C., "Digital Optical Computing", IEEE Proc. Vol. **72**, July 1984, pp. 758–779.
43. Sawchuck, A. A., Jenkins, B. K., Raghavendra, C. S., and Varma, A., "Optical Crossbar Networks", Computer, **20**, 6, 1987, pp. 50–60.
44. Taylor, G. W., Simmons, J. G., Cho, A. Y., and Mand, R. S., "A New Heterostructure Optoelectronic Switching Device using Molecular Beam Epitaxy", J. App. Physics, Vol. **59**, Jan. 1986, pp. 596–600.
45. Taylor, F. J., "Residue arithmetic: a tutorial with examples", Computer, **17**, 5, 1984.
46. Van Rosendale, J., "Minimizing Inner Product Data Dependencies in Conjugate Gradient Iteration", IEEE Parallel Processing Conference Proc., 1983.

CHAPTER 6

SEISMIC DATA PROCESSING ON A SIMD ARRAY PARALLEL SUPERCOMPUTER

by
SHUKI RONEN
Colorado School of Mines
Golden, Colorado 80401
and
ROBERT SCHREIBER
Rensselaer Polytechnic Institute
Troy, New York, 12180

Seismic exploration challenges require computer performance that is not achievable without either a revolution in computer technology or parallel processing.

Two basic control architectures for parallel computers have been suggested and built, Multiple Instruction Multiple Data (MIMD) and Single Instruction Multiple Data (SIMD). SIMD is cheaper to produce, easier to program and develop compilers for, but less general than MIMD. This is because SIMD allows only identical tasks to proceed in parallel, while with MIMD the tasks can be different.

An important question is whether seismic data processing can be done in parallel with SIMD. We claim that the answer is yes. To support this claim we describe a specific SIMD parallel computer, the Saxpy Matrix-1, and show how a variety of geophysical tasks can be addressed by that machine.

INTRODUCTION

The physical basis for seismic exploration is far ahead of what one can do on available computers. Seismic data processing is therefore the art of choosing the right currently calculable approximations. As faster computers are introduced, the approximations used will move closer to reality.

Today, in areas of mild lateral velocity variations and not-too-strong near surface anomalies, 2-D reflectivity of complex structures can be imaged. The traditional challenges have been to improve the ability to image 2-D reflectivity in the presence of higher complexity, stronger velocity variations, near surface anomalies, and multiples. Now interest is shifting from those traditional challenges to milestone projects like imaging 3-D structures (3-D prestack migration) and to obtaining more than reflectivity—physical parameters such as density, elasticity, viscosity (inversion). This inversion challenge probably implies integration of interactive processing and interpretation. Interactivity drastically raises the performance requirement to allow acceptable turnarounds.

Very few, if any, of these milestones can be reached without suitable computers.

We will first define what performance is required, in the areas of floating point speed, memory, I/O capacity, and graphic display—requirements that are beyond the limits of vector computers but are achievable in parallel. We then review a few parallel computers that have been produced, describe a specific SIMD parallel architecture, and illustrate how it can be used in seismic processing.

PERFORMANCE REQUIREMENTS

A three-dimensional seismic data set may contain anywhere from 10^8 to 10^{10} words. The number of operations per word varies between about 10^2 in a minimal processing sequence of normal moveout, stacking, and migration after stack, to about 10^4 with prestack migration. This gives a maximum number of 10^{14} operations, which would take 28 hours with a sustained rate of 10^9 operations per second. Supercomputers today have sustained rates of a few times 10^8 operations per second, which is therefore not enough.

The precision requirements in conventional processing are roughly 90% single (32 bit) precision, 10% double (64 bit) precision, so double precision slower than single by a factor of 10 may be tolerated. However, possible future processing techniques and other applications such as reservoir modeling require double precision more often, so a factor of 1-2 in floating point performance is desired.

To visualize results, movies are created from sequences of variable intensity snapshots. Movies of hundreds of two dimensional snapshots, 1-2 million pixels each, 1 byte per pixel (256 variable intensity color or gray levels) are required. A rate of 10^7 bytes a second from the memory to the graphic display will enable loading 5-10 snapshots a second, which is required for the perception of motion.

A combination of memory hierarchy and I/O capacity is needed to sustain the floating point performance. In addition to mass storage on tapes, disks, and main random access memory (RAM), we also need local RAM in each processor. The requirement is that the memory size and bandwidth sustain the floating point and graphics requirement. We expect that this implies storing about 10^9 words in random access memory, and mass storage transfer rates of a few times 10^7 bytes a second.

Last, but not least, suitable compilers and software are required to enable programming and usage of old and new algorithms for implementation of geophysical processes in parallel.

FROM ARRAY PROCESSORS TO PROCESSOR ARRAYS

The Limits to Serial and Pipelined Computers

The path to faster machines has traditionally been that of the faster clock. But progress has been slow. Between 1976 and 1986 the cycle time of Cray computers improved only by a factor of three. Because light travels so slowly (only 30 centimeters in a nanosecond) it is necessary to make the linear dimensions of machines smaller in direct proportion to the cycle time. In the absence of any reduction in the energy required to do a computation, the heat density grows as the cube of the clock rate. Cooling these machines becomes harder and harder, and every reduction necessitates a new cooling technology. Thus, to improve the speed of supercomputers by an order of magnitude or more will require either a revolution in component technology or parallel computers.

Serial machines with a few functional units, such as a multiplier and one or two adders, that may operate in parallel can run faster than nonoverlapped machines. Early supercomputers (CDC 6600, IBM 370/195) use hardware instruction pipelining to do this. Array processors avoid the expensive and complex control unit through a wide instruction word, but programming them has proved to be difficult, because of the memory mapping between the host and the AP, and because of the need to synchronize the functional units. Newer fast scalar machines (Multiflow for example) use a wide instruction and a sophisticated compiler to do this. But speedup by this method is quite limited.

The Parallel Computer Design Space

For the most part, multiple-instruction/multiple-data (MIMD) architectures have been proposed and built. These allow any parallel computation to be performed. Several of these machines (for instance the Cray XMP, the Sequent Balance 21000, and the Masscomp 5700) are *shared memory multiprocessors* that have a few independent computational units sharing a global memory. In these machines, some hardware synchronization mechanism is provided. If there are more than three or four processors, some local data caching is needed to limit the demand for data from the shared memory. Because cached data can become invalid when a cached variable is altered by another processor, the problem of cache coherence arises and must be solved with appropriate hardware. Even with caching, memory traffic limits the number of processors to ten or twenty.

Machines with logically shared but physically disjoint memories (such as the BBN Butterfly or the IBM RP3) may have hundreds or thousands of processors. In these machines, a substantial fraction of the hardware is devoted to the network that connects processors to memories. Advantageous placement of data in the memory can have a large impact on performance, so the programmer must pay attention to it.

The third class of MIMD machines, *message-passing multicomputers* (such as the Intel iPSC or the FPS T Series) employ fully independent computers, each with its own private memory. They communicate data and synchronize when necessary by sending messages to one another through a network, rather than by referring to shared global variables. The trend in these machines is to provide hardware support to speed up the message passing, which can otherwise become a bottleneck. Again, careful placement of the data in the machine is important for high performance.

In all these MIMD machines, the programmer is responsible for decomposing the problem into parallel processes. He then writes *node code* that runs in the individual processors; usually it can be written in Fortran or C, with extensions to pass messages (using SEND and RECEIVE primitives) or to read and write shared variables in critical sections protected by locks or semaphores. Various high level synchronization instructions (such as a barrier at which every process stops until all have arrived there) can be provided in software using the basic hardware tools. Their efficient realization is a subject of current research (Brooks, 1986). There are no compilers for "dusty decks."

In our experience, one program is almost always used by all the nodes. But different nodes can take different paths at conditional branches, so MIMD hardware is still needed. Thus, Single Program, Multiple Data (SPMD) is a more apt term than MIMD.

Other machines (such as the Connection Machine, the IBM GF11 and the Saxpy Matrix-1) operate in a tightly coupled, single-instruction, multiple-data (SIMD) fashion. Within this group there is also tremendous architectural diversity. Massively parallel SIMD architectures such as the Connection Machine employ thousands of simple processors; moderately parallel machines such as the Matrix-1 use fewer (32) fast processing elements. The GF11 is intended for a specific application, the others are general-purpose computers. The Connection Machine features a distributed memory and a near-hypercube interconnection network and the Matrix-1 has a shared memory connected by a high-bandwidth bus to the array of 32 processors with a simple linear connection in which processor i is connected to processor $i \pm 1$ (mod 32).

The virtue of MIMD machines is that they allow arbitrary, dissimilar tasks to proceed in parallel. For instance, in solving partial differential equations on irregular domains or with irregular grids, it may be impossible to decompose the problem in a SIMD manner, but a MIMD machine can be used. But MIMD machines are inherently more expensive and more complex for given speed. So it is an interesting question whether in seismic processing, the computations are regular enough for SIMD, or whether MIMD is really needed. Of course, there are two schools of thought on this question. It has been argued that in migration, due to lateral velocity variations, the flexibility of MIMD is needed. In this paper we take the opposite view: a SIMD Processor array supported by a large global memory is appropriate for a large and important part of seismic signal processing. Furthermore, the use of fast local memory and algorithms that make most of their memory references to the local memory allows for the use of inexpensive main memory. SIMD can provide a supercomputer with an outstanding performance to cost ratio.

SIMD ARRAY SUPERCOMPUTERS

We shall describe a class of SIMD array supercomputers, of which the Saxpy Matrix-1 is a notable example (Foulser and Schreiber, 1987). The key features of the architecture are these: the SIMD control approach; the use of a large global memory; the use of both systolic and global data paths in the systolic array; the choice of a small number of fast processing elements; the use of Fortran as the programming language; the emphasis on block algorithms and the provision in the hardware of double-buffered, software-managed local memory in the systolic array to support these algorithms. We then show by giving several examples how this architectural class is very well matched to seismic computing.

The Architecture

The five principal components of the Matrix-1 supercomputer (Figure 1) are:

(i) the System controller, a general-purpose computer that executes the application program and allocates resources (including the other components);

(ii) the Matrix processor, a linear SIMD array of 32 floating-point processors with systolic and global interconnections;

(iii) the System memory, which stores all data arrays for use by the Matrix processor;

(iv) the Mass storage system, an I/O interface for access to high-speed peripherals; and

(v) the Interconnect bus, a control bus and a data bus, running at 8×10^7 words a second, linking the other four units of the system.

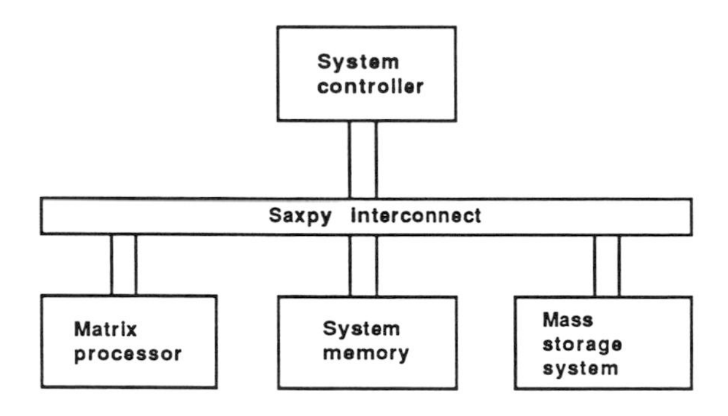

Fig. 1 System block diagram.

System Controller

The system controller compiles, links, and executes the application program, sending control information across the system bus to the other components. Since data are stored in the system memory and the Matrix processor performs practically all floating-point computation, the system controller's job is to coordinate resources.

Matrix Processor

The Matrix processor consists of a linear array of processing elements, called computational "zones" (Figure 2). Each zone has a 32-bit floating-point multiplier and a 32-bit floating-point adder with logic capabilities (ALU). The clock period is 64 nanoseconds. Reciprocal and square root calculation are also supported. Local memory is 4096 words per zone. The zone memory allows indirect addressing, in which elements of one vector are used as indices into another vector. Vector "scatter" and "gather" operations are implemented with this hardware and operate at half speed, one word every two cycles.

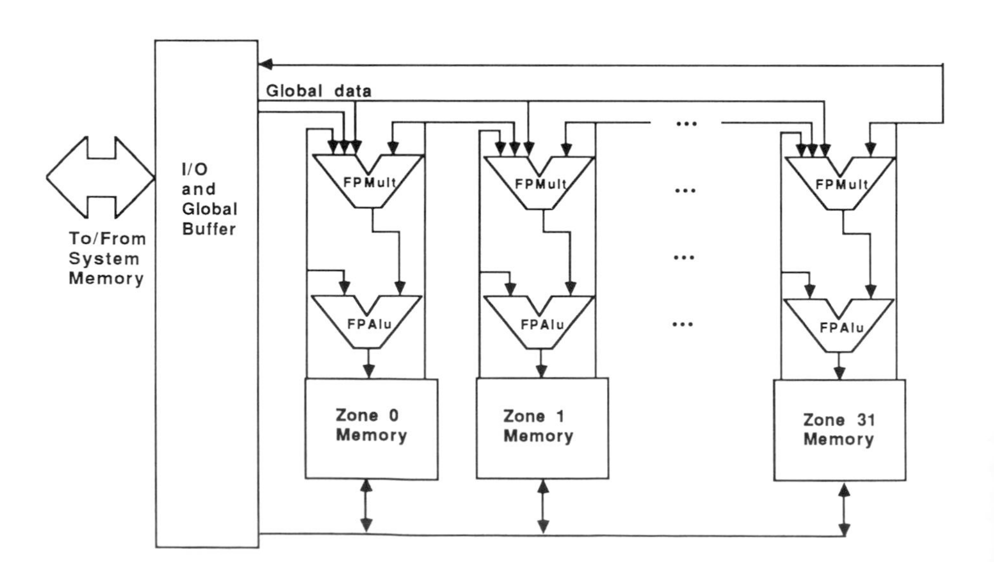

Fig. 2 The matrix processing unit.

We may view the zone memories as a set of vector registers, where the vector length is equal to the number of zones. On every clock, a vector multiply and a vector add are performed. 32 zones in parallel, each doing two operations every 64×10^{-9} second gives the Matrix-1 a peak rate of 10^9 floating point operations per second.

Within the Matrix processor there are three types of data interconnections. First, the systolic data path connects each zone to its neighbors; the last zone can connect to a systolic output buffer (the global buffer) or in a circular fashion to the first zone. A buffer provides data to the first and last zones over the systolic path, thereby acting as a systolic input buffer. The global buffer also sources a global data path for broadcast to all the zones. Third, the local memory of each zone has several connections, both to the zone's processing elements and to the system bus interface. All data paths are one word wide and operate at the clock speed of the Matrix processor. Thus the computational zones can communicate through a systolic shift or through a direct zone-to-zone path.

The Matrix processor can function in systolic mode, in which data are transferred between neighboring zones, or in block mode, in which all zones operate independently using only local data. All the zones execute the same operations in parallel. However, there are two mechanisms that allow the programmer more freedom. First, any subset of the zones may be disabled by setting the bits of a program-controlled mask. This allows for single zone computation or for computation in a subset of the zones. The second mechanism is the scatter/gather operations in which the index vectors can vary from one zone to another.

SIMD architectures have some important advantages. The control unit, control store and program memory need not be replicated. The programming model is simple. An SIMD design is synchronous. With a synchronous multiprocessor, it is possible to send data to another processor without incurring any overhead due to operating system software and its asynchronous hardware interface. The latency for communication with a neighboring processor is only one clock. This is two to three orders of magnitude faster than what is achieved by typical asynchronous MIMD multiprocessors.

The arrangement of arithmetic elements in the Saxpy Matrix-1 zone is shown in Figure 2. The output of the multiplier is cascaded into one of the ALU inputs in order to reduce the peak demand on the zone memory from 6 to 4 operands and results per cycle. This cascading allows peak performance when computing a sum of products, as happens during matrix-matrix multiplications and correlations.

Use of the global path and the shared global buffer has several benefits. Since shared data are not replicated, storage in the zone buffer is free to be used for locally varying operands. Furthermore, when one operand is received from the global path, less bandwidth is required of the zone memory.

Memory

While distributed memory is very suitable for some tasks, global memory seems to be imortant for general seismic data processing. It is very convenient for transposition (for frequency processing for example), for multidimensional Fourier transforms, for sorting (during surface consistent deconvolution for example), and for the display of movies.

The ability to store very large data-sets in main memory can reduce significantly the amount of intermediate data transfer to auxiliary storage devices. Thus from a few times 10^8 words to a few times 10^9 words of global memory is highly desirable. At about $256 per million words for NMOS dynamic RAM chips, this is now economically feasible. Unfortunately, large and inexpensive memory is slow.

At the peak rate of the Matrix-1 there are twice as many memory access operations as floating point operations, (three reads and one write, one add and one multiply). Therefore, to support computation at 10^9 floating point operations a second, data must be accessed at a rate of 2×10^9 words a second. Were these data to be provided directly by the system memory, its speed would have to be increased by a factor of 25 from the given rate of 8×10^8 words a second. To accomplish this, a far more expensive technology would be needed; in fact, no memory this large and fast have ever been built. Thus, a memory hierarchy is needed.

Most vector computers employ vector registers to buffer data between main memory and the computational unit; many scalar machines employ cache for the same purpose. We take a very different approach to the memory hierarchy. We use the zone memory as a block store and exploit it with block and systolic algorithms.

Multiplication of two $N \times N$ matrices takes $2N^3$ floating-point operations on $3N^2$ words of data access—an N-fold reuse of data. Furthermore, matrix multiply can be done by a block algorithm. By choosing the block size, the programmer (or the compiler) can obtain the desired ratio of computation to data and keep the data being used in the fast end of the memory hierarchy. Systolic algorithms also achieve reuse of data and require less memory bandwith. By choosing systolic and block lgorithms, the memory and data bus bandwidth needed to support the Matrix rocessor is effectively reduced by the necessary factor of 25.

The zone memory is explicitly managed by the program. This approach allows he programmer (or the compiler) to make near-optimum use of the zone memory. n cache memory systems, the local memory is under control of the hardware, which fetches data from main memory after "cache misses," or requests for data not esiding in the cache. We feel that for seismic processing, this software-managed ocal memory is better than hardware-managed cache because data may be reloaded by the software before it is needed.

nput/Output

In seismic processing, the data is moved through the levels of a memory ierarchy that spans the range from tape to disk to large, shared memory to small nd fast local memory to cache or registers and finally into the registers of the ALUs and multipliers. Unfortunately, the ultimate source of input and destination or output is magnetic tape. The realistic limit of tape I/O bandwidth is about 0.125 imes 10^6 words per second per tape drive. Thus, to sustain compute rates of 10^9 perations per second, one needs to have 80 tape drives running in parallel at full peed for simple processing (at 10^2 operations per input word), but can get by with ne drive if doing complex processing (10^4 operations per input word). This ptimistic assessment assumes that the disks can hold the whole data set so that it s enough to read from tape only once.

oftware

Another inescapable fact is that seismic processing is done with Fortran software. hus, machines that accept Fortran and can utilize existing software are much asier to use in seismic processing. A compiler that hides the hardware from the rogrammer is a big advantage. To date, progress on vectorizing compilers has een excellent. These compilers essentially detect SIMD parallelism at the inner oop level.

Compilers do not exploit the memory hierarchy very well, but we believe that here will be a major advance in this regard. By analysis of data dependences among statements, especially those contained in loops, the compiler can determine a way to break computation into subcomputations operating on blocks that can be held at the top level of the memory hierarchy. The compiler will look for compute intensive loops. These are loops nested more deeply than the dimensionality of the data; matrix multiply is a triply-nested loop with doubly-subscripted data, for instance. The compute intensive loops are the best candidates for this type of blocking.

While compilers for SIMD are achievable, compilers that detect and exploit MIMD parallelism appear to be much more difficult to achieve. The main problem is that large-grained MIMD parallelism requires consideration of the program in the large, a much more forbiding task than the analysis of inner loops.

GEOPHYSICAL PROCESSES IN PARALLEL

SIMD hardware is here, and compilers are achievable and are now here, too Meanwhile, it is worth-while to see if SIMD hardware can do enough geophysical processing so that continued development deserves the effort and attention.

The following are examples of geophysical data processes suitable to a SIMD parallel architecture.

Stack Optimization Residual Statics Estimation

Static time shifts are often an effective way to undo the effects of near surface anomalies on reflection seismic data. The estimation of the static shifts is a compute intensive process. The stack "power" maximization method (Ronen and Claerbout 1985), uses cross-correlations between long vectors to estimate the optimal static shifts implied by an objective to maximize the power in the common midpoint stack. The cross-correlations are between shot or receiver profiles, arranged one trace after another in a "super trace," to the relevant parts of the estimated stacked section, arranged in another "super trace." The method proceeds in a cyclic coordinate ascent:

```
Program (1)
  Repeat {
    Loop over shot and receiver stations {
      Form the two super traces
      Cross-correlate them
      Pick the maximum
      Update the stack
    }
  } until convergence
```

This algorithm spends almost all of its operations doing time-domain cross-correlations, which are performed in parallel as follows. Let the two vectors $\mathbf{x}=\{x_i\}$ for $i=0, 1,..., n,$ and $\mathbf{y}=\{y_i\}$ for $i=0, 1,..., n+m,$ be the data to be correlated. Their correlation $\mathbf{z}=\{z_k\}$ for $k=0, 1,..., m,$ is then defined by

$$z_k=\sum_{i=0}^{n} x_i y_{i+k}.$$

A copy of the vector $\mathbf{y}$ is stored in each of the local memories. The vector $\mathbf{x}$ is stored in a buffer that can broadcast to all the processors by a global data path. At time i, x_i is broadcast to all the processors. At that time, processor k calculates

$$z_k=z_k+x_i y_{i+k}.$$

The overall computation takes $2mn$ operations with traffic of $2m+2n$ words to and from the main memory, giving an $mn/(m+n)$ fold reuse of the data. The reuse is approximately m since in this case n is much larger than m. The number of correlation lags, m, has to be large enough for the data reuse to reach 25.

When $m+n$ exceeds the processor's memory or when m is larger than the number of processor's (32 for the Matrix-1) the algorithm is decomposed to subvectors and eventually recomposed at some small extra cost.

Crosscorrelation and estimation of surface consistent statics are important in seismic data processing but are not central. Migration is central. The following sections contain examples of SIMD application of migration methods.

Log-Stretch Dip Moveout

Dip moveout can be applied in shot profiles using two dimensional log-stretching (Biondi and Ronen, 1987). This method uses fast Fourier transforms (FFT) to perform a two dimensional frequency domain convolution. The algorithm is:

```
Program (2)
  Calculate and tabulate the operator
  Calculate and tabulate log-stretch interpolation coefficients
  Loop over shot profiles {
    Read a shot profile
    2-D log-stretch
    2-D FFT
    Apply the DMO operator (by complex vector multiply)
    2-D inverse FFT
    2-D inverse log stretch
    Add the processed profile to the stacked section
  }
```

This algorithm exploits the time-space invariancy of the DMO operator after two dimensional log-stretch transform. The operator is a two dimensional convolution and is applied as a multiplication by a data independent transfer function in the frequency wave-number domain. The recording geometry is assumed to be invariant from one shot to another; otherwise the operator has to be recalculated whenever the geometry changes.

The operations in this example are mainly two dimensional FFT and two dimensional stretch. The stretch operation is a variable mapping operation, similar to NMO only using a log function instead of a square root (Owusu *et al.*, 1983). Other less intensive operations are complex vector multiply (to apply the operator), and vector add (to stack).

The two dimensional transforms (FFT and stretch) are done in parallel by working on different columns, then on different rows in the separate processors. In a two dimensional FFT (or stretch transform), first the columns are processed in parallel, each of the SIMD processors performs a one dimensional FFT on one column, then the rows are processed in parallel, a row per processor. Internally, each processor is like an array processor, performing pipelined vector operations like FFT on the data in its local memory.

The floating point operations in an N-long one-dimensional FFT are $5N \log_2 N$, the data transfers are $2N$, thus leaving a reuse factor of $2.5 \log N$. The length N should be 1024 to reach a target data reuse of 25 fold.

The transposition associated with the row operations is of no cost because the data are transferred from the central memory (which contain the multidimensional array) to the local processor memories (which contain one column or one row each) and can be transposed "on the fly."

The one dimensional operations (the vector multiply and the vector add) are done in parallel by partitioning the vectors into chunks and assigning each chunk to a processor. However, the ratio of operations to word in the stretch, vector add, and vector multiply, is low so data reuse is not enough and the data transfers between the central memory and the local memory in the processors take more time than the computation itself. Thus a vector add may take about the same time as an FFT. We expect that SIMD compilers will chain vector operations to reduce the traffic between the global and the distributed memory.

Notice that in every operation in this algorithm all the processors run the same program, be it a one dimensional FFT, a one dimensional stretch, vector add, or vector multiply. Therefore, in this algorithm, SIMD is adequate.

Bucket Brigade Finite-Differencing Migration

A known disadvantage of Fourier and DMO methods is their limited ability to process data accurately in the presence of strong velocity variations. A computer company cannot expect a geophysicist to trade in his ability to run finite differencing programs, not even for a lightning fast FFT machine.

The partial differential equation,

$$\frac{\partial^2 u}{\partial t\, \partial z} = -\frac{v}{2}\frac{\partial^2 u}{\partial x^2}, \tag{1}$$

arises in finite-differencing (post-stack and prestack) migration. The equation can be obtained from the scalar (hyperbolic) wave equation, by the so called "15-degree" approximation, $\sqrt{1-\varepsilon} \approx 1 - \varepsilon/2$. The ε is the sine of the dip angle, and so the approximation above is accurate for small dips, 15 degrees or so.

Equation (1) can also be obtained by a linear transformation of the wave equation (Li, 1986), without any small dip approximation. When using the linear transform and also in prestack migration of shot profiles, the imaging condition is different to that in poststack migration, but the wave extrapolation is the same, using equation (1).

In the case of migration after stack, using the 15-degree approximation, the input stacked section is $u(z=0, t, x)$ and the output migrated section is $u(z, t=z, x)$, as shown in Figure 3. The migrated section is obtained by using the

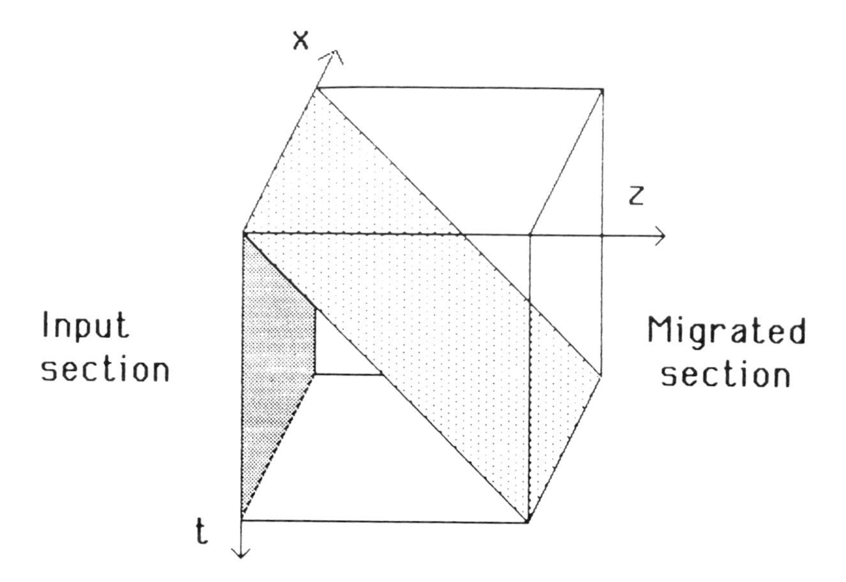

Fig. 3 Finite differencing grid for 15-degree migration. The input section is the $z = 0$ plane. The migrated section is the $t = z$ plane.

15-degree equation to extrapolate in the three dimensional grid of Figure 3, from the $z = 0$ side to the $z = t$ plane, using the difference equation

$$\mathbf{T}_1 \mathbf{u}(t, z) = -\mathbf{T}_1 \mathbf{u}(t + \Delta t, z - \Delta z) + \mathbf{T}_2[\mathbf{u}(t, z - \Delta z) + \mathbf{u}(t + \Delta t, z)]. \tag{2}$$

The $\mathbf{u}$'s are vectors along the x axis. $\mathbf{T}_1$ and $\mathbf{T}_2$ are symmetric tridiagonal matrices. They depend on the fixed sampling interval Δt, Δx and Δz, and on the velocity model, which may change laterally and vertically. For any velocity model the symmetric $\mathbf{T}_1$ and $\mathbf{T}_2$ are such that the extrapolation is stable (Claerbout, 1984).

The conventional migration algorithm is:

```
Program (3)
   Read the input into u(t, z=0)
   Loop over increasing depth z{
      Loop over decreasing time t{
         Solve equation (2) for u(t, z)
      }
   }
   u(t=z, z) is the migrated section
```

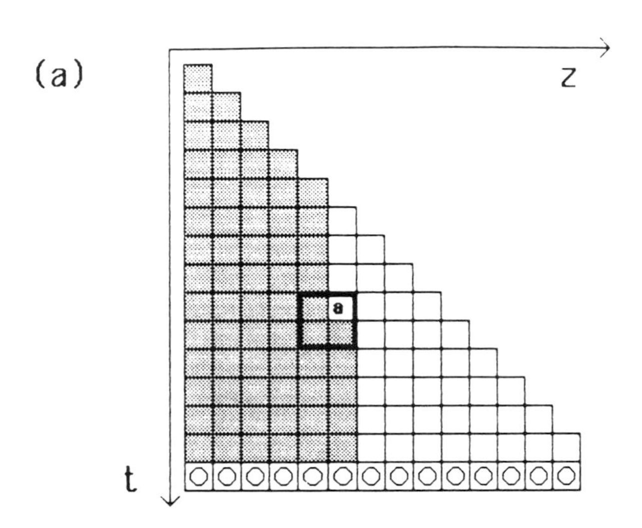

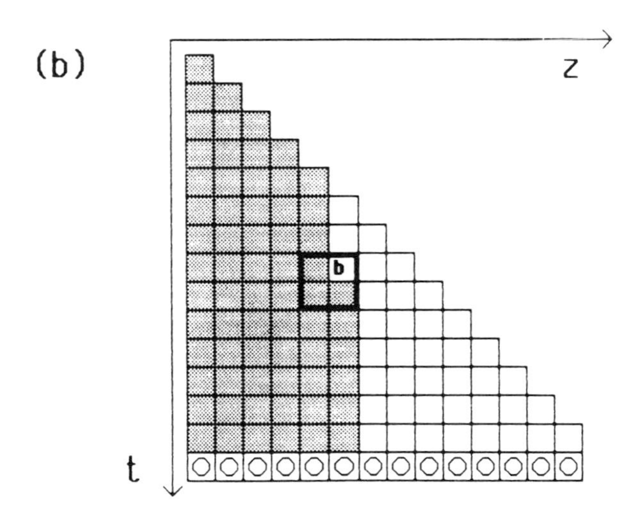

Fig. 4 15-degree migration on a non parallel computer. Each square represents a vector in the x direction. The shaded area denotes the known wave-field. The thick sided square is the two by two finite differencing star. (a) A single vector **a** is calculated. (b) A moment later, the vector **b** is calculated recursively as the differencing star moves up.

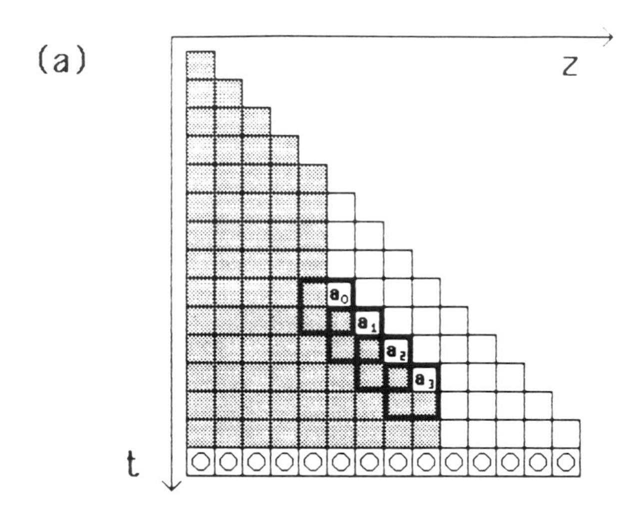

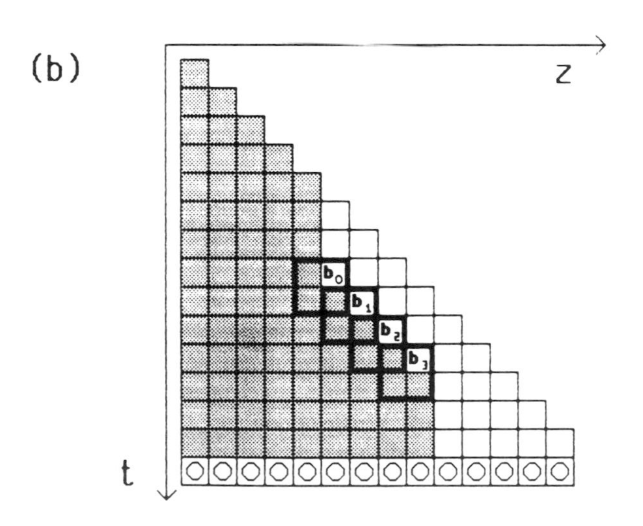

Fig. 5 15-degree migration on a four processor parallel computer. (a) Four vectors $\mathbf{a}_i$ are calculated in parallel. (b) A moment later, four vectors $\mathbf{b}_i$ are calculated recursively as the four overlapping differencing stars, one in each processor move up.

The wave extrapolation of Program (3) is illustrated in Figure 4. At any t and z a single vector is calculated.

The above migration algorithm adapted to an N fold parallel computer is:

```
Program (4)
  Read the input into u(t, z=0)
  Loop over depth z increasing in increments of N{
    Loop over decreasing time{
      For u(t+i Δt, z+i Δz), i=0, N−1
        Solve N equations (2), in parallel,
    }
  }
  u(t=z, z) is the migrated section
```

Program (3) is a special case of Program (4) when the number of processors is $N=1$. The difference between the programs is that Program (4) is incrementing the depth in steps of N, and in the inner loop, there is parallel computation of N vectors of $\mathbf{u}$ simultaneously! This is illustrated in Figure 5. What makes it possible is the independence of the unknown vectors of $\mathbf{u}(t+i\,\Delta t, z+i\,\Delta z)$ for different i's. The downward continuation in equation (2) is recursive in z and t but still can be done in parallel, because the recursion is between neighboring processors. To avoid a bottleneck between the local memories of the processors and the central memory, communication between processors is necessary. As can be seen in Program (4) and equation (2), processor i uses $\mathbf{u}(t+i\,\Delta t, z+(i-1)\,\Delta z)$ which has just been calculated in processor $i-1$. This one-way moving of information is like moving the bucket (containing $\mathbf{u}(t+i\,\Delta t, z+i\,\Delta z)$) in a brigade of N fire-fighters.

The velocity may vary both vertically and laterally. The lateral variations are expressed in the $\mathbf{T}_1$ and $\mathbf{T}_2$ matrices, internally in each processor. Vertical velocity variations are expressed in differences between those matrices from one processor to another. As can be seen in Figure 5, a processor stays in the same depth z for many t grid points, so there is no need to refresh the $\mathbf{T}$ matrices in a processor every step. (Running the program with z the inner loop would produce such a bottleneck.)

CONCLUSIONS

Seismic data processing is regular enough so that SIMD parallel computers are potentially very useful in seismic data processing.

To fulfill this potential, special algorithms for the parallel application of geophysical processes have to be developed and programmed. The main goals are to decompose processes into identical parallel tasks, and to avoid memory-access bottlenecks by reuse of data within the distributed memory of the parallel processors.

The software associated with the programming of these algorithms constitute the difference between the theoretical claim "the hardware can do it," which we make here, and actually doing it. Development of this software is a challenge whose rewards will include a significant improvement in our ability to explore the earth.

REFERENCES

Biondi, B. and Ronen, J., 1987. Shot profiles dip moveout. To appear in Geophysics *52*, (November issue). In press.

Brooks, E. D., 1986. The butterfly barrier. International Journal of Parallel Programming *15*, 295–307.

Claerbout, J. F., 1984. Imaging the Earth's Interior. Blackwell Scientific Publication.

Foulser, D. E. and Schreiber, R., 1987. The Saxpy Matrix-1: A general purpose systolic computer. Computer *20*, 35–43.

Li, Z., 1986. Wave field extrapolation by the linearly transformed wave equation. Geophysics *51*, 1538–1551.

Owusu, K., Gardner, G. H. F., and Massel, W.F., 1983. Velocity estimates derived from three-dimensional seismic data. Geophysics *48*, 1486–1497.

Ronen, J. and Claerbout, J. F., 1985. Surface consistent residual statics estimation by stack power maximization. Geophysics *50*, 2759–2767.

CHAPTER 7

HYPERCUBE SUPERCOMPUTING

by
W. D. MOORHEAD
Computer Science Department
Bellaire Research Center
Shell Development Company
Houston, Texas

1. INTRODUCTION

The purpose of this chapter is to familiarize the geophysicist with some of the issues of parallel computing that are relevant to geophysical data processing and modeling with almost exclusive concentration on the parallel processing engine that has come to be known as the hypercube. In the following text, sections are numbered with Arabic numerals, subsections with capital letters.

Parallel computing is not a new idea. It has been studied for a number of years and, as a result, there is an almost overwhelming diversity of opinion on the subject [24]. Many ways to connect a number of processors together into a coherent unit have been proposed but, until recently, no true multiprocessor computing engines were generally available from commercial sources. Within the past few years, a number of parallel processing engines have come to market. In particular, the hypercube multiprocessor computer is presently available as a commercial product from several manufacturers. Hypercubes are currently installed in universities, government laboratories, and industrial facilities all of which have active programs of both research and development.

Considerable understanding of hardware issues and algorithms for the hypercube architecture has been generated over the past few years [7, 12]. It is the purpose of this article to communicate some of these developments with an eye toward geophysical needs. To that end, we will consider several topics which include a brief discussion of multiprocessor computers along with a more detailed depiction of the hypercube architecture, typical system software, programming considerations, commercial products, and representative applications.

Many of the topics considered in this chapter have been the subject of numerous technical articles, while some have entire books devoted to them. Although no one topic can be discussed in any appreciable depth, the references should provide the interested reader with a starting point for further investigation.

Because the material contained in this chapter is largely devoted to elucidation of various achievements, it is well to include the following caveat. Parallel multiprocessor computing is not a fully developed field with respect to the availability of either automatic programming tools for the development of large production quality codes or of comprehensive software packages. In this regard, the field is still in an early stage of development with all of the promise and risk that can be attached to any rapidly unfolding area of technology.

2. DESIGN ISSUES FOR MULTIPROCESSOR COMPUTERS

There are a host of issues that must be regarded when contemplating the design of a parallel computer [24]. In particular, for the algorithm that will be run on a given machine, all processors must be kept busy doing useful work for a large portion of the time during the calculation. This is a factor of overriding concern in parallel computation for, without an ensemble of busy processors, no parallel speedup will result. Thus, there must exist a match between the algorithms of interest and the architecture in the sense that high parallel efficiency can be achieved. Of course, this is the case for most of the currently accepted parallel architectures. In this section we will address some of the principal design considerations for parallel computers and their relation to the hypercube architecture.

A. *Distributed vs Shared Memory*

In general, there are two ways in which memory can be apportioned in a multiprocessor. The first, called the shared memory scheme, has a single memory to which all processors have access. Perhaps the most ambitous shared memory device is the CEDAR machine [15] which is under construction at the University of Illinois. The second, the distributed memory scheme, is one in which each processor has access to only its own local memory. The hypercube is a distributed memory machine.

Each design has its own peculiar advantages and disadvantages. The shared memory machine must have a fast way with which to distribute data to the processors. This resource can become overloaded if it is too much in demand. For this reason, most shared memory machines have some local memory to hold often referenced data items. Also, a single data pathway or some particular area of memory may have many processors attempting to utilize it at one time during a calculation. This oversubscription of resource can cause a degradation in performance if either hardware or software is not implemented to overcome the situation. Proposed solutions to the memory "hot spot" problem that do not incur a large additional hardware expense are currently an area of active research.

An advantage to the shared memory machines is that it appears to be simpler to devise an automatic compiler for them than it is to automatically generate code for distributed memory architectures.

An advantage to the distributed memory machines is that data references are only made to the local memory under control of the processor that is executing the code. Of course, this implies that data that must be shared among processors is obliged to move along the communication lines that link the nodes together. Thus, for the distributed data machine, the problem must be apportioned among the processors in a manner that will avoid message paths that incur high transmission overhead along with the attendant degradation in performance.

Although the decomposition problem for parallel processing is still an active area of research, it is well understood how to decompose a rather large class of problems for the hypercube architecture. The latter sections of this chapter will be devoted to examples and discussion of some of the progress that has been made in this area.

B. *SIMD vs MIMD*

The acronyms SIMD, MIMD stand for Single Instruction Multiple Data, Multiple Instruction Multiple Data respectively. In a SIMD machine, every processor executes the same instruction at the same time. On the other hand, in the MIMD machine, the processors all run independent instruction sequences or, indeed, even completely different programs. Decomposition of a problem for SIMD versus

MIMD can be contrasted as separation in time as opposed to separation in space. This semi-aphoristic statement should not be taken too literally since, of course, each method can have components of the other.

C. *Interconnection Networks for Distributed Memory Machines*

The distributed memory machines have a network of data lines that interconnect the processors to each other. The method in which the data is physically moved and actually stored in the local memory is not important at this stage of the discussion. What is important, however, is the match of the necessary data movement among the processors with the topology of the network. Also of importance is the hardware cost in terms of the number of wires that must be paid for the construction of a particular interconnection stratagem.

There are many opinions as to which interconnection network is desirable for a multiprocessor computer [24]. The following is a short list of interconnection schemes which are used in distributed memory machines that are either available now or will be constructed in the near future.

The most obvious way to interconnect a set of processors is to join each to every other one. This is also the most expensive, requiring $N*(N\text{-}1)$ wires. A proposed solution to the wire problem utilizing a holographic mask that focuses a laser beam from a given processor onto the others is under investigation at J. P. L. in Pasadena.

Another possibility is to connect the processors in a linear chain, and, indeed, this has been done in the early systolic arrays of H. T. Kung [16] and in the WARP machine developed at Carnege Mellon University [17].

A third method is to connect the processors in a square grid of two or more dimensions. The grid is sometimes wrapped on a torus so that cyclic boundary conditions are imposed on the connectivity.

The hypercube interconnection subsumes many of the popular networks including the three mentioned above at a cost of $N*(2**N)$ connections among $2**N$ processors and can be implemented either as a physical network [22, 26] or as a switch sitting between the processors [21].

In order to perform computation in parallel on a set of processing nodes, a program for the execution of an algorithm must be broken into pieces and parceled out to the individual processors. Each processor will then perform the calculations necessary to execute its piece of the work. Given a particular algorithm which has been split into subprocesses, it is simple to draw a graph of the data dependencies among the various nodes that represent the procedures. The vertices of the graph represent computational work to be performed and the edges of the graph correspond to data pathways among the processors. The objective of such an exercise is to achieve a decomposition of the problem and to place the subprocesses represented by the vertices of the graph onto the individual processors. The next step is to assign the vertices of the graph to the processors in a way that will minimize an objective function that characterizes the utilization of the resources of the multiprocessor. This objective function is usually taken to be a combination of computational effort and communication cost. Although an algorithmic discovery of the optimal map of a data dependency graph onto a particular interconnect topology is typically a problem of combinatorial difficulty, it is easily seen that some interconnection schemes have a natural mapping of the data movement of a particular problem onto them. There is a growing understanding in a number of relevant cases [7] of how to choose algorithms that match the hypercube topology well enough to achieve efficient data movement. Some of these will be discussed in the sections on applications.

D. *Chip Count*

The mean time to failure of a computer should be large enough to accommodate a reasonable amount of useful computing time. The mean time to failure is an increasing function of the chip count so that roughly one duplication of a machine which thus doubles the number of chips will decrease the expected time between failures. This consideration is of importance for machines that could contain hundreds to thousands of processing elements. Thus attempting to keep the chip count per processing node as small as possible is very significant.

E. *Power and Cooling*

Some large vector machines require a special environment to accommodate their power and cooling requirements and incur a proportional cost for installation and upkeep. These requirements can restrict the circumstances in which it is feasible to install a supercomputer. Most commercial hypercube models have their physical size, power and cooling requirements at a more reasonable level than the large vector machines.

F. *Input Output*

For seismic processing, data handling at high speed is of primary importance. Any consideration of a multiprocessor computer for seismic computing must account for the way in which the data will be fed to the machine. Most commercial hypercube machines have hardware for direct injection of data into the nodes. This hardware can attach to a variety of *I/O* interfaces such as the DR11w or a disk or a tape controller.

3. THE HYPERCUBE MULTIPROCESSOR COMPUTER

This section gives a short history of the hypercube architecture, a brief description of some of the commercial models, and a view of a possible scenario for the next generation of machines.

A. *The History*

Driven by design criteria such as communication structures necessary to perform the FFT or sorting, number of communication wires, distance between processors, and uniformity of data movement, several proposals to build large hypercubes were made twenty or more years ago [21, 27, 28, 30, 31]. None of these machines were manufactured and it was not until the advent of VLSI that the density of packaging was sufficiently high to make construction of a large scale multiprocessor feasible.

A working 64 node hypercube was completed at Caltech in 1983 [26]. Each node processor consisted of the Intel 8086–8087 cpu, floating point chip combination, local memory, communication chips, and "glue" chips for the necessary control. Each node occupied an entire board whose communication channels were wired to those of other boards to form a hypercube interconnection network. An operating system was written to control the machine and numerous scientific applications were coded and run [6]. In many cases, the machine and its successors [6] provided an impressive performance advantage over comparably priced machines.

Motivated primarily by the work at Caltech, a number of commercial vendors have placed hypercubes in the marketplace since 1985. The hardware and performance of these machines is in a state of constant upgrade so that, by the time the reader is in possession of this article, any description of detail is bound to be out of date. With this caveat we shall list some of the commercial products along with their major features.

B. *Commercial Products*

Although other machines were installed in beta test sites at the time, Intel was the first to announce a commercial hypercube in 1985. Their machines come in a variety of configurations all of which are based on a combination of Intel chips along with floating point processors of variable power up to and including a vector processor on each node. The Intel iPSC (Intel Personal Supercomputer) can be configured to accommodate up to 128 nodes.

In 1985 Amatek Inc. NCUBE Corp. also announced their hypercubes as commercial products. The Amatek hypercube can have up to 256 nodes which employ Intel chips similar to those of the Intel machine and have an additional processor to control communication. The NCUBE machine makes strong use of the density provided by VLSI by packing 64 hypercube nodes and memory on a single board. Each NCUBE node chip has approximately the power of a VAX 750 processor [8], a hardware square root, hardware floating point, and 11 1-Mbyte/sec bidirectional communication channels. The current NCUBE machine can be configured with up to 1024 processors. In 1986 Floating Point Systems announced a hypercube which had as its node processor the INMOS Transputer coupled to a vector processor. Other developments include the Connection Machine and the non-commercial Mark-III under construction at Caltech/JPL [23].

C. *Second Generation Hypercubes*

One major advantage to the hypercube architecture is its scalability to greater size and speed. The number of communication links between nodes is $N^* \log(N)$ which implies that doubling the number of nodes does not violently increase the number of communication lines. The next generation of node processor could run with a floating-point speed of 10 times that of the node processor in today's machine. If the node-to-node communication bandwidth is increased proportionally, the present machine should scale well. Assuming this to be the case, a 2048-node second generation hypercube could be capable of producing 10 billion floating-point operations per second. At this computational rate, we would be approaching the speed necessary to do some of the more complex 3-D seismic data processing tasks.

4. HARDWARE, OPERATING SYSTEM, SOFTWARE

This section will describe the hardware configuration of a commercial hypercube along with a characterization of its operating system and software environment. As an example we shall choose the NCUBE hypercube computer manufactured by NCUBE Inc. of Beaverton, Oregon.

A. *Hardware*

The NCUBE machine consists of two enclosures one of which contains a backplane with up to 24 boards, while the second holds the disk/tape subsystem and power supplies. The cabinet for the machine measures 51 inches in width, 33 inches in depth, and 33 inches in height, approximately the size of an office desk. The entire machine is air cooled. Every node processor board contains 64 processing elements, each of which has its own operating system and memory. The entire machine can hold up to 16 node processor boards for a total of 1024 computing nodes. Each node element consists of a processing chip and one half megabyte of memory. The processing element is a single VLSI chip that combines communication control, memory access and management, logic, and floating point hardware. Each node processor has approximately the floating-point speed of a DEC VAX 11/750 and is capable of executing its own program independently of the others. Each processor is linked to a selected subset of the others via bi-directional communication lines. The communication network among the processors is such that the message lines constitute the edges of an N-dimensional hypercube with the processors as the vertices. The operating system running on the nodes has a set of communications procedures that may be called by the application program as subroutines. Thus, data is exchanged between processors as messages that are regulated by programs running in the processors in question.

B. *Operating System*

The operation of the hypercube system and disks is governed by the "host board" that contains its own multi-user, multi-tasking operating system. In addition, the host system provides a number of programming tools such as editors, FORTRAN and *C* compilers for both the nodes and host, and sundry UNIX utilities for file manipulation and searching.

The host system is responsible for loading the hypercube programs and for management of the hypercube resources. In order to execute a program on the hypercube, a user initiates execution of his host program, which in turn causes the node programs to be loaded from disk onto the appropriate processing nodes. A program, having been loaded onto a processing node, begins execution under its own control. Messages passed between the processors and the host program provide either input-output data or additional control.

The system also has a high speed graphics device that allows data residing on the host or in the hypercube nodes to be displayed directly on an attached high resolution color graphics screen.

A multichannel high speed input-output device that will allow attachment to any combination of tape, external disk, Hyperchannel, Ethernet or satellite computer is available. This hardware provides direct input-output with the hypercube processor nodes.

C. *The Programming Environment*

The host board of the NCUBE is a self-contained computer with disk and tape drive facilities, a UNIX-like operating system, and capabilities for Ethernet attachment. Up to eight users can operate on the system at one time in a multi-tasking environment when one host board is installed.

The NCUBE EMACS-style "NM" or vi editors offer the user a choice of full screen editing capabilities. The NCUBE programming environment also includes a number of the most useful of the UNIX tools for file management and manipulation.

The system has both FORTRAN 77 and C compilers for the host and nodes, and all of the host system calls are callable from either C or FORTRAN as subprograms. In addition, there are several FORTRAN- or C-callable functions for control of access to the hypercube itself and for management of message passing among the nodes.

5. THE HYPERCUBE INTERCONNECTION NETWORK

The hypercube is a generalization of an ordinary three dimensional cube. The three dimensional cube has the property that each of its corners (vertices) is attached to three other vertices by lines that represent the edges of the cube. Of course, the three dimensional cube has $2**3 = 8$ vertices. A hypercube of N dimensions has $2**N$ vertices with each vertex attached to N others. Thus, for the N-dimensional hypercube, each of the $2**N$ vertices has N edges attached. In a hypercube multiprocessor, the edges are physical data paths along which information can be communicated between the processor nodes (the vertices of the cube).

For applications concerned with data movement among the nodes, it is convenient to number the nodes in a binary representation. Each node of an N-dimensional hypercube is assigned a number whose binary digit representation is a sequence of N digits each of which is either zero or one. A point in N-space is located by giving N numbers that are the components of the vector location of the point along each of N linearly independent axes. The components of the node's location in N-space are defined to be the binary digits of the node number in question. Thus, we can visualize a particular node in N-space as located at a position described by its binary number representation. It should be clear that two nodes whose numbers differ in only one of their respective binary digits may communicate with each other over a one step path. Two nodes whose numbers differ in p places in their binary representation can communicate with each other with a minimum of p steps (p is called the Hamming distance between the nodes) and there are $p!$ distinct paths over which the message can travel.

A. *Meshes, Gray Codes*

For most hypercube machines, the least expensive communication is that which occurs between directly connected processors. Therefore it is desirable to decompose the data set such that whenever possible nearest neighbor communication is the dominant pathway. In some algorithms, a massive exchange of data among the processors is necessary at one or another stage during the calculation. In this case, there are efficient methods for performing data exchange that make use of the interconnection properties of the hypercube. This subsection will discuss some of the interconnection properties of the hypercube that are useful in understanding efficient data exchange among computational nodes.

The hypercube interconnection network subsumes a number of other connected graphs. By this we mean that there is a mapping of the connected graph in question onto the hypercube graph such that the edges of the connected graph map to edges of the hypercube and the vertices of the connected graph map to vertices of the hypercube. In particular, the linear chain, rectangular grids, and the binary tree [7] are examples. The linear chain and rectangular grids are discussed below.

Some systems, such as those described by partial differential equations, have a natural description as a linear, rectangular or volumetric set of grid points when the equation for the evolution of the system is discretized for digital computation. In order to achieve efficiency in the data transfer between hypercube nodes that will occur during the course of the computation, it is desirable to assign the hypercube nodes such that the nearest neighbor grid points in the spatial discretization of the problem are nearest neighbors in the hypercube. A mapping that has the above properties is called a Gray code, the properties of which will now be discussed.

Let us begin with a description of the map from a linear chain onto the hypercube since it is perhaps the most fundamental. We number a set of points in a linear chain as 0, 1, 2 ... N and assign these numbers to hypercube nodes so that a step in the linear chain from p to $p+1$ is a hop between nearest neighbors in the hypercube. We must do this in such a fashion as to never return to an already used hypercube node. The requirement for the assignment is thus simply to reorder the numbering of the hypercube nodes so that consecutive nodes differ by one in their binary representation. The way to accomplish the reordering has been known for some time to be the Gray code [10]. In particular, the following is called the binary reflected Gray code for reasons that will become apparent.

Start with the sequence 0, 1 and double its length by prepending a 0 to each member in the original order and then prepending a 1 to each member of the sequence but in reverse order to produce 00, 01, 11, 10. We now have produced the binary reversed Gray code for the numbers 0, 1, 2, 3 from that for 0, 1. Clearly, we can continue to produce the gray code for 0, 1,..., 7 as 000, 001, 011, 010, 110, 111, 101, 100 et cetera. Notice that the last and first members of the sequence differ in a single bit.

If an integer is represented as a column vector of its binary digits with the most significant in the top position, the Gray code of that integer can be obtained by multiplying this column vector by a matrix whose principal diagonal and first lower subdiagonal are all ones with the remainder of the entries being zeros. The multiplication must be performed in the ring of integers MOD(2). The matrix inverse of this operation is one wherein the principal diagonal and all subdiagonals are all occupied by ones, again in the same ring as above.

Given the number of a node in a hypercube, the node's linear chain number is found by applying the inverse Gray code to the node number. The left or right neighbor's node numbers in the chain are found by adding or subtracting 1 from the linear chain number and applying the direct Gray code to the result. These encodings are easily performed on a computer using bit manipulation.

For the case of regular rectangular mesh, one can also use the Gray code to make nearest neighbors in the mesh nearest neighbors in the hypercube by a simple extension of the linear chain code described above. Indeed, up to an N-dimensional grid can be embedded in an N-dimensional hypercube preserving the nearest neighbor property.

B. *Symmetry of the Hypercube*

The following discussion is given to indicate that there exists a tidy way to describe the interconnection properties of the hypercube and its subcubes. The usefulness of this description is found in the fact that it can be encoded easily and used to produce arrangements of nodes in any other region of the hypercube that will have the same interconnection properties as a starting set. The material in this section requires a rudimentary knowledge of group theory.

The hypercube has a high symmetry in that there are a number of mappings of the vertices onto themselves that preserve the physical interconnections of the nodes. A succinct description of this symmetry is obtained by mapping the nodes onto an abelian group generated by a set of order 2 generators $g0\ g1 \ldots g(N\text{-}1)$. Each node is an element of a group G of order $2^{**}N$ determined by a product of the generators each raised to a power equal to the corresponding binary digit of the node number. A generator, gj, can be visualized as a displacement along dimension j of the hypercube coordinate system. Further, displacements can be added componentwise if the scalar field associated with the N-dimensional hypercube vector space is the field of integers mod (2). Nodes n and m are physically connected if the group product $n * m$ is equal to a single one of the generators. A connection preserving map K of nodes to nodes must obey

$$K(n) * K(m) = gj \qquad \text{if} \quad m * n = gk$$

for some k.

Since the connection preserving maps must not produce new connections, the only automorphisms of the group that are allowed as candidates are those in the subgroup, Q of degree $N!$, of automorphisms produced by a permutation of the set of generators. Further, since the map of the group onto itself, $n - > p * n$, does preserve connections, we see that the full group of connection preserving mappings is the holomorph of G under the automorphisms in Q [32]. Because there are 2**N elements in G there are (N!) (2**N) elements in the group of connection preserving mappings of the hypercube onto itself. The composition of these maps is conveniently represented as the composition in the semi-direct product [32] of G and Q.

$$(g, R)(h, S) = (gR(h), RS)$$

where g and h are members of G, R and S are members of Q.

6. PROGRAMMING THE HYPERCUBE

Programming the hypercube is generally accomplished by breaking up the data set of the problem at hand and distributing the data over the node processors. In order for the whole machine to run efficiently, it is desirable to keep interprocessor communication "cost" at a minimum and to keep the work performed by the processors evenly distributed throughout the course of the entire computation.

In this section we will give a criterion for the efficiency of a parallel scheme along with a general discussion of some of the techniques for achieving it.

A. *Computational Efficiency*

The speedup, S, for execution of a task in parallel is defined to be the time for a single processor to perform the entire task divided by the time for N processors to perform the same task. The efficiency, E is defined to be the speedup per processor $E = S/N$. The commonly used upper bound for E is 1.

For the class of problems that have nearest neighbor communication on a static mesh of dimension d, it is possible to obtain a rather general expression that gives the efficiency in terms of a few node dependent parameters and n, the grain size of the decomposition [7, 9]. Let T-comm be the time to communicate a data item between neighbors and T-calc the time to perform a calculation on data items such as $c = a * b$, then the efficiency is given by

$$1/E = 1 + \text{const} * (T\text{-comm}/T\text{-calc})/(n)^{**}\ (1/d)$$

where const is usually between 1 and 10. The fact that the efficiency E is independent of the number of processors, N is indicative of the possibility for large performance gains on large problems by scaling the hypercube size with the problem size. The fact that the efficiency is only a function of the local properties of the processor is in conformity with the locality of data movement among the processors that was assumed in the derivation of this estimate. The validity of this estimate has been confirmed in a number of concrete examples that have been tested on the Cal-Tech, JPL hypercube machines. A more general discussion of these points can be found in the book by Fox *et al.* (7)

B. *Distribution of a Problem Over a Multiprocessor Computer*

A number of problems can be solved by using an algorithm such that the data generated in intermediate stages of the calculation is exchanged among sites which occupy points (computational nodes) located on a mesh. These algorithms are embedded in the hypercube by assigning a certain number of computational nodes to each of the processors in the hypercube. The computation then proceeds with the necessary data exchanges occurring among the processors. To achieve optimal utilization of the available computing resources two things are helpful. First, in order to minimize communication overhead, the assignment of computational nodes should be accomplished in a way that preserves the locality of data exchange. Second, in order to keep all of the processors busy doing useful work at all times during the execution of the program the work load should be distributed evenly among the processors.

If it is possible to maintain the work load balance throughout the calculation without reassignment of computational nodes, we have the case of a static mesh. If not, the mesh must be dynamically reallocated during the computation in order to maintain the desired load balance.

7. PARTITIONING TECHNIQUES

Several partitioning methods will be discussed in this section. We start with an ad hoc technique and discuss its effectiveness. Then we take up some of the procedures for semi-automatic decomposition.

A. *The Happy Guess Method*

If one knows the amount of work that must be performed in a particular computation, and it is observed for a given decomposition that the processors are busy doing useful work at all times, then one has as good a method in hand as any other. Thus, drawing on experience, the programmer can make a guess at how to decompose a problem, test the conjecture, and be pleasantly surprised to discover that high efficiency has been obtained. We call this technique "the happy guess method." It is a cousin of the familiar method of guessing the solution to a differential equation and then verifying its correctness. Of course there is the unhappy guess method which yields a very poor performance whcn tested. These decompositions usually do not appear in print! The happy guess method has been quite effective in simple cases involving matrix manipulation and decomposition of other sufficiently regular problems.

B. *Automatic Computation of Problem Decomposition*

There is a need to decompose problems which are too complex to be approached with the happy guess method. These can arise in finite element problems with irregular regions or other general situations such as dynamic multigrid algorithms.

Since we know that the decomposition problem is combinatorially hard insofar as it relates to graph mapping to a hypercube [14], it may appear that there can be no automatic method for finding an optimal decomposition. Fortunately, there are reasonably general methods yielding decompositions, that, although not truly optimal in their performance, are close enough to be acceptable.

We now turn to a discussion of such techniques. The first two methods start with an objective function that is to be minimized. This function accounts for communication overhead and computational load balance both of which can be estimated either before the program is run or during execution. If the program is constant in its demands for resources at all times during execution, a single decomposition that can be determined before runtime will suffice for the duration of the program. On the other hand, if the resource demands are time varying, the decomposition will have to keep pace as the computation proceeds. This situation is clearly more complex than the static case and will be touched upon again in the section on dynamic multigrid methods. Of course, the computation necessary to achieve resource balance must not utilize any appreciable fraction of the total time needed to compute the whole problem.

The objective function can be anything reasonable that tends to minimize the communication cost among nodes, groups local computations into single node processors and distributes the work evenly among the nodes. In the case for which the communication and computation costs can be estimated for each of the processors, the objective function, H, may be taken to be the sum of the communication overhead for each node plus the variance of the computation costs taken over all of the nodes.

B.1 *Decomposition by Simulated Annealing*

One method that has proven to be useful is to minimize H in stages as follows. We generate a random change in the decomposition scheme and find the change in H. If the change in H is negative we accept it, thus producing a new, currently accepted scheme. If the change in H is positive, we accept the new scheme with probability $\exp(-H/T)$. Thus the value of H is minimized over possible decompositions at a fixed value of T using the Metropolis Monte Carlo algorithm. The value of T is then lowered thereby reducing the number of statistically accessible states and the minimization is performed again. When a small enough value of T is reached, a decomposition approximating a minimum of H will have been attained. This technique, called the method of simulated annealing, has been successfully applied to a variety of problems of combinatorial difficulty including the one under discussion [7, 9].

B.2 *Decomposition by Neural Net Optimization*

It is possible to cast the decomposition problem into a form in which the neural net optimization scheme can be used. This method is dynamical as opposed to the statistical method used in the simulated annealing approach just discussed [7, 9].

B.3 *The Scattered Decomposition*

The scattered decomposition distributes the processors over the entire space of the problem in an attempt to equalize the work load while keeping the communications limited to nearest neighbors on the grid [7, 9].

In order to outline this method, let us assume for simplicity that the data space for the problem is two dimensional. The scattered decomposition is generated by first overlaying the computational grid with a large rectangle that covers the entire grid. The large rectangle is then covered with identical smaller squares that are called templates. Each of the identical templates will contain some number of computational nodes. Finally, the template is decomposed into a set of squares each of which corresponds to an available processor in the hypercube. For some problems it is convenient to Gray code the template decomposition so that nearest neighbors in the template correspond to nearest neighbors in the hypercube. We now have generated an assignment of computational nodes to hypercube nodes so that each processor handles computational nodes that are scattered throughout the entire grid. The scatter into small templates will tend to produce a better balance of the computational load across the processors, while the communication load will be increased as the template size decreases. Thus, there is some optimal size for the template that must be determined for each problem. This method has been shown to be useful in problems that have irregular boundaries and also for the cases wherein dynamic resource requirements are incurred [7, 9].

8. APPLICATION EXAMPLES

In this section we will illustrate how to use the hypercube for some familiar computing tasks. The examples in this section are not complete applications in themselves but rather are useful subtasks that could appear in a full blown application program. These simple examples should serve to illustrate some of the principles discussed earlier in the chapter with respect to communication and load balance.

A. *Transpose of a Matrix*

A matrix transpose occurs as an intermediate step in a multidimensional FFT. A method will be discussed in this subsection that will transpose a matrix using the hypercube in a manner that is easily understood and coded. A more detailed discussion of techniques to transpose a matrix on the hypercube is given in [11].

Consider a matrix whose columns have been distributed across nodes and, for the sake of simplicity, let us assume that there is one column of the matrix per node. The column number of a given element of the matrix is then the same as its node number. We now desire to transpose the matrix so that whole rows will occupy the nodes. Thus the number of the destination node is the same as the row number of an element of the matrix. In order to accomplish the transpose, sequentially compare the bits of the destination node number with the bits of the node number of the node that the data item currently occupies. Send the data item to the node where the first difference occurs in the bit patterns keeping all other bits constant. If correctly implemented, this procedure will produce a regular flow of all data items and will result in each item arriving in the proper place after at most a number of steps equal to the dimension of the hypercube. For the case of more than one column per node, a recursive block matrix transpose can be implemented [13].

B. *2-D FFT*

The two dimensional fast Fourier transform is used for velocity filtering, F-K migration and a variety of other well-known algorithms for seismic processing or modeling.

A two dimensional fast Fourier transform (FFT) can be computed on the hypercube in a number of different ways. The following method has the virtue of simplicity and speed. As indicated above, we apportion the data among the processors with an appropriate decomposition and then perform the computation using the interconnection network to achieve the necessary data movement. For clarity, assume that one seismic data trace has been loaded into each hypercube processor. It is helpful to visualize the entire data set as a matrix whose columns are the data traces. Thus, the node numbers correspond to the column numbers of the matrix. Of course, if a frequency-wavenumber (F-K) transform of a panel of seismic data is desired, the columns of the matrix should be loaded into the processors in a manner that represents the spatial order of the traces. The calculation proceeds as follows. First, perform an FFT on the data trace that is resident in each processor. We now have the F-X transform of the data with the frequencies numbered by the rows of our matrix while the spatial coordinate, X, corresponds to the column or processor number. Next, the matrix is transposed so that the data items for a single frequency and all X reside in an individual processor. The transpose is realized as described in the section on data exchange by swapping along consecutive dimensions of the hypercube. Finally, an FFT is performed in each processor on the entire data trace that resides there. We now have the F-K transform of the seismic panel such that a single column of the data matrix contains the various wave numbers and corresponds to a single frequency. At this stage, a filter can be applied to the data and the inverse F-K transform achieved by reversing the above processing steps using the appropriate inverse FFT in the processors.

C. *Matrix Multiplication*

Next, we describe one of several methods for matrix multiplication. This example will treat the product of two square matrices. Again, we decompose the data sets of the matrices to have a convenient and efficient method available for data movement among the various processors. This technique and its generalizations are described in detail in [7, 13]. The reader may want to diagram this procedure step by step in order to test its validity.

To begin, recall that the matrix multiplication $C = A * B$ can be accomplished by first partitioning A, B and C into square blocks and then performing the operation blockwise. The processors are numbered so that they occupy the nodes of a square grid in a nearest neighbor manner that matches the block decomposition of A and B. We assume that the data items of the matrix already reside in the processors with the mapping as just described. It is convenient to visualize the grid as wrapped on a torus with the implied periodicity and that a rolling of the grid data items can occur in each toroidal direction. The multiplication then proceeds as follows. First copy the contents of the diagonal blocks of A to locations corresponding to their respective rows. Thus, copy $A(0, 0)$ to locations corresponding to $A(0, 1) \ldots A(0, N)$ et cetera. Perform the block multiplication between the locally stored A and B blocks and buffer the result in C. Next "roll" the B subblocks around on the torus by one notch upward. Then broadcast the A subblocks that are one position to the right of the principal diagonal to their row partners as in the first step, and multiply and add as before. (The implied periodicity on the torus means that $A(N, 0)$ is broadcast in this step.) Continue to roll and broadcast until the rows of the B matrix have made one complete circuit on the torus. The multiplication is now complete.

9. APPLICATION PROGRAMS

In this section we will review a few algorithms that are full application programs and are representative of the activity in the field at present. The discussion will necessarily be brief but will contain sufficient references so that the reader can pursue the topics in greater depth if his interest so warrants.

A. *Finite Element and Preconditioned Conjugate Gradient*

Finite element calculations on the hypercube have been carried out by several groups [1, 18]. These consist mainly of a static grid that has been assigned to the problem at hand in combination with a pre-conditioned conjugate gradient method of solution [19]. A feature of this work is that the overall shape of the system may be irregular in which case the grid decomposition is nonobvious. For such systems, either the scattered decomposition or the method of simulated annealing as described above can be used to achieve communication and load balance.

B. *Adaptive Grid Methods*

Adaptive grid methods can have two dynamic aspects. The first is a changing grid density associated with a fixed region of space, the second is a change in the spatial location of the dense grid. Thus, the domain decomposition may or may not be time varying according to the particular problem to be solved. Of course, apart from stability and convergence considerations, the main problem to be solved with respect to the multiprocessor environment is the familiar one of load balance and communication cost.

Multigrid methods [29] are distinguished by a hierarchy of grid densities that are used at different stages of the computation. Many applications of multigrid methods are contained in the proceedings of the Copper Mountain Conferences on Multigrid methods. Applications of the multigrid method on the hypercube are discussed in [2, 5, 29] along with further references.

Dynamic adaptive grid methods are used for problems in which the computational action moves in space as the computation progresses. Some examples of this type of behavior are time varying problems such as wave propagation, shock propagation, or oil reservoir modeling. Other examples include fluid dynamics and aerodynamic modeling. Procedures for load balancing this type of computation on the hypercube are a very active area of research at this time. An example of one of the newer schemes can be found in the work of Berger [3].

C. *Further Applications*

The book by Fox *et al.* [7] contains a large collection of applications that have been adapted to the hypercube computers at Caltech and JPL. The list of topics already implemented there is of great breadth and we mention a few categorized by discipline.

Biology: brain models, protein dynamics.

Engineering: fluid dynamics, graphics ray tracing, chaos, earthquake engineering.

Geophysics: seismic wave modeling, geodynamics, ray tracing and tomography.

Physics: Computational astrophysics, Monte Carlo studies of condensed matter, granular physics, lattice gauge calculations.

10. SEISMIC DATA PROCESSING EXAMPLE

This section will discuss a simple example of seismic data processing along with a possible method for embedding it into a hypercube. The application is a standard frequency, wavenumber (F-K) filter which is part of the demonstration suite that has been shown by NCUBE corporation at a number of exhibitions during the past few years. Hopefully, the usefulness of the techniques that were given earlier with respect to hypercube programming and data space decomposition will be apparent in their application to the following example.

A. *Data Space*

The data space will be taken to a standard two dimensional, pre-stack marine seismic line. We will assume that there are 120 receiver stations per shot and that each data trace has 2400 amplitudes associated with it. Also, we assume that a shot is fired every 100 feet and that the line is 10 miles in length. Given these assumptions, we see that the entire line is a data set that consists of 152 million words and that each shot panel contains 288 thousand words of data. The data set is conveniently diagrammed according to either the shot, receiver or common midpoint, offset coordinates. For each of the examples, we shall decompose the data set over the hypercube node processors in a manner that will tend to achieve computational efficiency.

B. *F-K Velocity Filter*

Let us suppose that an F-K filter is to be applied to a shot panel, and that the transform is to be 128 traces in the K-domain and 3096 traces in the F-domain. Then, each panel will contain 793 thousand words of complex data in the transform domain. If each processor can hold 100 thousand words of trace data, there is an available store of 12.8 million words in a 128 node hypercube. This amount of storage will accommodate 15 shot panels that have been transformed into the F-K domain. So, we decompose the data domain into groups of 15 shot panels and load shot panel groups into a 128 node hypercube. The hypercube nodes should be Gray coded to correspond to a linear chain and a one to one correspondence made between the receiver station number and the linear chain processor number. Thus, each processor will take care of one receiver station with a few processors left over for zero padding. Traces from the receiver stations are then loaded into their corresponding processors and a time-frequency transform is performed on each trace. Then the hypercube network is activated and each of the panels is transposed as described in the section on data manipulation. Next, the transform in the station coordinates is performed in the individual processors which will produce the fully transformed panels. At this stage, the K-F filter will be applied, and the inverse transform executed on the panels.

11. CONCLUSIONS

The multiprocessor computer field has experienced a growth spurt in the last several years. In particular, several commercial versions of the hypercube architecture have come to market since 1985. Hypercube machines offer an environment for implementation of parallel algorithms and the prospect for high price performance in comparison to high speed vector machines. In addition, the hypercube architecre can be scaled up in both size and performance concomitant to improvements VLSI and communication technology. These advantages must be tempered with e realization that, at this time, the hypercube domain is not at a mature stage ith respect to either automatic parallelization of code or large numbers of full-task riented software packages.

EFERENCES

[1] C. Aykanat, and F. Ozguner, "Large Grain Parallel Conjugate Gradient Algorithms on a Hypercube Multiprocessor," in 1987 Proceedings of the International Conference on Parallel Processing, p. 641, ed. S. K. Shani, IEEE Computer Society Press in Cooperation with the Association for Computing Machinery, 1987.

[2] Mark E. Bassett, "An Implementation of Multigrid on a Hypercube Multiprocessor," in Hypercube Multiprocessors 1986, Michael T. Heath, Editor, SIAM, Philadelphia, 1986.

[3] Marsha J. Berger, and S. H. Bokhari, "A Partitioning Strategy for Nonuniform Problems on Multiprocessors," IEEE Transactions on Computers, Vol. C-36, May 1987.

[4] A. Brandt, "Guide to Multigrid Development," in "Multigrid Methods," Lecture Notes in Mathematics, Vol. 960, Springer Verlag, Berlin 1982.

[5] Tony F. Chan, Youcef Saad, and Martin H. Schultz, "Solving Elliptic Partial Differential Equations on Hypercubes," in Hypercube Multiprocessors 1986, Michael T. Heath, Editor, SIAM, Philadelphia 1986.

[6] Geoffery C. Fox, "The Performance of the Caltech Hypercube in Scientific Calculations," Caltech Report CALT–68–1298, April 1985.

[7] Geoffery C. Fox, Mark A. Johnson, Gregory A. Lyzenga, Steve W. Otto, and John K. Salmon, "Solving Problems on Concurrent Processors," Prentice Hall, Englewood Cliffs, 1987.

[8] Geoffery C. Fox, and Paul C. Messina, "Advanced Computer Architectures," Scientific American, vol. 257, p. 66, October, 1987.

[9] Geoffery C. Fox, and Steve W. Otto, "Concurrent Computation and the Theory of Complex Systems," in Hypercube Multiprocessors 1986, Michael T. Heath, Editor, SIAM, Philadelphia 1986.

10] E. N. Gilbert, "Gray Codes and Paths on the *n*-Cube," Bell System Technical Journal, vol. 37, p. 915, 9158.

[11] Chien-Tien Ho, and S. Lennart Johnson, "Algorithms for Matrix Transpositions on Boolean n-cube Configured Ensemble Architectures," in 1987 Proceedings of the International Conference on Parallel Processing, p. 621, ed. S. K. Shani, IEEE Computer Society Press in Cooperation with the Association for Computing Machinery, 1987.

[12] "The Characteristics of Parallel Algorithms," ed. Leah H. Jamieson, Dennis Gannon, Robert J. Douglas, The MIT Press, Cambridge Mass, 1987.

[13] S. Lennart Johnson, "Data Permutations and Basic Linear Algebra Computations on Ensemble Architectures," Research Report YALEU/DCS/RR-367, Feb. 1985.

[14] D. W. Krumme, K. N. Venkataraman, and George Cybrenko, "Hypercube Embedding is NP-Complete," in Hypercube Multiprocessors 1986, Michael T. Heath, Editor, SIAM, Philadelphia, 1986.

[15] D. J. Kuck, E. S. Davidson, D. H. Lawrie, and A. H. Sameh, "Parallel Supercomputing Today and the Cedar Approach," Science, v. 281, Feb. 28, 1986.

[16] H. T. Kung and C. E. Leiserson, "Systolic Arrays for VLSI," in Sparse Matrix Proceedings 1978, p. 256, ed. I. S. Duff, SIAM, 1979.

[17] H. T. Kung, "Systolic Algorithms for the CMU WARP Processor," p. 570, Proc. 7th Int'l Conf. Pattern Recognition, July 1984.

[18] G. A. Lyzenga, A. Raefsky, and G. H. Harper, "Finite Elements and the Method of Conjugate Gradients on a Concurrent Processor," in Asme International Conference on Computers in Engineering, p. 401, 1985.

[19] Oliver A. McBryan, and Eric F. Van de Velde, "Hypercube Programs for Computational Fluid Dynamics," in Hypercube Multiprocessors 1986, Michael T. Heath, Editor, SIAM, Philadelphia, 1986.

[20] L. Ni, C. King, and P. Prins, "Parallel Algorithm Design Considerations for Hypercube Multiprocessors," in 1987 Proceedings of the International Conference on Parallel Processing, p. 717, ed. S. K. Shani, IEEE Computer Society Press in Cooperation with the Association for Computing Machinery, 1987.

[21] M. C. Pease, "The Indirect Binary N-Cube Microprocessor Array," IEEE Transactions on Computers, vol. C-26, p. 458, May, 1977.

[22] J. P. Hayes, T. N. Mudge, Q. F. Stout, S. Colley, and J. Palmer, "Architecture of a Hypercube Supercomputer," in 1986 Proceedings of the International Conference on Parallel Processing, p. 653, ed. K. Hwang, IEEE Computer Society Press in Cooperation with the Association for Computing Machinery, 1986.

[23] Hayes J. C. Peterson *et al.*, "The Mark III Hypercube-Ensemble Concurrent Processor," Proc. Conf. on Parallel Processing, p. 71, Aug. 1985.

[24] (1971–1987) Proceedings of the International Conference on Parallel Processing, IEEE Computer Society Press in Cooperation with the Association for Computing Machinery, various Editors, ISSN 0190–3918.

[25] P. Sadayappan, F. Ercal, and S. Martin, "Mapping Finite Elements Graphs onto Parallel Processor Meshes," in 1987 Proceedings of the International Conference on Parallel Processing, p. 192, ed. S. K. Shani, IEEE Computer Society Press in Cooeration with the Association for Computing Machinery, 1987.

[26] C. L. Seitz, "The Cosmic Cube," Comm. of the ACM, vol. 24, p. 300, Jan. 1985.

[27] J. S. Squire and S. M. Palais, "Physical and Logical Design of a Highly Parallel Computer," Tech. Note, Dept. of Elec. Engin., Univ. of Michigan, Oct. 1962.

[28] J. S. Squire and S. M. Palais, "Programming and Design Considerations for a Highly Parallel Computer," Proc. Spring Joint Computer Conf., pp. 395–400, 1963.

[29] K. Stuben, and U. Trottenberg, "Multigrid Methods, Fundamental Algorithms, Model Problem Analysis and Applications," in "Multigrid Methods," Lecture Notes in Mathematics, Vol. 960, Springer Verlag, Berlin 1982.

[30] H. Sullivan, and T. R. Bashkow, "A large Scale, Homogeneous Fully Distributed Parallel Computer I," Proc. Computer Architecture Symp., p. 105, 1977.

[31] H. Sullivan, T. R. Bashkow and D. Klappholz, "A Large Scale, Homogeneous Fully Distributed Parallel Computer II," Proc. Computer Architecture Symp., p. 118, 1977.

[32] H. J. Zassenhaus, "The Theory of Groups," Chelsea Publishing Company, New York, 1958.

CHAPTER 8

LARGE SCALE ELASTIC WAVEFIELD INVERSION

by
PETER MORA,
Stanford University,
Stanford, CA, 94305, U.S.A.,
and
ALBERT TARANTOLA,
Institut de Physique du Globe,
4 place Jussieu, Paris, France

SUMMARY

Seismic recordings depend on the seismic source, the properties of the Earth, the location of the seismic receiver stations, and the physics of seismic wave propagation. It has always been a dream in seismology to predict the Earth properties directly from the seismograms using our knowledge of how seismic waves are affected by the rocks. Thanks to theoretical developments and advances in computer technology, this dream is on the verge of being achievable. The Earth properties can be estimated using a least squares conjugate gradient algorithm to solve for the Earth model which predicts seismograms that best match the observed data. A new theory puts the gradient direction required by this algorithm in terms of wave simulations. In the past, wave simulations in realistic Earth models were too CPU intensive for this formulation to be practical but this no longer appears to be the case.

A well understood method to do wave simulations in media of arbitrary complexity is by directly solving the discretized wave equation using the method of finite differences. The Earth is parametrized as a grid with each node of the grid associated with the elastic properties governing seismic wave propagation. Finite differences are used at each node to propagate the seismic waves from one instant of time to the next. At any instant of time, the calculations at a given node are independent from the calculations at other nodes (though data stored at a few nearby nodes must be accessed). Therefore, the calculations at an instant of time can be done at all nodes simultaneously. Hence, the method is ideally suited to fine grain parallel computer architectures such as that of the Connection Machine®[1] which is capable of rapid parallel communications between a large number of nodes. Results suggest that by using such fine grain parallel computers, realistic sized inverse problems can be solved for the first time!

THE SEISMOLOGISTS' DREAM

Earth images using traditional seismic processing methods

Traditionally, interpretation of reflection seismograms has been based on signal processing methods. These methods modify the seismograms to obtain a picture of the reflectors in the Earth. It has always been dissatisfying that most of these methods require interpretive steps, approximations and oversimplifications.

The processing methods consist of sequential steps to modify the seismic records to produce an image of the Earth. In oil exploration, the most common processing methods are called velocity analysis, NMO stacking and migration. Normally, each of these steps requires considerable interpretive input from an experienced seismologist. Even then, several attempts at processing may be necessary before the seismologist is satisfied he has obtained a good image of the Earth.

Because of the large quantity of data involved in seismic experiments, the processing steps had to be fast and so the simplest approximations to describe seismic wave propagation were used. For instance, many methods are based on the assumption that seismic waves can be approximated as acoustic waves but the Earth is not a liquid even to first order! Both compressional and shear waves are observed! Even if the waves were acoustic, the most common methods in oil exploration still make simplifying assumptions that restrict the applicability of the different techniques. For example, velocity analysis methods normally assume that

[1] The Connection Machine is a registered trademark of Thinking Machines Corporation.

there is no refraction of seismic waves in the Earth. This assumption would only be true if the Earth were homogeneous but the Earth has structure and is not homogeneous!

The dream of obtaining Earth properties from seismic observations

Seismologists dream of the day when it will be possible to automatically obtain the Earth's physical properties directly from the recordings of seismic waves with no approximations or oversimplifications. In principle, this can be achieved by inverting the equations of physics describing seismic wave propagation. Instead of computing the data observations (seismograms) from known Earth properties using the wave equation, the Earth properties are computed from a set of seismic observations using an inverse equation. This can be done by finding the Earth model which predicts seismograms that best match the observed seismograms. The measure of match depends on the statistics of the noise in the data and the statistics of the Earth properties.

This inverse problem is not easy to solve because of its immense dimensions. If a $4 \times 4 \times 4$ kilometer cube of the Earth is discretized every 10 meters then there are $3 \times 400^3 \approx 2 \times 10^9$ parameters required to define an isotropic elastic solid. Even more parameters are required if the Earth is anisotropic in this volume. The size of the data space, the digitized seismic records, for such a volume of the Earth is about 10^{10}. These are the size of model and data spaces corresponding to seismic surveys used for oil exploration. Full Earth seismic studies using Earthquake seismograms involve comparable sized spaces.

Even if an Earth model were obtained by solving the seismic inverse problem, its meaning may not be clear because of non-uniqueness. The dream to use the huge volume of seismic observations to reconstruct a single picture of the Earth can be compared to the dream of the pan-dimensional creatures (mice) in the "Hitch Hiker's Guide to the Galaxy" of finding the answer to the meaning of "life, the universe and everything". They built a huge computer to solve it, waited a few eons and voilà! The answer was "42" but they didn't know what this meant (c.f. the unclear meaning of the solution to the seismological inverse problem).

Head in the cloud dreams

Rather than a single Earth model, a better answer would be the probability of each possible Earth structure (i.e. a range of answers and their meanings). This range of answers and associated meanings could be represented as a multidimensional function giving the probability of every possible distribution of rocks in the Earth. The probability map would tell the seismologists what their answer means! As if the inverse problem to find a single answer was not hard enough, now we need all possible answers to know what the single answer means! The mice in the "Hitch Hiker's Guide" tackled this daunting task by building an even bigger computer consisting of a biological/physical system (the Earth) to solve for the meaning to the answer. Unfortunately for them, the Earth was destroyed to make way for a hyper-spaceway just prior to the time the answer was due. In a sense their computer suffered a mega-crash. The moral of the story is that when a problem takes on a large magnitude, its difficulty is compounded by bureaucracy and computer reliability.

Back to reality

The dream of probability maps of every possible Earth model is of such immense dimensions that it is considered impossible by most humans. To see why, consider how to compute a function giving the probability of each possible Earth model. One way is to generate synthetic seismic data for each realization of our discretized Earth and subsequently measure the probability of each realization. The probability would be measured by comparing the synthetic data with the data observations. When the two data sets look alike, the probability of the corresponding realization of Earth properties is high and when they are dissimilar the probability is low. For seismic inverse problems with 10^9 Earth parameters and say 100 realizations of each parameter then we would require 100^{10^9} forward modeling runs. Considering wave simulations take from seconds to hours on most computers, it is infeasible to compute probability maps using this brute force approach.

How about trying to solve for the single most likely solution? Is this smaller dream realizable? The answer appears to be yes provided a few assumptions are made. The most crucial assumption is that we can make educated guesses of the Earth properties that are accurate enough that the more generally applicable Monte Carlo and probability map methods are not required. In that case, we can obtain the most probable Earth model by doing only a few seismic wave simulations. Each simulation would determine the probability of the current Earth model contained in the computer. Some other calculations would determine how to change this Earth model to improve the probability.

Maximum probability inverse solution

Statistical knowledge is required in order to derive expressions for the most probable Earth properties. We assume Gaussian probability density functions for the Earth parameters and the noise in the seismic data observations. This corresponds to the least squares criterion to measure the fit between the synthetic and observed seismograms. Then the most probable Earth model can be found by iterative least squares which updates an Earth model iteratively until the best fit solution is obtained. Mora (1987a) used the preconditioned conjugate gradient algorithm which updates the Earth model as a linear combination of the model perturbations at the current and previous iteration. The current perturbations are a function of the least squares steepest descent direction which is the set of perturbations that most rapidly decrease the sum of squared difference between observed and synthetic seismograms.

The elastic forward problem

Least squares theory requires forward and adjoint calculations. The forward calculations consist of seismogram synthesis by modeling the propagation of seismic waves. The adjoint calculations consist of solving for the model perturbations that most rapidly decrease the square error sum (i.e. solving for the "steepest descent" direction). Tarantola (1984, 1987) and Mora (1987a) have shown that for elastic waves, the adjoint calculations can be formulated in terms of the forward calculations.

If the Earth is assumed to be perfectly elastic, then the seismic forward problem, that of computing seismic data (seismograms) $\mathbf{d}$ from Earth properties $\mathbf{m}$ denoted $\mathbf{d}(\mathbf{m})$ may be computed by solving the elastic wave equation (Freeman Press. Aki and Richards, 1980. Quantitative Seismology: Theory and Methods),

$$\rho \partial_{tt} u_i - \partial_j c_{ijkl} \partial_l u_k = f_i, \tag{1a}$$

$$n_j c_{ijkl} \partial_l u_k = T_i, \tag{1b}$$

$$u_i = 0, \; t < 0, \tag{1c}$$

$$\partial_t u_i = 0, \; t < 0, \tag{1d}$$

where $u_i = u_i(\mathbf{x}_S, \mathbf{x}, t)$ is the ith component of displacement resulting from shot S (i.e. body force $\mathbf{f}$ and/or surface traction $\mathbf{T}$) located at $\mathbf{x}_S$. If the receivers are located at $\mathbf{x}_R$ then digital data recorded every Δt seconds can be represented as:

$$d(i, S, R, J, \mathbf{m}) = u_i(\mathbf{x}_S, \mathbf{x}_R, J\,\Delta t, \mathbf{m}), \text{ and } J = 0, 1, ..., \tag{2}$$

so, for given $\mathbf{m}$, $\mathbf{d}(i, S, R, J, \mathbf{m})$ is an array of size $\mathrm{n}_i \times \mathrm{n}_S \times \mathrm{n}_R \times \mathrm{n}_t$, the discrete representation of the abstract vector $\mathbf{d}(\mathbf{m})$. n_i is the number of components recorded by the receivers (e.g. $\mathrm{n}_i = 2$ if the ground displacement in both the x and z directions is measured by the receivers).

The elastic inverse problem

The adjoint calculations allowing the Earth model to be iteratively updated are of form

$$\delta \hat{c}_{ijkl} = \sum_S \int dt\, \varepsilon_{ij}(\mathbf{x}, t)\, \tilde{\varepsilon}_{kl}(\mathbf{x}, t), \tag{3}$$

for the elastic moduli c_{ijkl} and

$$\delta \hat{\rho} = \sum_S \int dt\, \partial_t u_i(\mathbf{x}, t)\, \partial_t \tilde{u}_i(\mathbf{x}, t), \tag{4}$$

for the density ρ where ε_{ij} is the strain $\frac{1}{2}(\partial_i u_j + \partial_j u_i)$ associated to the background wavefield u_i computed using Earth model $\mathbf{m}$ and $\tilde{\varepsilon}_{ij}$ is the strain associated with the "back propagated residual wavefield" $\tilde{u}_i$ which will be defined shortly. Of course, not all of the 21 c_{ijkl}'s must be computed. For instance, assuming the Earth's is isotropic, only the Lamé moduli λ and μ or P- and S-wave velocity are computed, but the corresponding results are easily obtained from the general formula (3). For more details, the reader may refer to Tarantola (1984, 1987) and Mora (1987a). Note that the choice of parameters is important, for instance, assuming isotropy, the P- and S-wave velocity are better resolved from one another than the Lamé moduli.

The wavefield $\tilde{u}_i$ is defined as the field whose sources are the data residuals, acting as sources, and with *final* (instead of initial) conditions at rest.

$$\rho\partial_{tt}\tilde{u}_i - \partial_j c_{ijl}\partial_l\tilde{u}_k = \sum_R \delta u_i(\mathbf{x}_S, \mathbf{x}_R, t)\,\delta(\mathbf{x}-\mathbf{x}_R), \tag{5a}$$

$$n_j c_{ijkl}\partial_l\tilde{u}_k = T_i, \tag{5b}$$

$$\tilde{u}_i = 0,\ t > T, \tag{5c}$$

$$\partial_t\tilde{u}_i = 0,\ t > T, \tag{5d}$$

Equation (3a) can be understood conceptually by observing that the time integral performs a correlation between the strain ε_{ij} of the synthetic background wavefield u_i and the strain $\tilde{\varepsilon}_{ij}$ of the "missing wavefield" $\tilde{u}_i$ generated from the part of u_i that does not match the seismic observations. When the estimate of the Earth parameters is good, the missing wavefield is weak and the correlation (computed model perturbations) are small. When the estimate of the Earth parameters is poor, the missing wavefield is strong and the computed model perturbations are large. For example, if the estimate of the Earth model did *not* contain a particular reflector, then the missing wavefield would contain the reflected wave and this would correlate well with the shot wavefield at the location of that reflector (i.e. a reflected wave intersects with a direct wave at the interface that generates the reflector). Similarly, if the estimate of the Earth model contained an incorrect interval velocity in some region, the missing wavefield would contain a phase shifted reflected wave that has a strong correlation with the reflected part of the background wavefield in the region that caused the traveltime distortion (c.f. reflection tomography). This discussion provide a clue as to how the inversion formula can update both interval velocities and reflector locations (i.e. low- and high-wavenumbers in the Earth parameters). See Mora, 1988 for details.

A Newton preconditioning of the adjoint may theoretically speed convergence. Newton preconditioning corresponds to space-varying deconvolution to resolve between the different Earth parameters and to remove the source signature. While Newton preconditioning decreases the number of iterations it increases the cost per iteration. Experience shows that a simple approximation to the Newton preconditioning leads to convergence after only a few iterations (Mora, 1988a).

The inverse calculations

Consider equations (3) through (5) which define the computation of the gradient direction for our optimization problem. We need to perform the following steps:

(i) Propagation of elastic waves through some Earth model **m** using equation (1) to solve for the background wavefield $u_i(\mathbf{x}_S, \mathbf{x}, t)$ and the synthetic seismograms $u_i(\mathbf{x}_S, \mathbf{x}_R, t)$.

(ii) Compute the residual seismograms $\delta u_i(\mathbf{x}_S, \mathbf{x}_R, t) = u_i(\mathbf{x}_S, \mathbf{x}_R, t) - u_i(\mathbf{x}_S, \mathbf{x}_R, t)_{obs}$. These seismograms measure the difference between the synthetic and observed seismic data.

(iii) Back propagation of the residual seismograms in order to compute the back propagated residual wavefield $\tilde{u}_i$ defined in equation (5). In practice, the problem defined by equation (5) is solved using the same computer code that is used to solve equation (1), but running time backwards, and using residuals $\delta u_i(\mathbf{x}_S, \mathbf{x}_R, t)$ as a forcing function.

(iv) Calculate the time integral and shot sum in equation (3). In practice, this step is done during the back propagation to avoid storage of the wavefield $\tilde{u}_i$.

The power of the inverse formulation

Mora (1987b, 1988) pointed out that these inversion formula are like a combination of traditional migration methods of seismic processing that obtain reflector locations and tomographic and velocity analysis methods that obtain wavespeed between the reflectors. All seismic waves would be accounted for using this method and the most likely Earth model obtained. For example, Rayleigh waves which are traditionally treated as noise would be useful in the inversion, especially to resolve the compressional and shear wavespeeds near the Earth's surface. If this worked well, one of the oldest problems of reflection seismology called "the statics problem" would be solved. Namely, to account for traveltime delays of seismic waves caused by near surface wavespeed fluctuations (these delays distort the image of deeper reflectors if they are not taken into account).

Note that due to incomplete observations (discretization, finite recording axes (both time and space) and band-limited source functions) there will be some part of the model that cannot be resolved. Convergence to a locally optimal solution is an extreme manifestation of data incompleteness. A less extreme case would be slow convergence. Careful selection of the input data (choice of discretization and axis lengths etc) overcomes some of these difficulties but more research is required to quantitatively understand these phenomena.

To summarize, the real power of the inverse formulation in this paper is that it can theoretically do everything that all the methods of seismic data processing could do combined. It is a unified approach to estimate the Earth properties.

SIMULATING SEISMIC WAVES

In principle, two methods can be used to model seismic waves propagating in a complex inhomogeneous medium like the Earth:

Cellular Automata

Cellular Automata (CA) simulate the physics of wave propagation using a set of rules describing particle interactions on a microscopic level. Therefore, this method can be compared to real physics where molecules interact at a microscopic level giving rise to seismic waves and other macroscopic phenomena. Although research on acoustic wave simulations (e.g. Rothman (1987), Muir (1987)) based on fluid flow models (see Wolfram, 1986) is progressing rapidly, it has not yet extended to cover elastic waves.

Finite differences

Finite differences simulates the continuous differential equations describing wave motion by using finite differences approximation to the true derivatives. In comparison to the Cellular Automata approach where a microscopic description of molecular interactions is used to statistically simulate the macroscopic phenomena, finite differences directly solves the macroscopic equations. This method is well understood and is routinely applied at the present time. Mora (1986) has derived a fast and accurate isotropic elastic modeling scheme based on Kosloff et al. (1984) but using convolutional operators to perform the spatial derivatives rather than Fourier transforms.

The finite differences are done using the algorithm:

for all time {

$u(\mathbf{x}, t) \rightarrow e(\mathbf{x}, t)$ *from the strain displacement relation*

$e(\mathbf{x}, t) \rightarrow \sigma(\mathbf{x}, t)$ *from Hooke's Law*

$\sigma(\mathbf{x}, t) \rightarrow \ddot{u}(\mathbf{x}, t)$ *from the elastic wave equation*

$\ddot{u}(\mathbf{x}, t) \rightarrow u(\mathbf{x}, t+\Delta t)$ *step in time using an explicit finite difference scheme*

}. (6)

Although easy conceptually, this algorithm requires considerable expertise to implement efficiently in terms of memory and computational effort. The different stress variables are carefully specified on grids separated by 1/2 a gridpoint along the space and time axes for improved accuracy. Also, we have found the use of an eight-point convolution first-derivative operator to be optimal in terms of memory versus speed of calculation (see Mora (1986) for details of the scheme).

The number of floating-point operation is

$$Nflop \propto nc * nx * ny * nz * nt = nc * X * Y * Z/\Delta h^3.$$

where nc is the number of points in the convolutional first-derivative operator (nx, ny and nz are the number of gridpoints in the x-, y-, and z-directions and nt is the number of points on the discrete time axis). For a stable explicit finite-difference scheme with a specified accuracy in a volume of the Earth (X^*Y^*Z) and for a recording time (T), the grid-spacing Δh, is required to be roughly proportional to nc for $nc \leqslant 8$. An eight-point operator is accurate to about 3/4 of the spatial Nyquist. Lengthening the operator beyond eight points long does not significantly improve the accuracy. Hence, we can approximate the number of floating-point operations as

$$Nflop \propto nc\,\Delta h^{-4} \propto \begin{cases} nc^{-3}, & nc \leqslant 8 \\ nc8^{-4}, & nc > 8 \end{cases}.$$

Note that the finite difference method requires a large memory. Currently, the "Connection Machine" (with up to 512 MW) has enough memory potential for anisotropic elastic finite-difference calculations on a $256 \times 256 \times 256$ block of the Earth (e.g. $5 \times 5 \times 5\ \text{km}^3$ block if $\Delta h = .02$ km or the whole Earth discretized at 50 km). While large enough for most problems in Exploration and Global Seismology, more memory is still desirable to allow for better resolution. Note that assuming 4 Gflops and an accurate 8-points first-derivative operator ($\Rightarrow$ approx 250 operations per node per time-step for 3D calculations), the CPU for this sized forward problem with $nt = 2048$ is approximately

$$T_{CPU} \approx nt * nx * ny * nz * 250/(4 * 10^9) \approx 1/2 \text{ an hour.}$$

In two-dimensions we would have

$$T_{CPU} \approx nt * nx * nz * 170/(4 * 10^9) \approx 6 \text{ seconds,}$$

which suggests that in 2D, our inversion is viable but in 3D it remains too expensive to be widely applicable.

One further comment, the finite-difference method implicitly assumes that there are no sharp boundaries in the Earth model (i.e. the Earth parameters are wavenumber limited at a Nyquist wavenumber of $\pi/\Delta h$). In practice, this wavenumber limitation on the Earth model does not detract from the validity of the inversion considering seismic source spectra are band-limited implying the resolvable Earth parameters are wavenumber limited.

PARALLELISM IN NATURE AND COMPUTERS

How can finite differences be done fast enough to be useful for inversion? First observe that nature is intrinsically a parallel process (i.e. particles may vibrate simultaneously in different locations of the Earth). Surely, we can build a computer that can simulate wave propagation as fast or faster than they propagate in the Earth! Then, the outlined inversion method using wave propagations would be feasible and an Earth image could be automatically computed in real time! What would be necessary to achieve this kind of speed of calculation? The most obvious answer is a computer that is built to look like the Earth with many particles (processors) that operate simultaneously (c.f. the mice's computer, the Earth, in "the Hitch Hiker's Guide to the Galaxy"). The most crucial feature of a fine grain parallel computer is the ability of nodes (processors) to communicate with adjacent nodes. This communication must be done about as rapidly as the processors do a floating point calculation or the parallel computer would be inefficient.

Physical processes such as wave propagations are easily simulated on massively parallel computers. To see this observe that the calculations at all $\mathbf{x}$ locations at an instant of time in the finite difference algorithm (equation (6)) can be performed simultaneously. The "Connection Machine" of Thinking Machines Corporation (Hillis, 1986) with 64,000 processors has the highest level of parallelism existing today and is well suited to solving the seismic inverse problem using equations (1) through (6). (The Connection Machine has 64K processors. An instruction enables it to be configured as almost any size multi-dimensional grid with each gridpoint assigned to a "virtual processor". For large grid sizes, a nearest-neighbor communication is about the same speed as a floating-point instruction.) It is interesting that the creation of the "Connection Machine" was motivated by another physical problem, that of simulating the brain. The brain consists of many

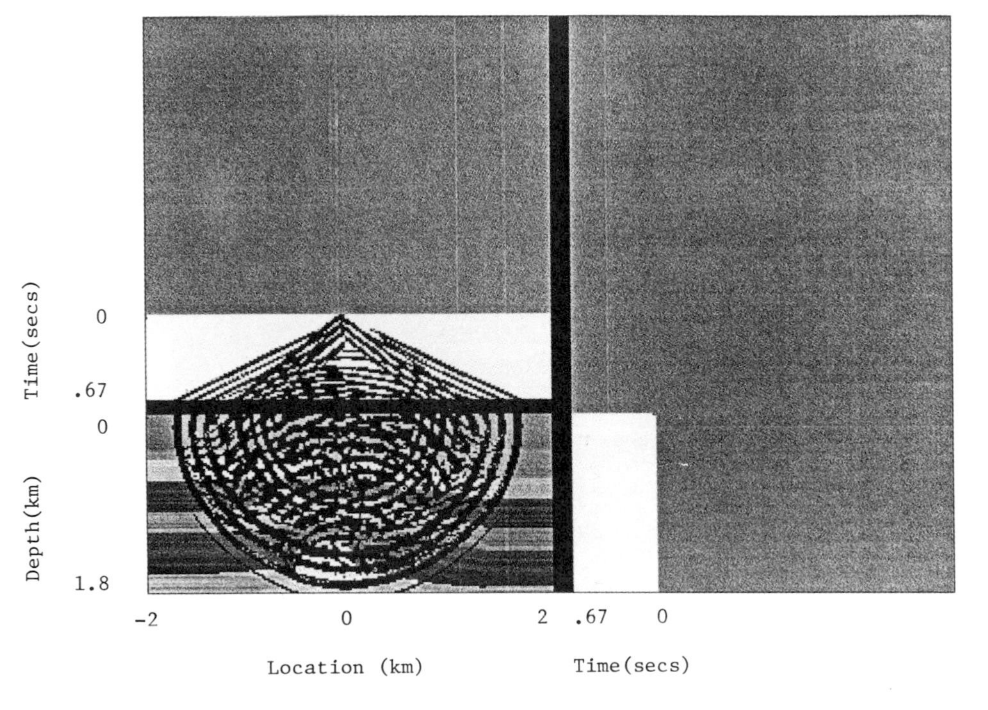

Fig. 1 A frame of an elastic wave simulation .67 seconds after a seismic source was activated. The velocity model with the seismic waves superposed is plotted in the lower left, the seismograms recorded at the Earth's surface are in the upper left plot and the seismograms recorded down a well on the right of the model are shown in the lower right plot.

interconnected neurons and so the creators of the "Connection Machine" put great effort into solving the important processor connectivity problem (hence the computer's name).

Figure 2 shows how some synthetic data were generated by simulating elastic waves propagating through an 2D Earth model using the method of finite differences (equation (6)). The Earth model in this figure represents a typical cross section of a sedimentary basin in an oil producing region. The different frames contain snapshots of waves propagating through the Earth model at an instant of time and the seismograms recorded thus far in the calculations (see Figure 1 for a description of one frame). The finite difference calculations over the entire Earth model are done in parallel so the computer time is proportional to the length of the

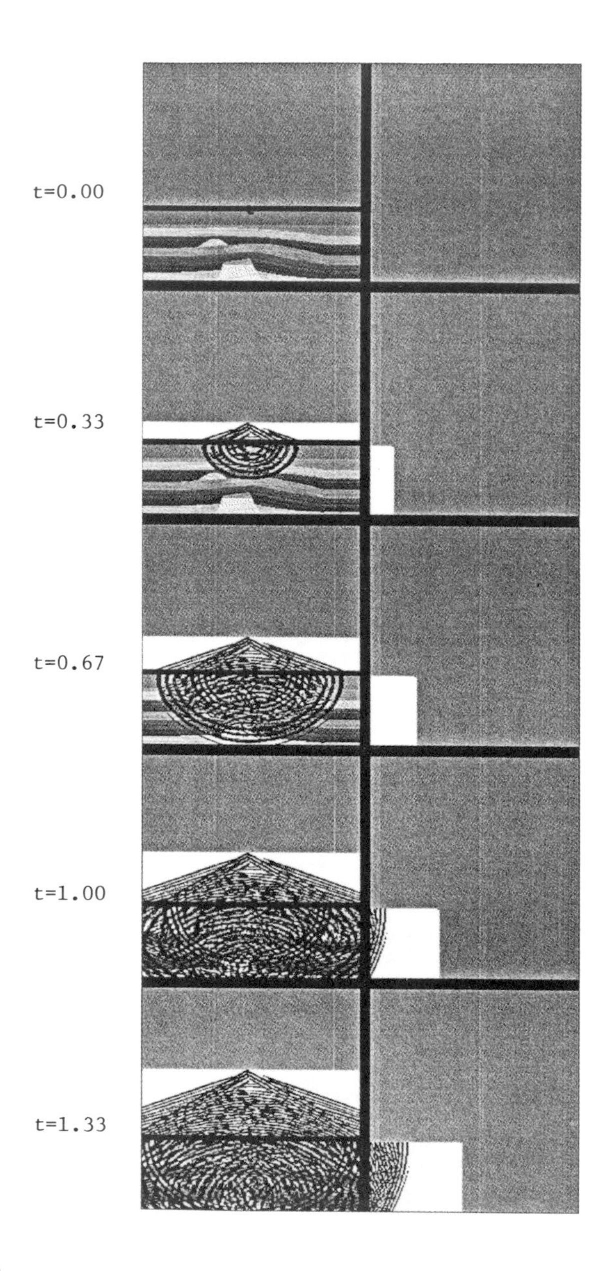

Fig. 2(a) Snapshot frames of an elastic wave simulation from $t = 0.00$ until $t = 1.33$ seconds. Refer to Figure 1 for a description of a frame.

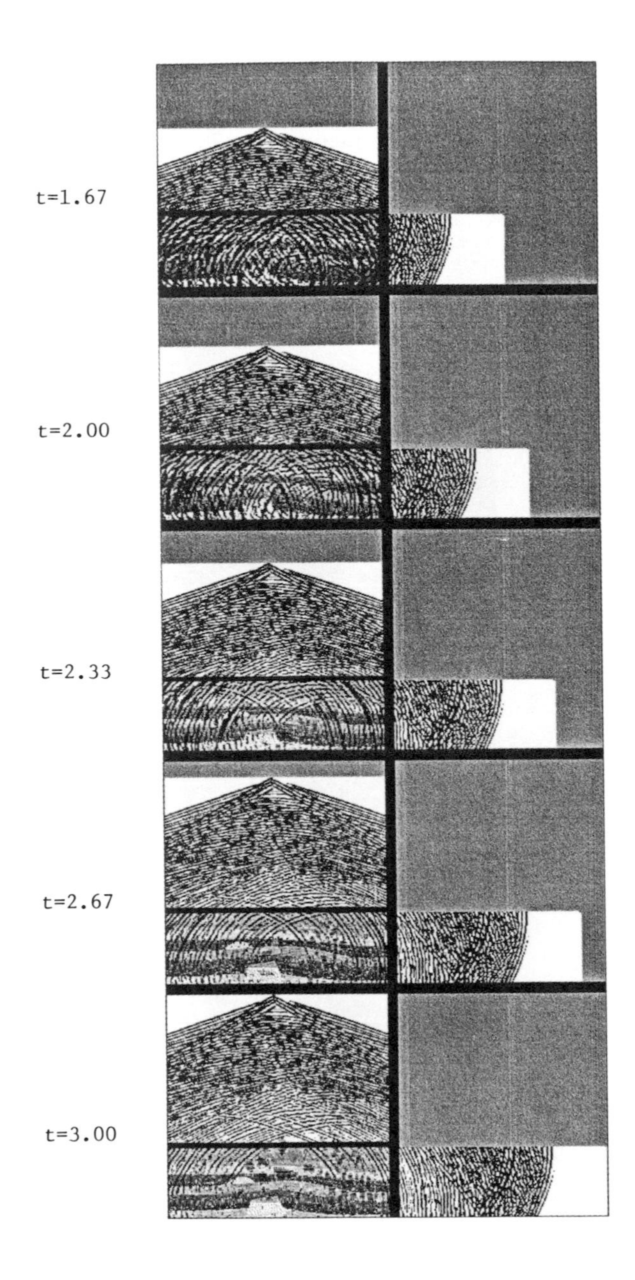

Fig. 2(b) Snapshot frames of an elastic wave simulation from $t = 1.67$ until 3.00 seconds. Refer to Figure 1 for a description of a frame.

seismogram time axis rather than the complexity of the Earth model. This is just the way real physics works with time progressing at the same rate whether or not the Earth has a complex structure.

The CPU time for this simulation on the Cm-2 "Connection Machine" is greater than, but the same order as, the time taken by waves to propagate through the Earth in reality. Because the simulation is slower than the physical experiment, real time inversions using equations (1) through (6) are not yet possible. However,

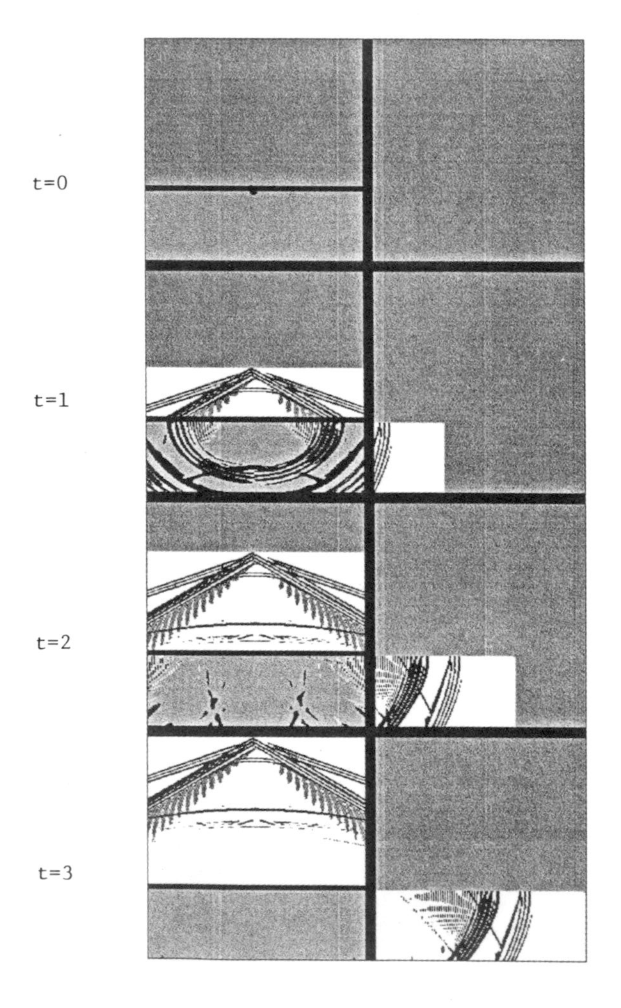

Fig. 3 Snapshot frames showing waves propagating through the initial velocity model to compute the background wavefield and synthetic data at iteration one.

the simulation is fast enough that the inversion process using recorded seismic data is feasible for the first time!

Assuming that each grid-point is assigned a processor (data level parallelism) the finite difference calculations are 100 % parallel at each time step. The conjugate gradient computations which are typically less than 1 % of the total inverse calculations can also be done in parallel. Hence, the inversion algorithm is virtually 100 % over the model domain so the number of sequential steps (and thus CPU time) is approximately proportional to recording time (c.f. normal physics). This

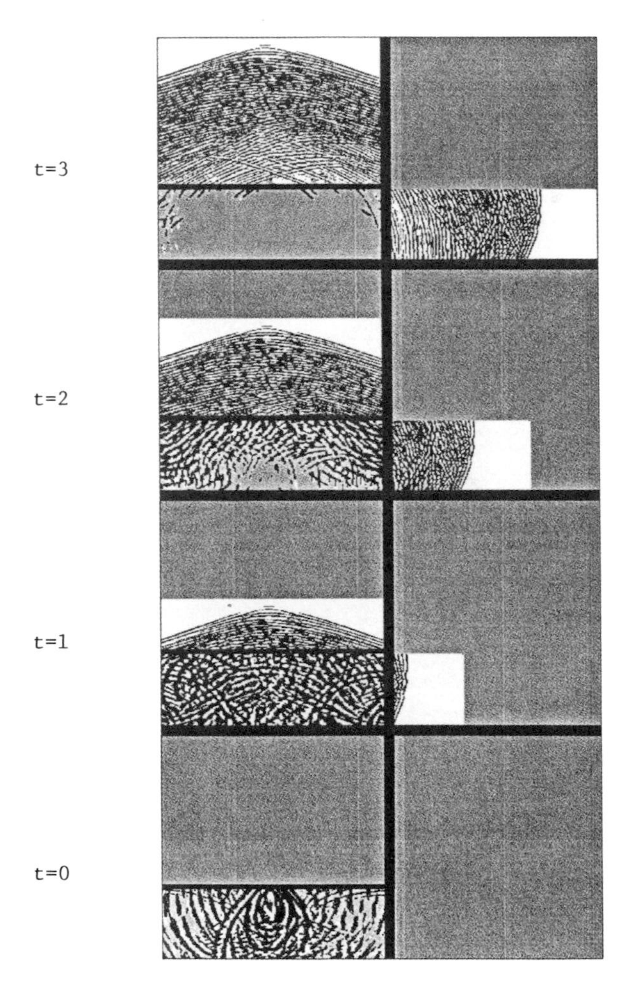

Fig. 4 Snapshot frames showing the computation of the back propagated residual wavefield at iteration one. The data residuals are used as a forcing function in reverse time.

gives hope that as parallel computers become faster, inversion will not only be feasible but will be commonplace and we will be able to invert the entire Earth!

INITIAL TESTS

Inversion tests done by Mora (1988a) are encouraging and fuel the dream and desire to exploit parallelism. They indicate that very good pictures of the Earth may

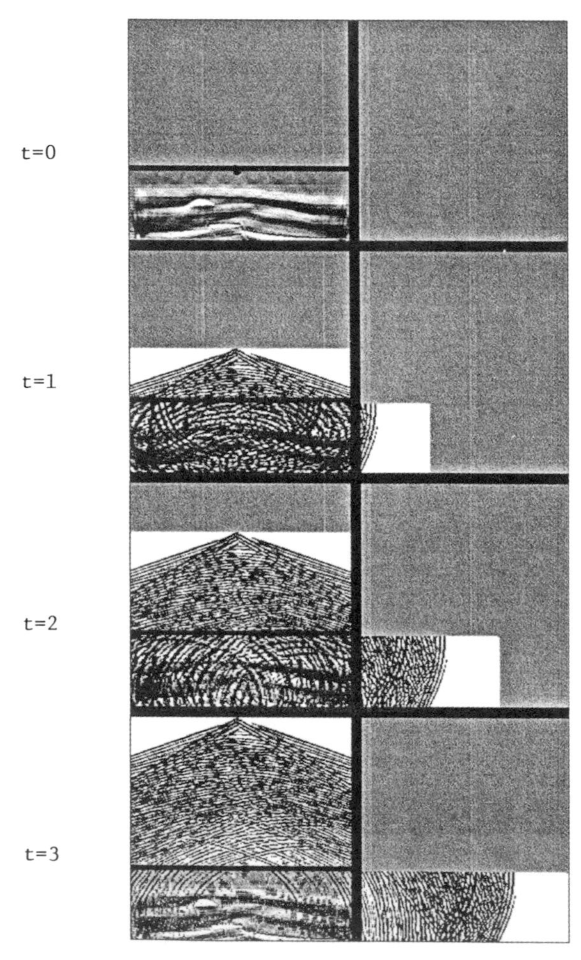

Fig. 5 Snapshot frames showing waves propagating through the ten iteration velocity model result to compute the synthetic data at iteration ten. This synthetic data matches well with the data being inverted shown in Figure 2 and the ten iteration velocity model looks like the true model of Figure 2 so the inversion was a success.

be obtained by the inversion process. Therefore, the dream of feeding seismograms into computers to obtain the Earth properties may soon become a reality as fine grain parallel architectures become more widespread.

The synthetic seismograms shown in Figure 2 were inverted with a linear with depth initial velocity model. Figures 3 through 5 depict the inversion process and demonstrate its dependence on the wave simulations. Figure 3 shows a forward modeling run using the initial velocity model to generate the background wavefield u_i. The data residual calculated by subtracting u_{obs} from u_i is used as a forcing function in reverse time to calculate the wavefield $\tilde{u}_i$ as shown in Figure 4. As the calculations proceed, the velocity and density perturbations are computed using equations of the form of equation (3). The new Earth model is computed by adding these perturbations to the current model. This summarizes one iteration of the inversion procedure.

In this example, the inversion algorithm converged to a solution Earth model that generated best matching synthetic data after 10 iterations (see Figure 5). The 10 iteration solution shown in Figure 5 looks like the Earth model of Figure 2 verifying that the inversion technique works at least under ideal circumstances.

CONCLUSIONS

Fine grain parallel computers are well suited to simulating physical processes. In seismology, the inverse problem to find the Earth properties using the seismic data observations can be formulated in terms of the physics of wave propagation and is hence suited to parallel computations. The overall level of parallelism of the seismic inversion algorithm is almost 100%, the only sequential steps being to propagate waves in time. Results from an implementation on the "Connection Machine" indicate that realistic sized oil exploration seismic inverse problems and whole Earth inversions can already be tackled. This brings the seismologists' dream of feeding seismograms into a computer and waiting for an Earth model to pop out one step closer to becoming a reality. As fine-grain parallel computer speeds increase through use of more processors, better chips etc, we can expect a revolution in geophysics. Sophisticated global inversions will be performed in real time as the seismic data is gathered!

ACKNOWLEDGEMENTS

Thanks to the sponsors of the Stanford Exploration Project, Jon Claerbout and Thinking Machines Corporation during this research.

REFERENCES

Hillis, W. D., 1986, The connection machine: The MIT Press.

Kosloff, D., Reshef, M., and Loewenthal, D., 1984, Elastic wave calculations by the Fourier method: Bulletin of the Seismological Society of America, **74**, 875–891.

Mora, P., 1986, Elastic finite differences with convolutional operators: Stanford exploration project report **48**, 277.

Mora, P., 1987a, Nonlinear 2D elastic inversion of multi-offset seismic data: Geophysics, **52**, 1211–1228.

Mora, P., 1987b, Elastic wavefield inversion for low and high wavenumbers of the *P*- and *S*-wave velocities, a possible solution: in Proceedings for the research workshop on deconvolution and inversion, September 1986, Rome, Italy.

Mora, P., 1988a, Elastic wavefield inversion of reflection and transmission data: Geophysics, **53**.

Mora, P., 1988b, Inversion = migration + tomography: Geophysics, submitted.

Muir, F., 1987, Three experimental modeling systems, Stanford exploration report **51**, 119.

Rothman, D., 1987, Modeling *P*-waves with cellular automata, Geophysical Research Letters, V. 14, p. 17–20, 1987.

Tarantola, A., 1984, The seismic reflection inverse problem, in Inverse problems of acoustic and elastic waves, edited by: F. Santosa, Y. H. Pao, W. Symes, and Ch. Holland, SIAM, Philadelphia.

Tarantola, A., 1987, Theoretical background for the inversion of seismic waveforms, including attenuation, submitted to Pageop.

Wolfram, S., 1986, Theory and applications of cellular automata: World Scientific Press.

CHAPTER 9

THE FUTURE OF ITERATIVE MODELING IN GEOPHYSICAL EXPLORATION

by
KURT J. MARFURT and C. S. SHIN

ABSTRACT

Supercomputers and massively parallel processors have and will continue to profoundly impact the application of seismic modeling. Most of the emphasis in the recent past and in this memoir has been devoted to either solving the same modeling problems faster or to solving the samc modeling problems bigger, as in 3-D modeling versus 2-D modeling. The authors of this chapter take a slightly different view and propose an efficient modeling scheme applicablc when only part of the model is subject to revision.

Anyone who has provided a modeling service to a geophysical operating division has certainly encountered iterative modeling. The thrust of this chapter is that by careful planning and the use of extremely large memory supercomputers, one can provide the suite of models that the acquisition geophysicist and interpreter really desire with only moderately increased cost and delay.

The four implementations of iterative modeling described in this chapter–substructuring, partial factorization, homotopy and Taylor Series expansions–all exploit properties of the numerical Green's function obtained by a frequency domain finite element formulation. In particular, perturbations of the model can be viewed as controlling equivalent sources that add or subtract energy from the total wave fields.

WHAT IS ITERATIVE MODELING?

Most structures of engineering and geologic interest can be broken into smaller components. It is obvious that in order to model a Fokker triplane aircraft that one needs to be able to model the wings, the tail, the engines and the fuselage, subject to surface air currents. We denote iterative modeling to be the process where one changes only one component or parameter at a time (such as the tail design on our jet aircraft), keeping the others fixed. One continues to change this parameter until one determines the model sensitivity (of airflow and stresses) and if desired an optimum design. One may continue the modeling exercise by subjecting the new structure to different applied fields (air speeds) or by modifying a second component (perhaps the wings).

Analogously, in order to model a seismic experiment, one needs to be able to model the target, the overlying rocks, and the underlying rocks (Fig. 1) subjected to a suite of applied seismic sources. One may wish to study the target response as a function of its porosity, thickness and fluid content. Alternatively, one may wish to optimize the illumination of the target given various overburden sequences.

Whether for cultural reasons or for the limited capacity of the human mind, it is rare that one would vary more than one model component at a time, even though computers are quite capable of doing so. We hope to show in this chapter

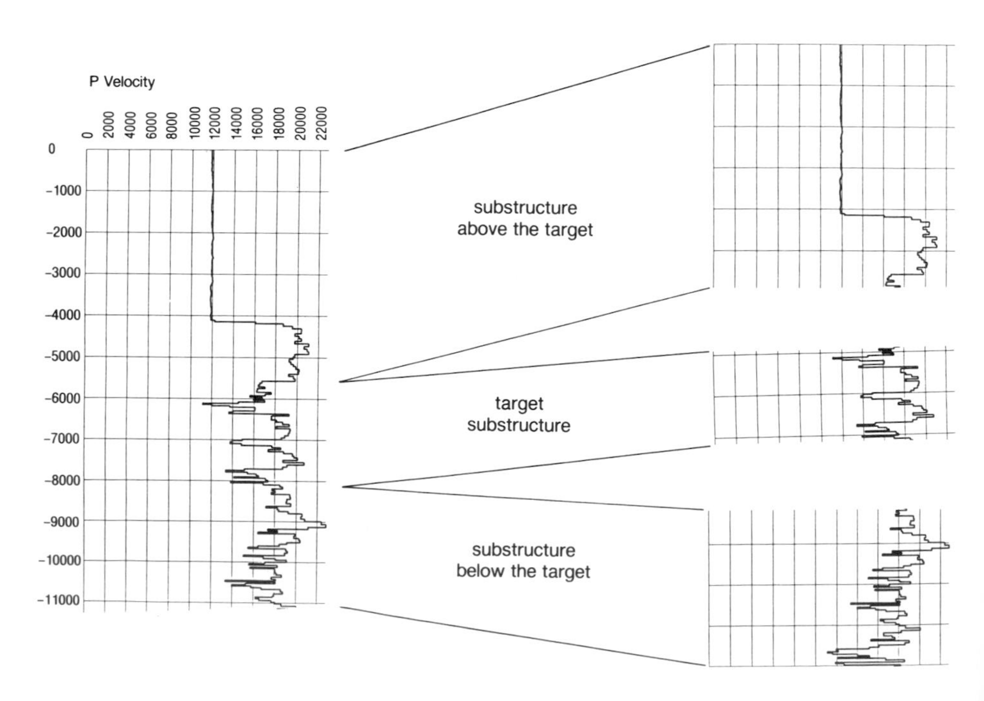

Fig. 1 Substructuring of a target gas sand.

that one can achieve significant computational savings by simultaneously solving a suite of problems that vary with only one or two parameters than by solving the same suite of problems independently.

ITERATIVE MODELING USING SUBSTRUCTURE ANALYSIS

Efficient iterative modeling has long been used in structural analysis. Since one of the major differences between finite element and finite difference techniques is the greater effort in model definition and bookkeeping effort required, it soon became clear that there was much to be gained by cataloguing libraries of major model components or substructures. In this way, one would need to explicitly define only one of six main wings and one of two tailwings on a Fokker triplane. These substructures (also called superelements because they may be composed of many hundreds or thousands of finite elements) can then be assembled to the fuselage using conventional finite element techniques. The fully assembled structure is then subjected to appropriate boundary conditions and applied fields to obtain the full solution. So far, one has only saved in the model definition and assemblage phase, not in the numerical factorization and solution phase. In general, savings in the factorization and solution phases is possible for neither nonlinear nor time marching schemes; but one *can* achieve significant savings in the linear seismic wave propagation problem solved in the frequency domain.

$$(-\omega^2 M + i\omega C + K)U = F \tag{1}$$

where

M is the mass matrix,

C is the damping matrix,

K is the stiffness matrix,

F is the applied source field vector,

and

U is the solution field vector.

For the remainder of this chapter, we will simplify the notation by rewriting Equation (1) as

$$SU = F \tag{2}$$

where the complex impedance matrix

$$S \equiv -\omega^2 M + i\omega C + K$$

is symmetric because of source/receiver reciprocity. Substructuring is particularly easy to implement in a flat layered hybrid finite element/spectral method algorithm such as those developed by Alekseev and Mikhailenko (1980), Orsag (1980) and others.

Here, one solves the problem of a fixed unperturbed geologic sequence by forming a substructure (Figure 2). Since any element interacts only with its immediate neighbors (because the interpolation functions are zero outside the element in question), one can eliminate the interior of a substructure before the total problem is solved. Labeling those degrees of freedom on the substructure boundary and along the output surface as "*a*" and those on the interior as "*b*" (Figure 2), one can formally partition Eq. (2) by rearringing rows and columns to obtain

$$\begin{bmatrix} S_{aa} & S_{ab} \\ S_{ba} & S_{bb} \end{bmatrix} \begin{bmatrix} U_a \\ U_b \end{bmatrix} = \begin{bmatrix} F_a \\ F_b \end{bmatrix} \tag{3}$$

Solving for $\{U_b\}$ using the second equation of (3)

$$[U_b] = [S_{bb}]^{-1}([F_b] - [S_{ba}][U_a])$$

one substitutes this value into the first equation

$$[S_{aa}][U_a] + [S_{ab}][U_b] = [F_a]$$

and obtains

$$([S_{aa}] - [S_{ab}][S_{bb}]^{-1}[S_{ba}])[U_a] = \{F_a\} - [S_{ab}][S_{bb}]^{-1}[F_b]$$

Defining

$$\begin{aligned} [\tilde{S}_{aa}] &= [S_{aa}] - [S_{ab}][S_{bb}]^{-1}[S_{ba}] \\ [\tilde{F}_a] &= [F_a] - [S_{ab}][S_{bb}]^{-1}[F_b] \end{aligned} \tag{4}$$

one has the equations for the substructure (related to the Schur complement in linear algebra):

$$[\tilde{S}_{aa}][U_a] = [\tilde{F}_a] \tag{5}$$

which can be assembled (added) to other elements or substructures to form a larger problem. All multiple events, transmission effects, and mode conversions still reside in the substructure. No short cuts have been taken, indeed the amount of computation required to arrive at Eq. (5) is slightly more than the direct solution of Eq. (2) since one needs to perform "a" times as many forward and back substitutions in the numerical solution of Eq. (4) and (5).

Figure 3 shows a suite of layered models run over a volcanic sequence of

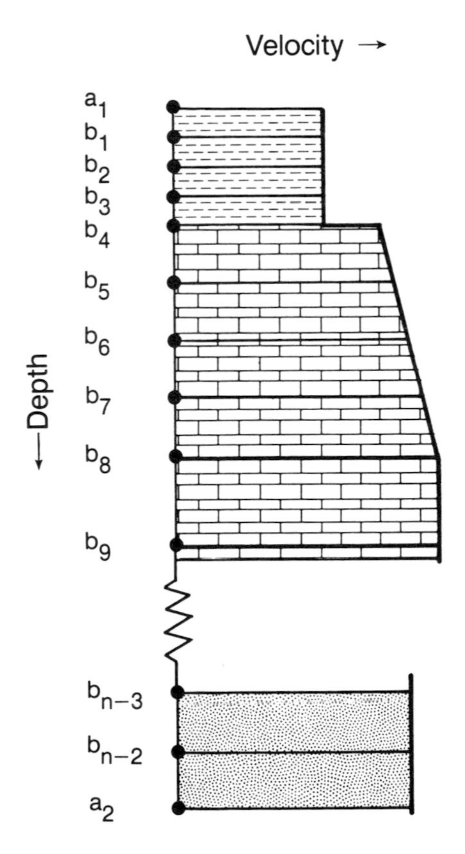

Fig. 2 Nodal numbering of a substructure. Nodes b_1 through b_{n-2} will be eliminated. Nodes a_1 and a_2 will be retained to interact with overlying and underlying substructures.

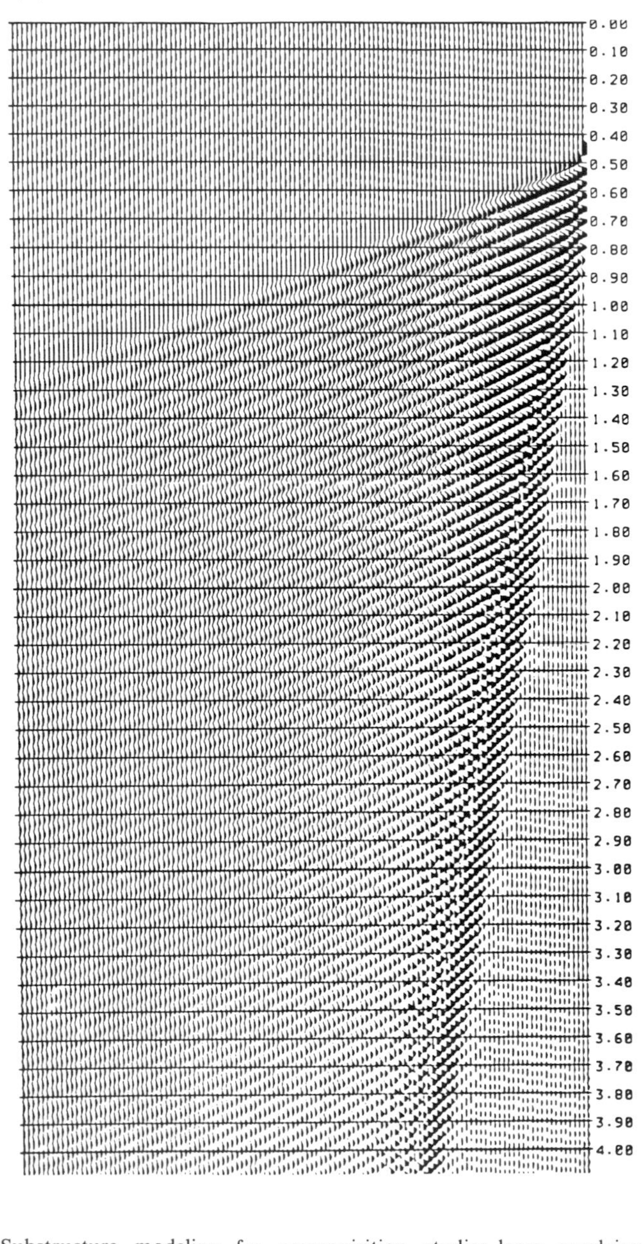

Fig. 3(a) Substructure modeling for preacquisition studies–loess overlying–thick volcanics overlying sediments: 12 ft loess with $Q_p = Q_s = 200$.

(b)

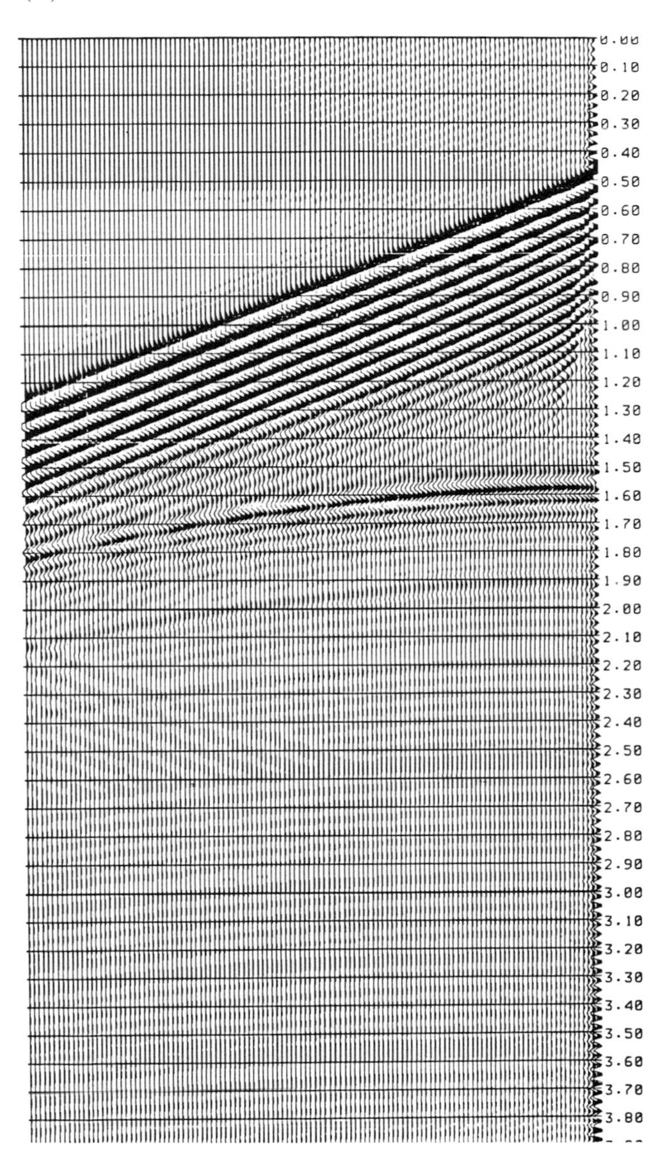

Fig. 3(b) 12 ft loes with $Q_p = Q_s = 10$.

(c)

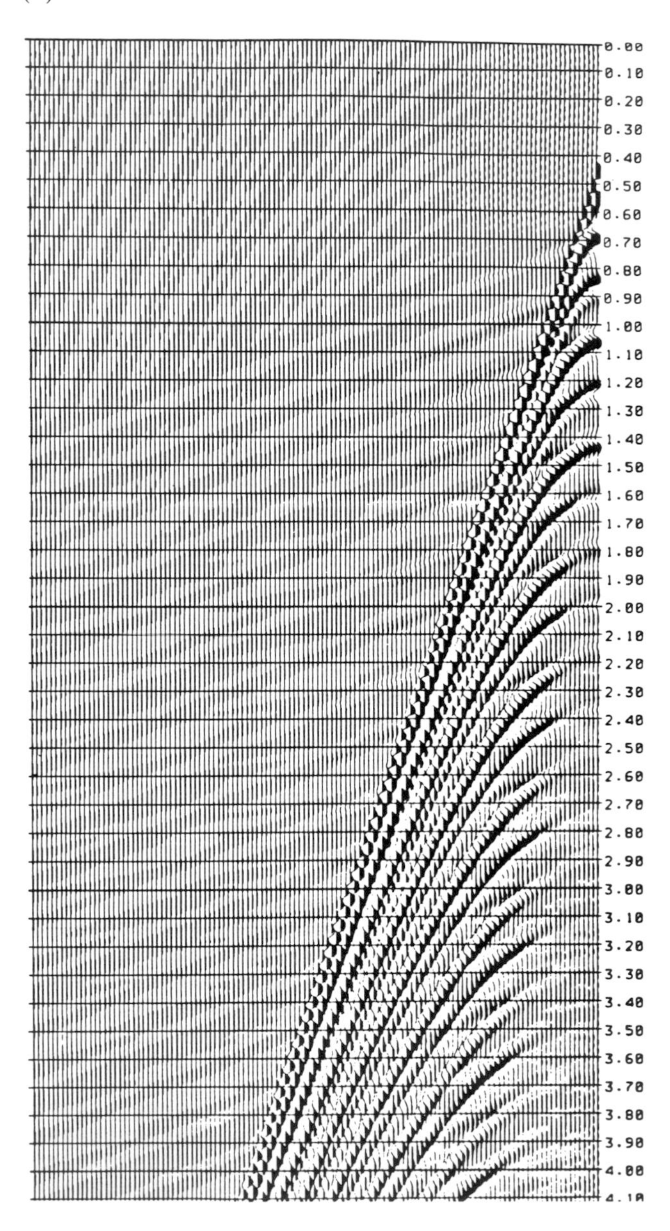

Fig. 3(c) 100 ft loes with $Q_p = Q_s = 200$.

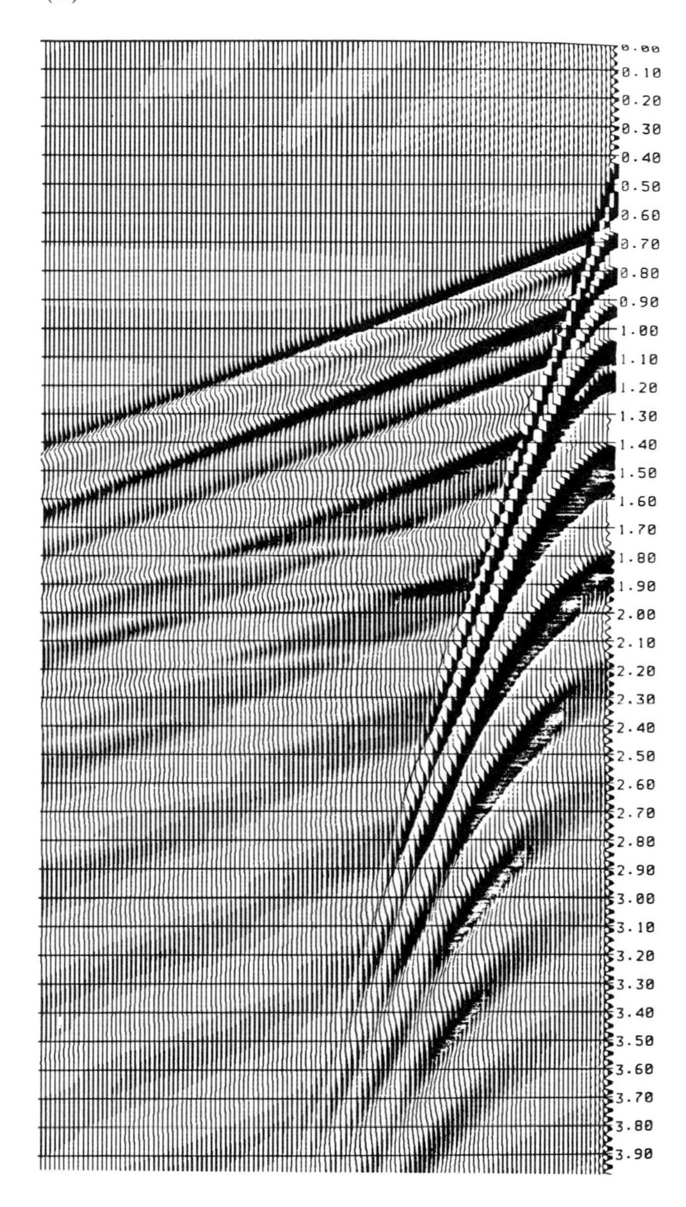

Fig. 3(d) 100 ft loess with $Q_p = Q_s = 10$. All figures are vertical component receivers due to a vertical source applied to the surface.

varying thickness. The cost of N_M models using substructuring compared to 1 model is

$$\frac{C(N_M)}{C(1)} \leqslant N_{DOF}(1 + N_M \Delta L/L)$$

where ΔL is the length of the substructure, L is the length of the total model (measured in the number of contained nodes) and N_{DOF} is the number of degrees of freedom (independent variables) per node (1 for the scalar wave equation and 2 for the elastic wave equation); if the objective is to model both horizontal and vertical sources in elastic wave propagation, the N_{DOF} factor disappears. Thus, it pays to use the substructuring technique when modeling more than two variations of the same model.

The substructuring of a general 2D model is considerably more difficult to implement. Before detailing this method, it is desirable to introduce the nested dissection solution technique.

ITERATIVE MODELING BY PARTIAL FACTORIZATION

A Lumberjack's Guide to Nested Dissection

Nested dissection as originated by Alan George (George and Lui, 1981) is numerically similar to substructure analysis.

If carried to the two dimensional extreme (Figure 4), one could assemble groups of four neighboring finite elements in the x direction (substructure level 1) and eliminate the internal degrees of freedom (of course, it simplifies the argument if one has powers of 2 finite elements in each direction). One then takes the level 1 superstructures and assembles them in groups of two in the z direction and eliminates internal degrees of freedom (substructure level 2 with 8 elements). This pattern of assemblage and internal node elimination continues until the complete mesh is reduced.

The nested dissection algorithm works in the opposite direction by breaking the fully assembled matrix into progressively smaller and smaller matrices (Figure 5). Although applicable to a much more general class of matrices, it successively cuts the finite element mesh in two until one obtains easily solvable 2×2 element submatrices. Our incomplete nested dissection algorithm stops at some

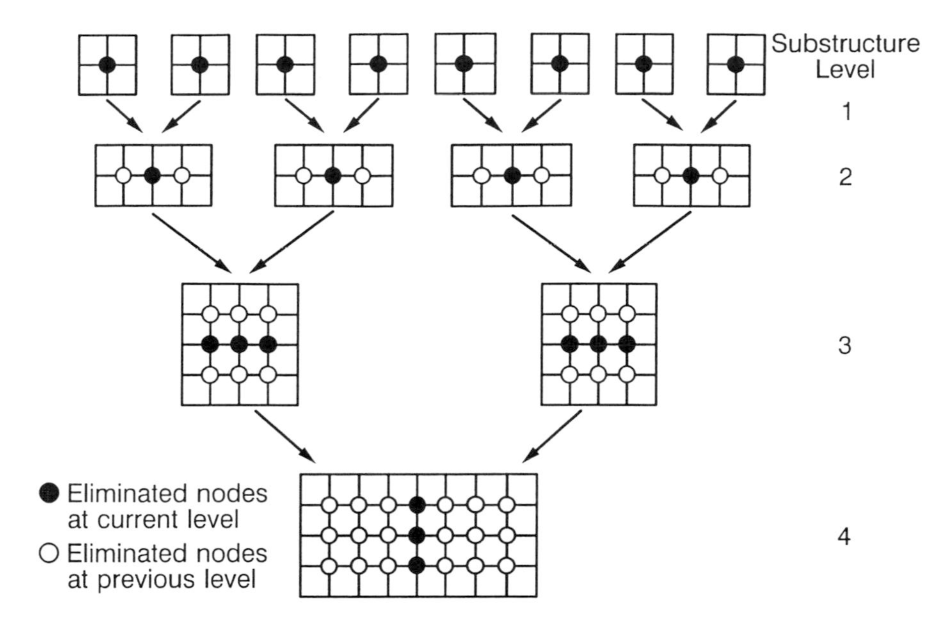

Fig. 4 Building a regular 2-D model using substructures.

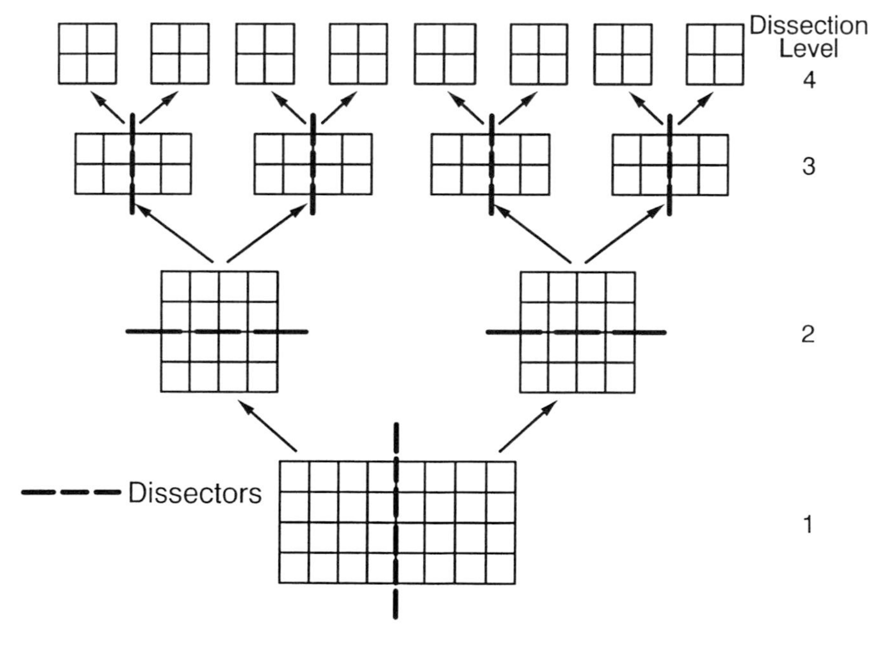

Fig. 5 Breaking apart a regular 2-D model using nested dissection.

machine dependent intermediate matrix size whereby we switch to a conventional band matrix solver. This results in a very simple graph tree whose leaf blocks (Figure 6) represent the banded matrices. Nested dissection is typical of divide and conquer matrix factorization schemes. Like substructuring, the process makes repetitive use of the Schur complement (Eq. 4) for updating the previous dissector level. The cost per level for a regularly numbered N_x by N_z finite element mesh is given in Table 1. George and Lui (1981) have calculated the cost of factoring an $N = N_x = N_z$ mesh which are compared in Table 2 against the conventional active column (Zienkiewicz, 1977, Chapter 24) and general dense factorization techniques. It is clear that nested dissection can produce a significant savings for typical problem sizes of interest.

A typical seismic model will contain on the order of one million equations, one hundred to one thousand sources and 10–15 levels of dissection. If one wishes to obtain the solution at every node for a great number of sources, the cost of forward and back substitution (column 2 of Table 2) dominates that of factorization and the total cost can become prohibitive.

Sparse Solutions and Pruned Trees

After factorization, matrix Eq. (2) is of the form

$$SU = (LDL^T)U = F.$$

Fig. 6 The nested dissection tree.

TABLE 1

Cost per Level in Complex Operations for the Incomplete Nested Dissection Scheme. Total cost can be figured by multiplying the number of dissectors at each level by the cost of factorization and updating, then adding all the levels. Levels 3 through n do not reflect small savings possible along the perimeter of the mesh.

level	number of dissectors	cost of factorization per dissector	cost of updating previous level per dissector
1(root-block)	1	$1/3(N_{DOF}N_z)^3$	0
2	2	$1/3(N_{DOF}N_x/2)^3$	$N_{DOF}^3(N_x/2)^2\,(N_z)$
3	4	$1/3(N_{DOF}N_z/2)^3$	$N_{DOF}^3(N_z/2)^2\,(N_x/2+N_z/2)$
4	8	$1/3(N_{DOF}N_x/4)^3$	$N_{DOF}^3(N_x/4)^2\,(N_x/2+2N_z/2)$
⋮			
$n=2m+1$	2^{2m}	$1/3(N_{DOF}N_z/m)^3$	$N_{DOF}^3(N_z/m)^2\,(2N_x/m+2N_z/m)$
leaf blocks	2^{2m+1}	$1/2N_{DOF}^3(N_z/m)(N_x/m)^3$	$N_{DOF}^3(N_z/m)(N_x/m)^2\,(2N_x/m+2N_z/m)$

TABLE 2

Cost in Complex Operations for Three Direct Methods of a Square N by N Finite Element Mesh. N_s is the number of sources (right-hand sides).

method	cost of factorization	cost of complex forward and back substitution
nested dissection	$10N^3$	$10N^2(\log_2 N)\,N_s$
active column	$1/2N^4$	$1/2N^3N_s$
general dense	$1/3\ N^6$	$1/3N^4N_s^s$

Traditionally, one solves for U, given F, by defining a temporary vector Y

$$Y = L^T U.$$

One begins by solving for Y by forward substitution starting at the top of the lower triangular matrix L:

$$LY = F.$$

Next, one scales Y by the diagonal matrix D to obtain

$$\tilde{Y} = D^{-1} Y.$$

Finally, one solves for U, given $\tilde{Y}$, by back substitution from the bottom of the upper triangular matrix L^T:

$$L^T U = \tilde{Y}.$$

In almost all seismic exploration problems, the applied sources are spatially limited to only a small subset of the nodes on the finite element grid. One can exploit this sparsity of the right hand side vector of Eq. (2) in the simplest matrix solution schemes by simple renumbering. Assume the scenario in Figure 7 with source (and receivers) placed near the top of the model. Number the nodes from lower right to upper left such that the degrees of freedom corresponding to sources and reeivers lie in the bottom of the right-hand side vector, F.

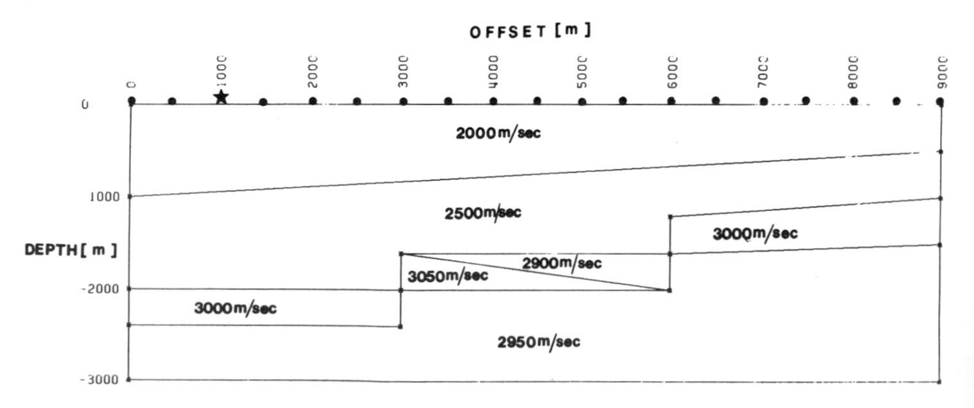

Fig. 7 Typical source and receiver distribution for seismic modeling of a gas sand model. Source denoted by star, receivers by circles.

Partitioning the forward substitution step for this sparse matrix such that all nonzero sources and their adjoining elements lie within the small subvector F_b:

$$\begin{bmatrix} L_{aa} & 0 \\ L_{ba} & L_{bb} \end{bmatrix} \begin{bmatrix} Y_a \\ Y_b \end{bmatrix} = \begin{bmatrix} 0 \\ F_b \end{bmatrix}$$

One notices that the submatrix Y_a is identically zero and that this (major) part of the calculation need not be done.

After scaling the nonzero portion of Y:

$$\tilde{Y}_b = D_{bb}^{-1} Y_b$$

one partitions the back substitution step

$$L^T U = \tilde{Y}$$

to obtain the only small solution subvector U_b:

$$\begin{bmatrix} L_{aa}^T & L_{ba}^T \\ 0 & L_{bb}^T \end{bmatrix} \begin{bmatrix} U_a \\ U_b \end{bmatrix} = \begin{bmatrix} 0 \\ \tilde{Y}_b \end{bmatrix}.$$

Here, although one notes that U_a is nonzero, all the desired receivers are contained (by the unconventional numbering scheme) within U_b. Thus, one can stop the black substitution phase after the U_b components are calculated, at a greatly reduced cost.

Such simple renumberings are not possible when sources and receivers lie in different parts of the mesh (Fig. 8a). More complicated renumbering schemes could destroy the efficiency of conventional banded and active column matrix solvers. The nested dissection algorithm, however, is quite amenable to arbitrarily located sparse source and receiver locations. After factorization, one merely “prunes away” those branches of the tree (Fig. 8a) with zero source loads (Fig. 8b) and unwanted receiver solutions (Fig. 8c). The cost of such a sparse solve is roughly one to two orders of magnitude less than a complete solve at all nodal points for typical seismic model problems. Similar savings in memory or peripheral storage are obtained by pruning the unnecessary portions of the tree.

Partial Factorization and Grafting

For most modeling problems, the cost of factorization is significantly greater than the cost of solution. One can easily renumber the traversal order on the tree

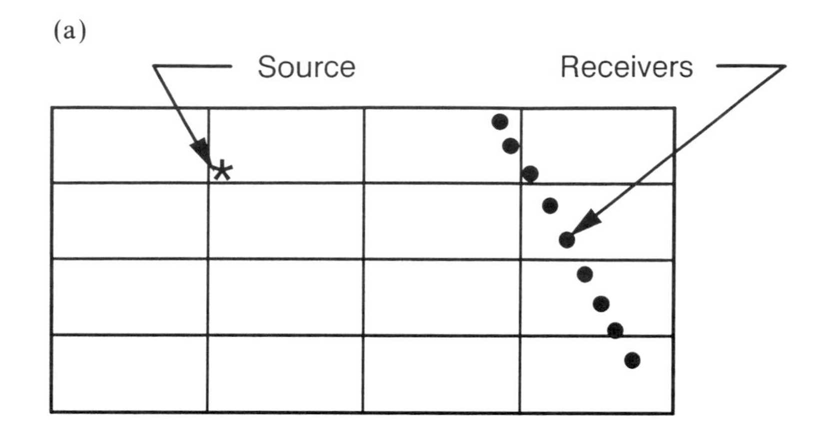

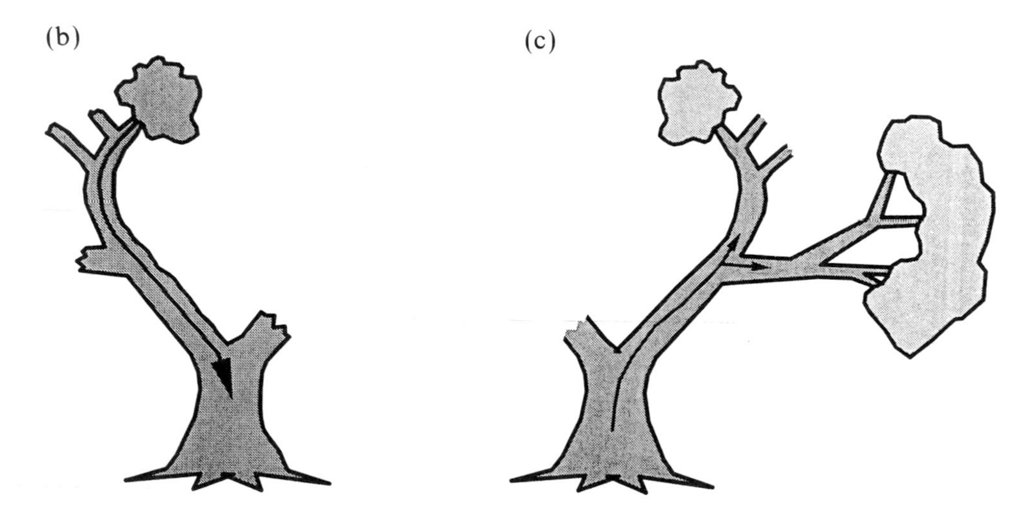

Fig. 8 (a) Distribution of sources and receivers for a VSP model. (b) Branches and leaves of the nested dissection tree necessary for forward substitution, and (c) for backsubstitution.

(what computer scientists call the stack) during the factorization step such that those limbs that are to be changed during iterative modeling will be treated last. Thus, in iterative modeling one starts with a partially factored tree (Fig. 9), grafts a new branch in the proper location, and completes the factorization. The cost of completing the factorization as compared to the total factorization is shown in Figure 10. Note that in cases where the blocks to be modified do not contain the

(a)

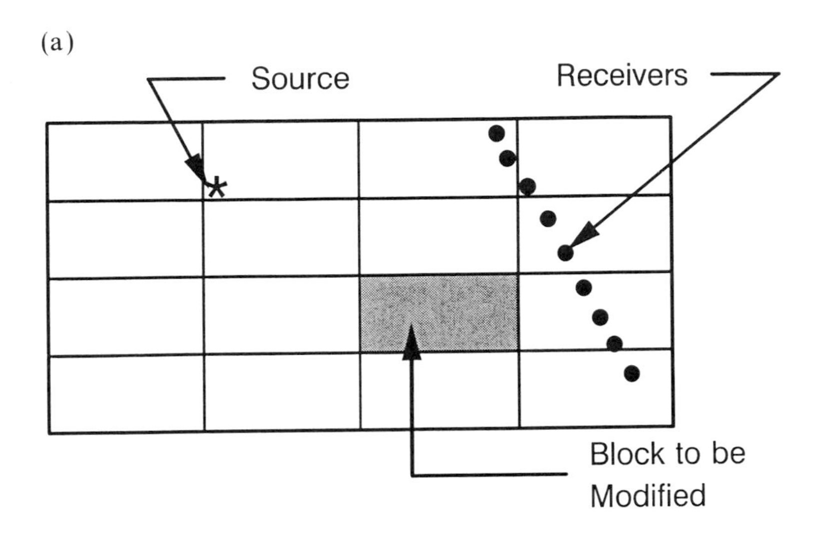

(b)

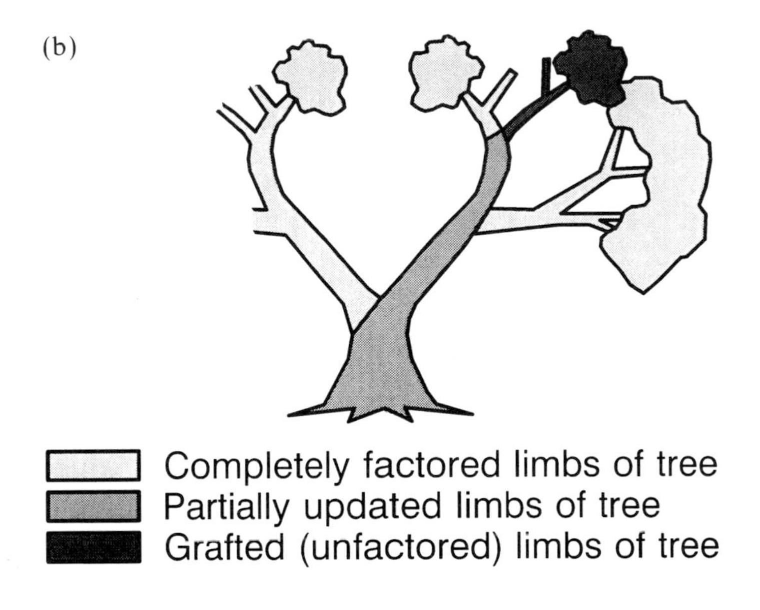

Fig. 9 Iterative modeling by partial factorization: (a) the block of the model to be modified, and (b) the corresponding, grafted limb on the nested dissection tree.

(a)

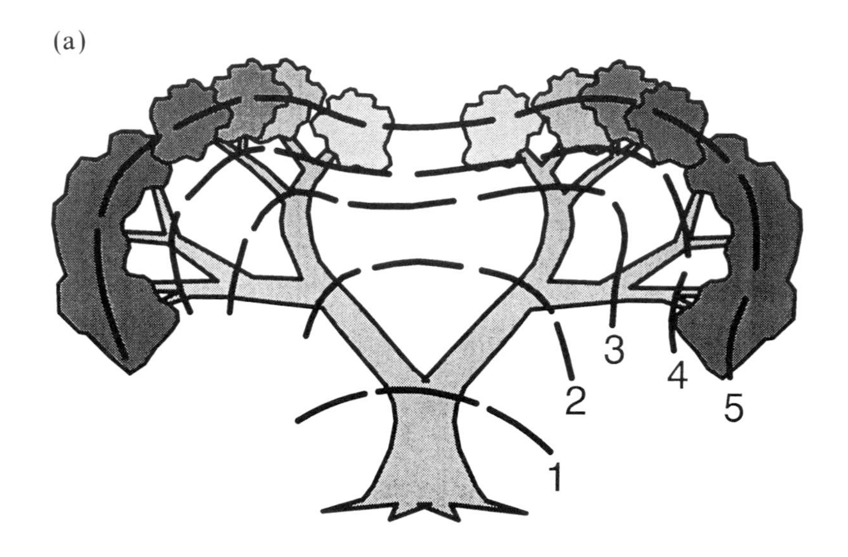

(b)

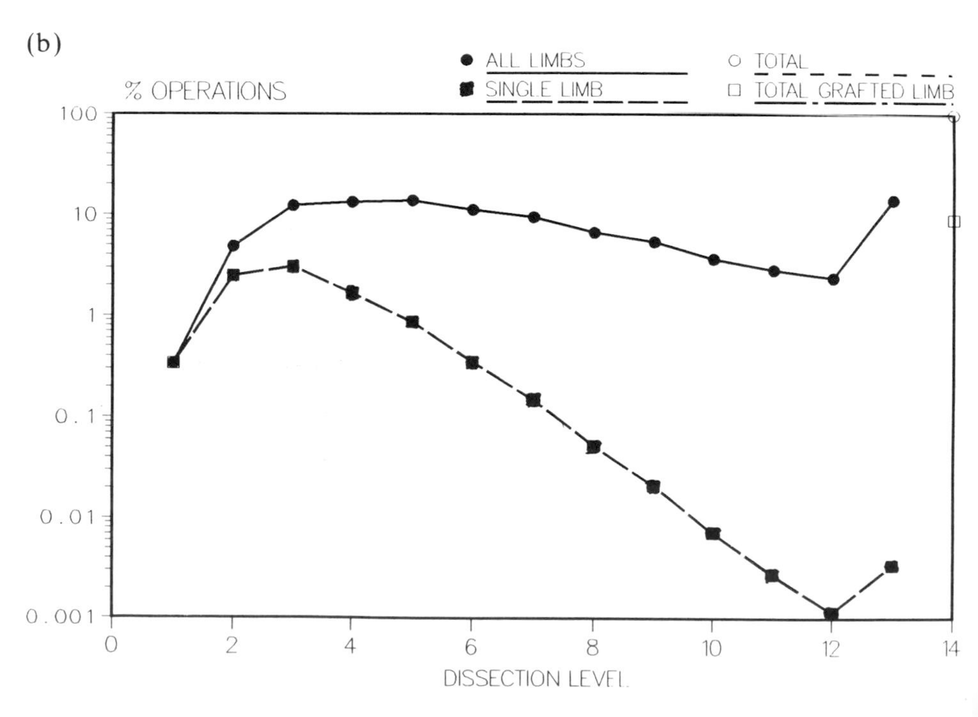

Fig. 10 (a) Levels of the numerical tree. (b) Cost per level to factor the matrix vs. cost for updating one limb.

source blocks, one may often perform a great part of the forward substitution step before encountering the modified block, thus sharing the overhead of the forward substitution among many iterations.

ITERATIVE MODELING BY HOMOTOPY METHODS

The homotopy method has enjoyed a considerable amount of success in ray theoretical modeling (Keller and Perozzi, 1983) and is well established in solving differential equations (Watson, et al., 1988). Here we show how Keller's philosophy can be extended to the more general finite element method.

Assume one changes the parameters p in one or more blocks of the model shown in Figure 9a. Assuming $p = p_0$ to be the parameter of some representative initial or average model, one defines the impedance matrix of the perturbed model as

$$S(p_0 + \Delta p) \equiv S_0 + \Delta S,$$

and the solution of the perturbed model as

$$U(p_0 + \Delta p) \equiv U_0 + \Delta U.$$

Given the equations for the unperturbed model

$$S(p_0)\, U(p_0) = F_0, \tag{6a}$$

and for the perturbed model

$$S(p_0 + \Delta p)\, U(p_0 + \Delta p) \equiv F_0, \tag{6b}$$

one can rewrite Eq. (6b) as

$$(S_0 + \Delta S)(U_0 + \Delta U) = F_0,$$

or

$$S_0 U_0 + S_0\, \Delta U + \Delta S\, U_0 + \Delta S\, \Delta U = F_0.$$

Assuming ΔU to be small compared to U_0 when ΔS is small compared to S_0 and using Eq. (6a) one obtains

$$S_0\, \Delta U \cong -\Delta S\, U_0 \equiv \tilde{F} \tag{7}$$

Having obtained the correction term ΔU one can approximate a new solution $U_1 = U_0 + \Delta U$ that is hopefully closer to $U(p + \Delta p)$ and further refine the solution recursively. It is very important to notice that the matrix S_0 has already been factorized for the first model, whose solution was U_0. In addition, if only a small portion of the model has been modified, then ΔS and hence the equivalent sources $\tilde{F}$ are correspondingly sparse. Indeed, only those parts of U_0 and ΔU corresponding to the desired receiver locations and the numerical support of ΔS need be calculated. The nested dissection algorithm described in the previous section is most effective in exploiting this sparsity.

(a)

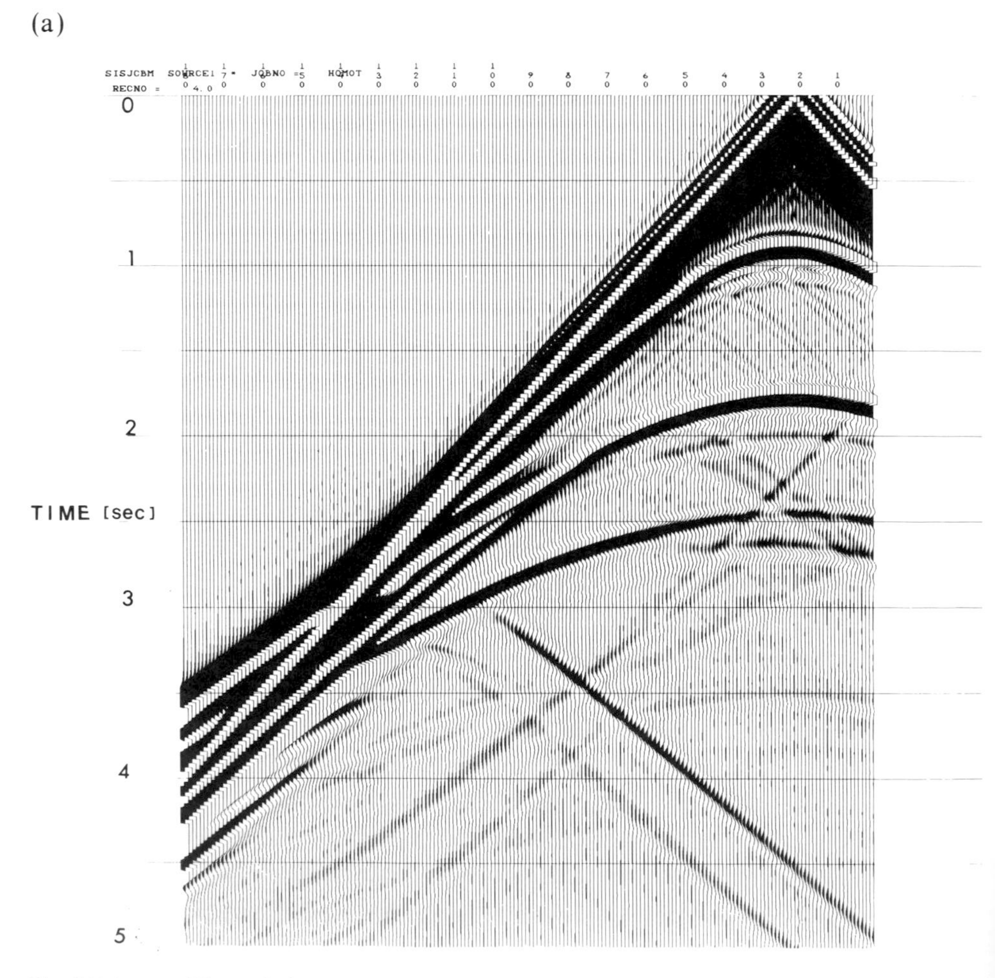

Fig. 11(a) Three of eleven models generated by the homotopy method for the gas sand model shown in Figure 7. U for $v = 2740$ m/s.

For the homotopy approach to work, the models should be quite similar to each other and S_0 and ΔS need to be regularized. We have found the technique of moving the temporal frequency ω off the real axis (Rosenbaum, 1974) to be quite adequate in regularizing S_0 and ΔS. We feel a suite of closely related models will be the normal mode of operation for iterative modeling where one may wish to know the seismic response due to a geologic structure for a finite range of porosity, fluid content and thickness. Figure 11 illustrates a suite of models for a variable porosity gas sand (Figure 7) obtained from a rock properties database.

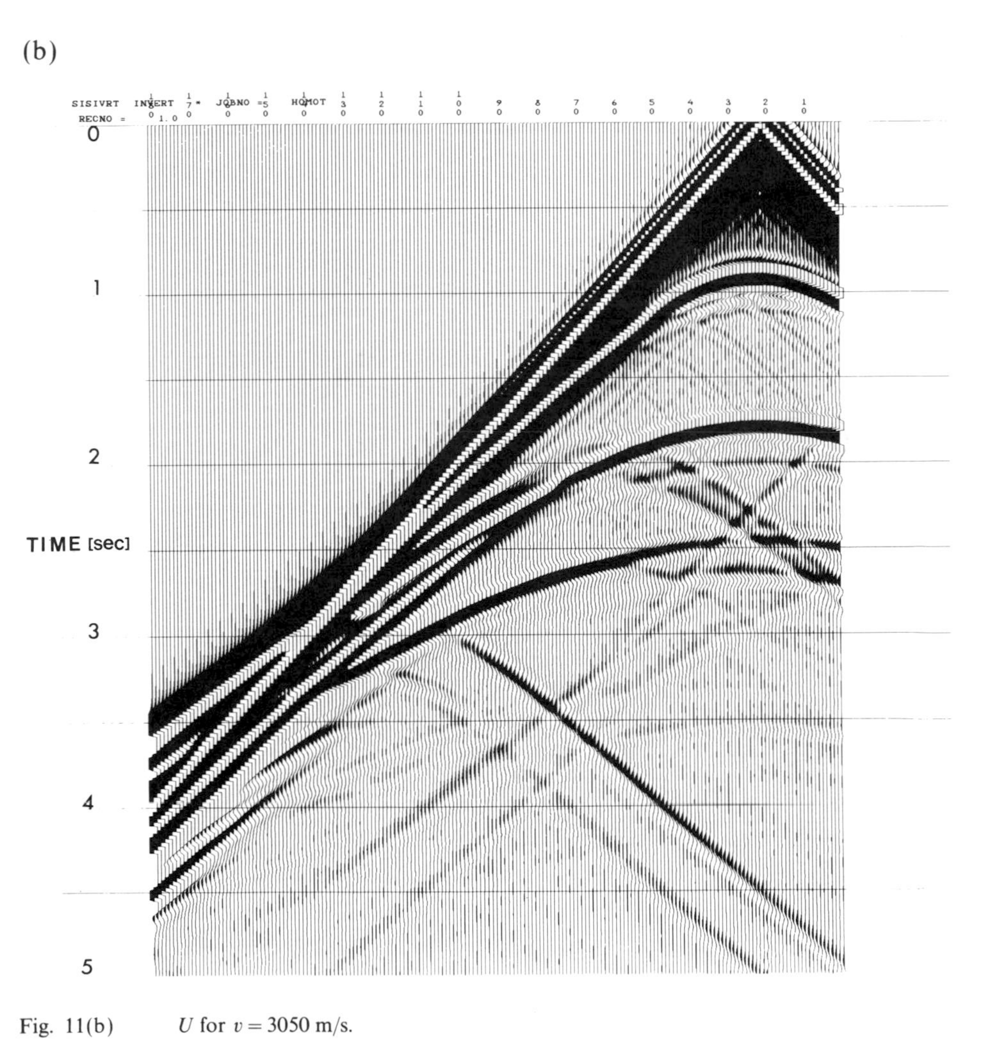

Fig. 11(b) U for $v = 3050$ m/s.

(c)

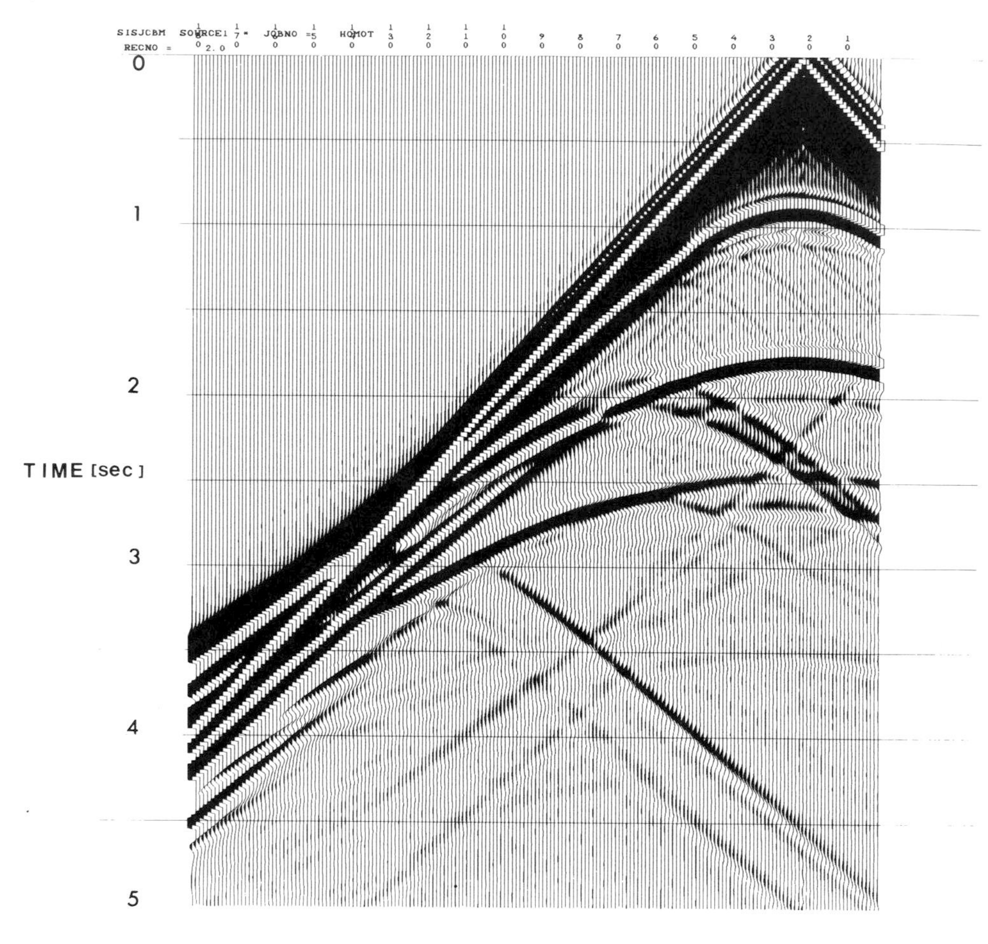

Fig. 11(c) U for $v = 3350$ m/s.

ITERATIVE MODELING USING TAYLOR SERIES EXPANSIONS

Often, one wishes to determine how sensitive the seismic section is to a given geologic parameter. Alternatively, one may wish to represent the seismic response as a continuous function of the parameter in question, perhaps for later interactive analysis. Both these goals suggest a Taylor Series expansion. Modeling of parameter sensitivity is not terribly new to brute force seismic modeling—indeed, it is the basis of full waveform seismic inversion in its most direct formulation (e.g. Shin, 1988) where one determines the Jacobian of the error function at each approximation step towards some local minimum in the parameter space.

The seismic response about some initial parameter p_0 can be expressed as

$$\begin{aligned} U(p_0+\Delta p) = U(p_0) &+ \Delta p \left.\frac{\partial U}{\partial p}\right|_{p=p_0} \\ &+ (\Delta p)^2/2! \left.\frac{\partial^2 U}{\partial p^2}\right|_{p=p_0} \\ &+ (\Delta p)^3/3! \left.\frac{\partial^3 U}{\partial p^3}\right|_{p=p_0} \\ &+ \cdots \end{aligned} \tag{8}$$

The calculation of each derivative in Eq. (8) has the same numerical structure and a similar computational cost to that of the homotopy method described earlier. Taking the partial derivative of Eq. (2) with respect to the parameter p:

$$\frac{\partial}{\partial p}[SU] = \frac{\partial}{\partial p} F$$

or

$$S \left.\frac{\partial U}{\partial p}\right|_{p=p_0} = -\frac{\partial S}{\partial p} U|_{p=p_0} \equiv \tilde{F} \tag{9}$$

Once again, $\partial S/\partial p$ has only limited numerical support giving rise to a highly sparse $\tilde{F}$, while $S(p_0)$ has already been factored to obtain $U(p_0)$. The next higher derivative is obtained by further operating on Eq. (9)

$$\frac{\partial}{\partial p}\left[S\frac{\partial U}{\partial p}\right] = -\frac{\partial}{\partial p}\left[\frac{\partial S}{\partial p} U\right],$$

or

$$\begin{aligned} S \left.\frac{\partial^2 U}{\partial p^2}\right|_{p=p_0} &= -2\frac{\partial S}{\partial p}\frac{\partial U}{\partial p} - \left.\frac{\partial^2 S}{\partial p^2} U\right|_{p=p_0} \\ S \left.\frac{\partial^3 U}{\partial p^3}\right|_{p=p_0} &= -3\frac{\partial S}{\partial p}\frac{\partial^2 U}{\partial p^2} - 3\frac{\partial^2 S}{\partial p^2}\frac{\partial U}{\partial p} - \left.\frac{\partial^3 S}{\partial p^3} U\right|_{p=p_0}. \end{aligned} \tag{10}$$

It is interesting to note that in the important case where one wishes to calculate U as a function of p where p stands for λ, μ, ρ, Q_p or Q_s of a given region, one need not retain the rightmost terms in Eq. (10) because S is a linear function of Lame's parameters, density and attenuation ($1/Q_p$, $1/Q_s$) for moderate to large values of Q. In contrast, the perturbation of the geometric boundaries is quite non-linear. The numerical support, however, is even more sparse than that of the material parameters (Figure 9). Details on the perturbation of boundaries can be found in Shin (1988).

Since the solution U is also quite nonlinear in parameter p, it is not obvious how many derivatives need to be calculated. We feel that one can use a nonrigorous

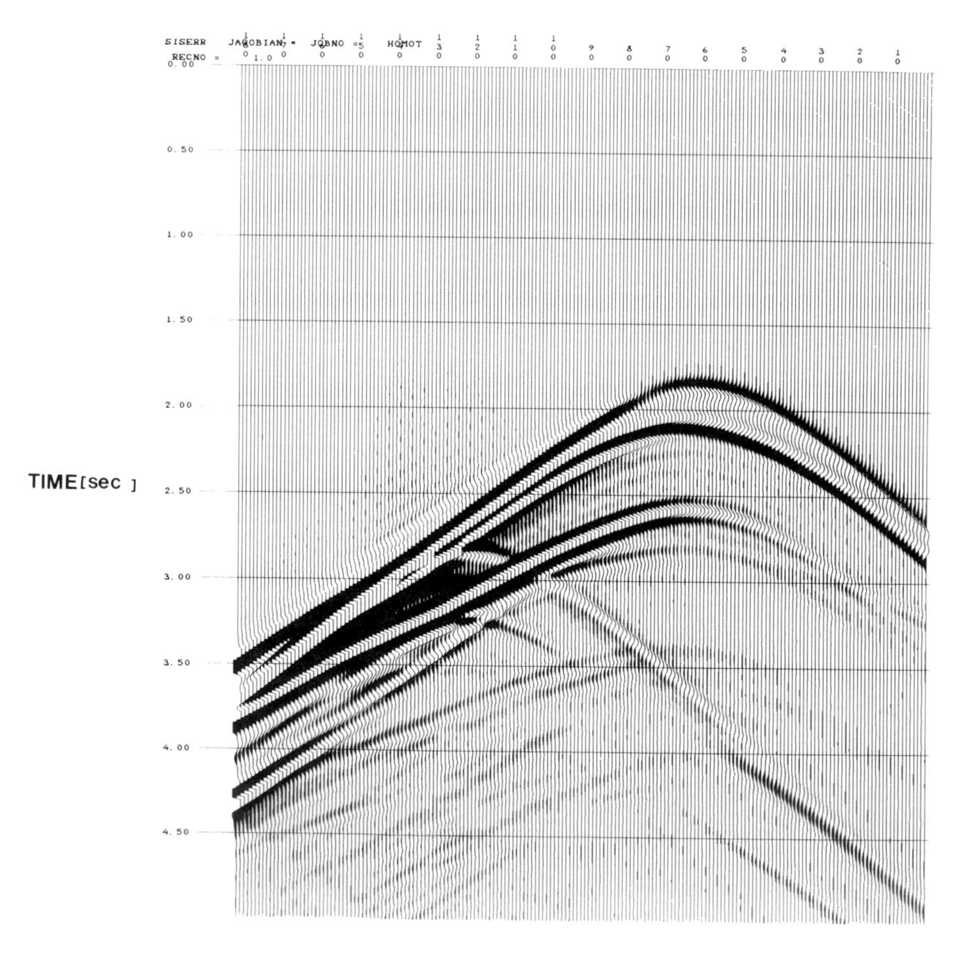

Fig. 12 $\partial U/\partial p$ for the model in Figure 7. p is the velocity of the gas sand.

application of the remainder theorem (Lanczos, 1956) to guide us in this choice. One needs to choose a priori the range of values of the parameter p

$$p_{\min} \leqq p \leqq p_{\max}.$$

Defining

$$U^n(p) = \sum_{k=0}^{n} \frac{1}{k!}(p-p_0)^k \frac{\partial^k U}{\partial p^k}$$

one can grossly estimate the remainder over the interval (p, p_0) by

$$\varepsilon = (p-p_0)^{n+1} \frac{\partial^{n+1} U}{\partial p^{n+1}}(\zeta)$$

where we set $\zeta = p_0$ instead of $p_0 \leqq \zeta \leqq p$. As in the homotopy method, regularization of the impedance matrix by moving ω sufficiently far off the real axis makes this problem tractable. If the remainder is too high, one has the choice of adding higher Taylor terms, limiting the range of interest or switching to the homotopy method.

The derivatives of the seismic response due to the velocity change of the model shown in Figure 11 is shown in Figure 12.

CONCLUSIONS

Cost effective iterative modeling can be achieved by several techniques: substructuring, partial factorization, homotopy, and Taylor Series expansion. Iterative modeling by substructuring and partial factorizing is as stable a process as forward modeling itself. The authors feel that partial factorization is a much more tractable approach than substructuring, as it results in fairly simple, but general modifications to the sparse matrix solver. Substructuring can be more cost effective in certain specific problems (such as flat layer modeling using a hybrid finite element–spectral technique or semianalytic reflectivity techniques) but is extremely tedious to implement and often more numerically intensive than partial factorization for the general 2-D problem.

The homotopy and Taylor Series expansion techniques are much easier to implement in that they require no modification to existing matrix equation solvers.

To be cost effective, they require sparse forward and back substitution capabilities in the matrix equation solver, which is easily and efficiently achieved by the nested dissection algorithm. The authors have not had enough experience with either of these "numerically iterative" schemes to quantitatively state when convergence could fail. It is not clear how such techniques would behave near singular perturbations, such as where a solid degenerates to a fluid or to a vacuum. At the time of this writing, however, convergence for a wide variety of models has been achieved in the homotopy method by making the parameter steps smaller, and/or by further smoothing the response spectrum by moving the frequency integration further off the real axis.

Since we know the model response to be nonpolynomial, we have little reason to believe that the Taylor Series expansion method will converge for an arbitrary model variation. The major advantage of the Taylor Series approach is that a user can generate a large suite of closely spaced models from a limited number of coefficients in an interactive, postprocessing stage.

REFERENCES

Alekseev, A. S. and Mikhailenko, B. J., (1980), The solution of dynamic problems of elastic wave propagation in inhomogeneous media.

George, A. and Lui, J. W. H., (1981) Computer solution of large sparse positive-definite systems. Prentice-Hall, Inc., Englewood Cliffs, N.J.

Keller, H. B. and Perozzi, D. J., (1983) Fast seismic ray tracing: SIAM J. Appl. Math, Vol. 43, No. 4, p. 981-992.

Lanczos, C. (1956), Applied Analysis. Prentice-Hall, Inc., Englewood Cliffs, N.J.

Orsag, S. A., (1980), Spectral methods for problems in complex geometries: J. Comp. Physics, v. 37, p. 70–92.

Rosenbaum, J. H. (1974), Synthetic microseisms–Logging in porous media: Geophysics, v. 39, p̂ 14–32.

Shin, C. S. (1988), Nonlinear elastic inversion by blocky parameterization, Tulsa Univ., PhD thesis, Tulsa, OK.

Watson, L. T., Billups, S. C. and Morgan, A. P., (1988) HOMPACK: A suite of codes for globally convergent homontopy algorithms; ACP Trans. Math Software (to appear).

Zienkiewicz, O. C. (1977), The finite element method: McGraw-Hill Book Co., Third Edition, NY, NY.

CHAPTER 10

APPLICATION OF SUPERCOMPUTERS IN THREE-DIMENSIONAL SEISMIC MODELING

by
IRSHAD R. MUFTI
Mobil Research and Development Corporation

INTRODUCTION

About a decade ago, the geophysical community launched a serious effort to develop two-dimensional finite-difference seismic models. The major motivation behind this endeavor was to be able to investigate more fully the seismic response associated with complex geological structures. It was a big leap forward as compared to the classical approach to modeling based on the ray theory approximation. Since then, the computers have become immensely more powerful, making it feasible to investigate the modeling problem in three dimensions.

There are a number of advantages to be gained by doing 3D modeling. Some of the more significant advantages are mentioned below.

1. *More dependable interpretation*

It is now widely known that the surface images as they appear on a seismic section do not, in general, correspond to their true subsurface location. Recent advances in the area of seismic migration have gone a long way to alleviate this problem; however, the major cause of this mislocation has its origin in lateral variations of velocity. In most cases, the current methods of migration fail to treat this problem satisfactorily. When we reognize that the structures that we have to deal with are actually three-dimensional, the problem of lateral variations in velocity turns out to be far more troublesome. Under these circumstances, one can treat the results of seismic data interpretation as an initial guess of the subsurface picture, and use this information to compute the corresponding synthetic seismic data. The departure between the observed and the synthetic sets of data can be utilized to modify the interpreted results.

2. *Better understanding of amplitude variations*

Anomalous variations in amplitude as a function of source-receiver distance are often indicative of the presence of hydrocarbons in the subsurface. The current investigations in this critically important area do not adequately account for the geometric effects on amplitude caused by topographic variations of the subsurface horizons. 3D models will play a vital role in attacking this problem.

3. *Velocity analysis*

The current methods of velocity analysis are based on the assumption that the various events identified in a seismic section originate from a single vertical plane section of the earth containing the source and receiver. This is a gross simplification which can lead to erroneous estimation of velocities. 3D models can be used possibly in an iterative fashion as a means to reduce such errors.

4. *Determination of data acquisition parameters*

The selection of a site for making seismic measurements and an optimum choice of data acquisition parameters which would maximize the influence of a subsurface target on the field data are accompanied by questions which are both crucial and difficult to answer. 3D model studies can provide valuable insight for answering such questions.

The various points mentioned above provide sufficient justification in favor of 3D seismic modeling. But there are also some drawbacks. Even on supercomputers such as the Cray X/MP, such models require a lot of CPU time and a huge memory for manipulation of data. In order to make such models cost effective, we shall base the following treatment on the acoustic wave equation. In the case of offshore data, the quantity that we actually measure is pressure and this equation is quite acceptable. In the case of onshore surveys, as long as we do not record three-component field data and continue to use this equation at important stages of data processing such as migration, such a choice is both reasonable and practical.

THE CONVENTIONAL APPROACH

Let us consider a 3D space in which the z-axis, positive downward, denotes depth below the surface of the ground which coincides with the plane (x, y, o). In this frame of coordinates, the acoustic wave equation can be expressed as

$$u_{xx} + u_{yy} + u_{zz} = c^{-2} u_{tt} + f(t)\,\delta(x - x_s)\,\delta(y - y_s)\,\delta(z - z_s) \tag{1}$$

where

$c(x, y, z) =$ velocity of the medium

$u(x, y, z, t) =$ pressure

$f(t) =$ a time-dependent source located at (x_s, y_s, z_s)

$t =$ time

The subscripts in (1) indicate derivatives of the wavefield wth respect to x, y, z or t. For the purpose of setting up a finite-difference model, it would be convenient to introduce a set of indices i, j, k and n such that

$$\begin{aligned} x &= i\,\Delta x \\ y &= j\,\Delta y \\ z &= k\,\Delta z \\ t &= n\,\Delta t \qquad i, j, k, n = 0, 1, 2, \ldots \end{aligned} \tag{2}$$

In (2), Δx, Δy and Δz denote uniform grid spacings along the x, y and z axes respectively and Δt means the time sampling interval. By using these indices, we can write

$$\begin{aligned} u(x, y, z, t) &= u^n_{i,j,k} \\ f(t) &= f_n \qquad n = 0, 1, 2, \ldots \end{aligned} \tag{3}$$

$u^n_{i,j,k}$ denotes the discrete value of the wavefield at the grid point (i, j, k) at time n. A similar notation can be used to indicate discrete values of related quantities such as u_{xx}.

By virtue of the central difference formula (see, e.g. Smith, 1965, p. 6), the first term on the LHS of (1) be approximated as

$$(u_{xx})^n_{i,j,k} = \frac{u^n_{i-1,j,k} - 2u^n_{i,j,k} + u^n_{i+1,j,k}}{(\Delta x)^2} + O[(\Delta x)^2] \tag{4}$$

The remaining wavefield derivatives in (1) can be treated in a similar fashion. For the source term, we can write

$$f(t) = \begin{cases} f_n & \text{at} \quad (i_s, j_s, k_s) \\ o & \text{elsewhere} \end{cases} \qquad n = 1, 2, \ldots \tag{5}$$

Substituting expressions such as (4) and (5) into (1), we get

$$\begin{aligned} u^{n+1}_{i,j,k} &= a_{i,j,k}(u^n_{i-1,j,k} - 2u^n_{i,j,k} + u^n_{i+1,j,k}) \\ &\quad + b_{i,j,k}(u^n_{i,j-1,k} - 2u^n_{i,j,k} + u^n_{i,j+1,k}) + e_{i,j,k}(u^n_{i,j,k-1} - 2u^n_{i,j,k} + u^n_{i,k+1}) \\ &\quad + 2u^n_{i,j,k} - u^{n-1}_{i,j,k} - (c\Delta t)^2 f_n \, \delta(i-i_s)\, \delta(j-j_s)\, \delta(k-k_s) \end{aligned} \tag{6}$$

where

$$\begin{aligned} a_{i,j,k} &= (c_{i,j,k}\, \Delta t/\Delta x)^2 \\ b_{i,j,k} &= (c_{i,j,k}\, \Delta t/\Delta y)^2 \\ e_{i,j,k} &= (c_{i,j,k}\, \Delta t/\Delta z)^2 \end{aligned} \tag{7}$$

It is usually feasible to set

$$\Delta x = \Delta y = \Delta z = h \tag{8}$$

In that case (7) reduces to

$$\begin{aligned} u^{n+1}_{i,j,k} &= g_{i,j,k}(u^n_{i-1,j,k} + u^n_{i+1,j,k} + u^n_{i,j-1,k} + u^n_{i,j+1,k} \\ &\quad + u^n_{i,j,k-1} + u^n_{i,j,k+1} - 6u^n_{i,j,k}) + 2u^n_{i,j,k} - u^{n-1}_{i,j,k} \end{aligned} \tag{9}$$

where

$$g_{i,j,k} = (c_{i,j,k}\, \Delta t/h)^2 \tag{10}$$

In (9) the source term is implied but we have dropped it in favor of notational convenience.

Figure 1 shows a simple example of a 3D finite-difference model consisting of two different materials with velocities c_1 and c_2. It extends along the x-axis from

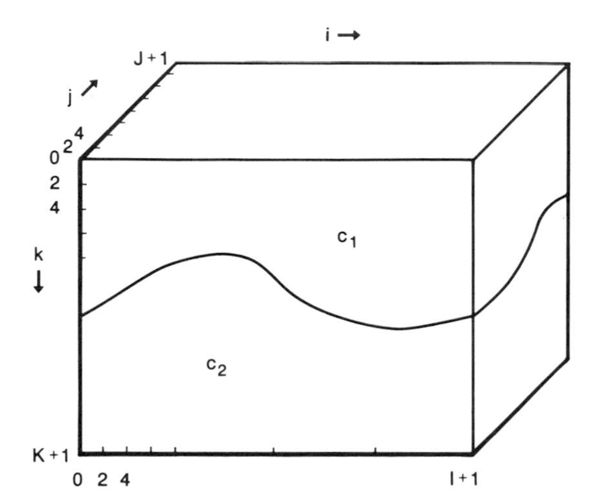

Fig. 1 A simple example of a three-dimensional finite-difference seismic model.

$i=0$ to $i=I+1$, along the y-axis from $j=0$ to $j=J+1$ and vertically from $k=0$ to $k=K+1$. The plane $k=0$ corresponds to the surface of the ground. The subsurface boundaries of the model such as the planes $i=0$ and $k=K+1$ are introduced to keep the size of the model finite. The value of the wavefield or its derivative is usually specified along the boundaries of the model. Then with the help of the initial conditions:

$$u^{o}_{i,j,k}=o \tag{11a}$$

$$(u_t)^{o}_{i,j,k}=0 \tag{11b}$$

one tries to evaluate the field successively for $n=1, 2,...$ at a set of grid points $\{(i,j,k),\ i=1, 2,..., I;\ j=1, 2,..., J;\ k=1, 2,..., K\}$.

A very simple way to initiate evaluation of the wavefield is to set

$$u^{1}_{i,j,k}=\begin{cases} f_1 & \text{at} \quad (i_s, j_s, k_s) \\ o & \text{elsewhere} \end{cases} \tag{12}$$

Now if we set $n=1$ in (9), then in view of (11a) and (12) all the quantities on the LHS of (9) are known. Consequently, the value of the field for the time step $n=2$ can be readily computed for the various grid points. This process can be repeated for subsequent time steps $n=3, 4...$ For a given value of n, the set of data

$\{u^n_{i,j,o}\ (i=1,..., I; j=1, 2,..., J)\}$ is called the time slice of the wavefield at time n. A more widely used information commonly referred to as shot record represents the set of data $\{u^n_{i,j,o}\ (n=1,..., N)\}$ for a fixed value of i or j.

SOME PRACTICAL PROBLEMS

The modeling algorithm outlined above is an established technique and its 2D version has been used for many years. Some of the most frequently reported problems regarding its use are

1. Huge memory for data manipulation
2. Prohibitive amount of CPU time
3. Approximate results

The first problem arises from the necessity to use extremely fine grid intervals. Any attempts to use coarser intervals causes dispersion of energy which degrades the quality of results to such an extent that the signal of interest may be totally obliterated. In order to grasp the magnitude of this problem, let us consider a 3D model with horizontal dimensions of 3000 m × 3000 m and a depth of 3000 m. In view of the usual range of velocities and the frequency of the source signal needed in such models, we may need a grid spacing of, say, 10 m. In terms of grid dimensions, the size of this model will be 300 × 300 × 300 which represents a grid population of 27 million! Another problem known as numerical stability makes the actual situation much worse. In order to keep the model numerically stable, the ratio of time sampling interval and grid spacing ($\Delta t/h$) must be kept very small. This necessitates the use of extremely fine values of Δt and results in a prohibitive amount of CPU time. Further discussion on grid dispersion and numerical stability will be taken up at a later stage.

Let us now consider the question of accuracy. A major portion of errors in the finite-difference data is contributed by the truncation errors introduced by replacing the wavefield derivatives by the corresponding difference relations. The magnitude of such errors is directly dependent on the value of grid spacing and time sampling interval. Since the evaluation of the wavefield necessitates extremely fine spatial and temporal sampling, the computed data are pretty accurate. Thus, the accuracy of results is mostly a concern of orthodox theoreticians who prefer analytical methods. The real problem is how to design a seismic model which would permit coarser sampling of the wavefield even if it amounts to sacrificing some accuracy.

The problem of numerical stability can be overcome to varying degrees by designing finite-difference models in the implicit mode. In a recent article, Mufti (1985) introduced an implicit method of seismic modeling which is unconditionally stable; consequently, one can use any arbitrarily chosen values of time samples as long as the desired resolution can be maintained in the computed results. It is obvious that in order to be able to use large grid intervals, one must evaluate derivatives of the wavefield to higher orders of accuracy. We shall now discuss higher-order algorithms for 3D modeling.

A SIMPLE HIGHER-ORDER ALGORITHM

Relation (4) is obtained by using the Taylor series expansion; $O[(\Delta x)^2]$ implies that all the terms in the series whose contribution is $O[(\Delta x)^n]$, $n \geqslant 2$ are discarded. By retaining more terms in the series, one can obtain a similar expression but accurate to $O[(\Delta x)^4]$. The result can be expressed in the form

$$(u_{xx})^n_{i,j,k} = \left[I - \frac{\delta_x^2}{12} \right] \frac{\delta_x^2}{(\Delta x)^2} u^n_{i,j,k} + O[(\Delta x)^4] \tag{13}$$

where I represents the identity matrix and

$$\delta_x^2 u^n_{i,j,k} = u^n_{i-1,j,k} - 2u^n_{i,j,k} + u^n_{i+1,j,k} \tag{14}$$

Therefore, the first term on the LHS of (1) can be expressed as

$$(u_{xx})^n_{i,j,k} = \frac{1}{12(\Delta x)^2} (-u^n_{i-2,j,k} + 16u^n_{i-1,j,k} - 30u^n_{i,j,k} + 16u^n_{i+1,j,k} - u^n_{i+2,j,k}) + O[(\Delta x)^4] \tag{15}$$

Similar expressions can be derived for the remaining spatial derivatives.

If we use fourth-order expressions for the spatial derivatives, but only a second-order expression for the time derivative, the resulting difference relation for the wave equation can be expressed as

$$\begin{aligned} u^{n+1}_{i,j,k} = -\frac{1}{12} \{ & a_{i,j,k}[u^n_{i-2,j,k} + u^n_{i+2,j,k} - 16(u^n_{i-1,j,k} + u^n_{i+1,j,k}) + 30u^n_{i,j,k}] \\ & + b_{i,j,k}[u^n_{i,j-2,k} + u^n_{i,j+2,k} - 16(u^n_{i,j-1,k} + u^n_{i,j+1,k}) + 30u^n_{i,j,k}] \\ & + e_{i,j,k}[u^n_{i,j,k-2} + u^n_{i,j,k+2} - 16(u^n_{i,j,k-1} + u^n_{i,j,k+1}) + 30u^n_{i,j,k}] \} \\ & + 2u^n_{i,j,k} - u^{n-1}_{i,j,k} \qquad i = 1,..., I;\ j = 1,..., J;\ k = 1,..., K. \end{aligned} \tag{16}$$

Finally, if we set $\Delta x = \Delta y = \Delta z = h$, (16) reduces to

$$
\begin{aligned}
u_{i,j,k}^{n+1} = \tilde{g}_{i,j,k}[u_{i-2,j,k}^{i} &+ u_{i+2,j,k}^{n} + u_{i,j-2,k}^{n} + u_{i,j+2,k}^{n} + u_{i,j,k-2}^{n} \\
&+ u_{i,j,k+2}^{n} - 16(u_{i-1,j,k}^{n} + u_{i+1,j,k}^{n} + u_{i,j-1,k}^{n} + u_{i,j+1,k}^{n} \\
&+ u_{i,j,k-1}^{n} + u_{i,j,k+1}^{n}) + 90u_{i,j,k}^{n}] + 2u_{i,j,k}^{n} - u_{i,j,k}^{n-1}
\end{aligned}
\tag{17}
$$

where

$$\tilde{g}_{i,j,k} = -(c_{i,j,k}\,\Delta t/h)^2/12 \tag{18}$$

Relation (17) is only slightly more involved than (9) and yet it permits the use of much coarser grid intervals. Its striking simplicity makes it a very powerful candidate for designing efficient 3D models.

GRID DISPERSION AND NUMERICAL STABILITY

The maximum value of grid spacing which can be used in a model without causing excessive dispersion of energy is governed by the relation

$$\max(\Delta x, \Delta y, \Delta z) = \frac{\min(c_1, c_2, \ldots, c_m)}{w f_{\max}} \tag{19}$$

where

m = number of velocity types defining the model
$f_{\max}$ = maximum frequency in the source wavelet
w = number of samples per wavelength corresponding to $f_{\max}$

In (19) we may set $\Delta x = \Delta y = \Delta z = h$. In that case, this relation yields the largest value of h, say $h_{\max}$, for a given model. However, if we choose different values of Δx, Δy, and Δz, each of these quantities may be less but none of them may exceed $h_{\max}$. Therefore, the use of equal grid spacings along all the three axes of coordinates minimizes the population of grid points and it represents the most efficient choice.

In the conventional scheme, it is usually necessary to use $w \geqslant 10$ in order to avoid grid dispersion. But for the models based on (17), this problem arises only when $w < 3$. This means that for a model of given physical dimensions, (17) requires only about 2.7 % of the grid points needed for setting up the corresponding model based on (9). This idea can be best comprehended with the help of a concrete example. Typical grid dimensions of a 3D model based on (17) with $w=3$ are in the neighborhood of $250 \times 250 \times 250$. This results in 15.625 million grid points. The same model based on (9) with $w=10$ will consist of 579 million grid points. These numbers explain why the development of seismic models had to be restricted to two dimensions in the past. The use of algorithms such as (17) coupled with the availability of supercomputers such as Cray 2 which has a core memory of 256 million words have opened the avenue of 3D modeling.

Let us consider the problem of numerical stability. The largest value of the time sampling interval which can be used in a given model without making the system numerically unstable is given by

$$(\Delta t)_{\max} = \frac{\mu \max(\Delta x, \Delta y, \Delta z)}{\max(c_1, c_2, \ldots, c_m)} \tag{20}$$

The quantity μ depends on the algorithm used for computing the wavefield; it can be determined by following von Neumann method (Smith, 1965, p. 70). In the case of (17), $\mu=0.5$ which is somewhat less than in the case of (9) which yields $\mu=0.57$. Thus, the advantage of using larger values of grid spacing is somewhat offset by the more restricted choice of μ. In practical terms, it is not a severe constraint.

PROBLEMS RELATED TO MODEL BOUNDARIES

Relation (17) indicates that for computing $u^{n+1}_{i,j,k}$, we must know the quantity $u^{n}_{i+2,j,k}$. This implies that for computing the wavefield along the grid plane $i=I$, we must know the field for time n along the plane $i=I+2$ which lies outside the model. A similar situation arises along grid planes such as $i=1$ and $j=J$ which are adjacent to the boundaries of the model and it gets worse along the grid lines which are located at the intersection of such planes. This problem can be easily avoided by using asymmetric operators for evaluating the wavefield at such grid locations.

Another problem which is much more troublesome is the unwanted reflections from the subsurface boundaries of the model. These boundaries correspond to five different planes, viz, $i=0$, $i=I+1$, $k=0$, $k=K+1$ and $j=J+1$. Thus, it is a much bigger problem than the corresponding problem which arises in 2D models. Several approximate algorithms are available for getting rid of such reflections (Clayton and Engquist, 1977; Reynolds, 1978; Korn and Stoekel, 1982) all of which lead to more or less similar results. Reynolds' algorithm seems to be the simplest among them and will be described here. In essence, it amounts to estimating the wavefield for the new time step $(n+1)$ by using the following relations along the boundaries:

$$\left.\begin{aligned}\frac{\partial u}{\partial x}-\frac{1}{c}\frac{\partial u}{\partial t}&=0 \qquad i=0\\ \frac{\partial u}{\partial x}+\frac{1}{c}\frac{\partial u}{\partial t}&=0 \qquad i=I+1\end{aligned}\right\} j=0,\dots,J+1;\ k=0,\dots,K+1$$

$$\left.\begin{aligned}\frac{\partial u}{\partial y}-\frac{1}{c}\frac{\partial u}{\partial t}&=0 \qquad j=0\\ \frac{\partial u}{\partial y}+\frac{1}{c}\frac{\partial u}{\partial t}&=0 \qquad j=J+1\end{aligned}\right\} i=0,\dots,I+1;\ k=0,\dots,K+1$$

$$\frac{\partial u}{\partial z}+\frac{1}{c}\frac{\partial u}{\partial t}=0 \qquad k=K+1 \Big\} \ i=0,\dots,I+1;\ j=0,\dots,J+1 \tag{23}$$

Derivation of difference relations for (21) to (23) involves first-order differencing. For (21), one gets

$$\begin{aligned}u^{n+1}_{o,j,k} &= u^{n}_{o,j,k}+\frac{C\Delta t}{\Delta x}\left(u^{n}_{1,j,k}-u^{n}_{0,j,k}\right)\\ u^{n+1}_{I+1,j,k} &= u^{n}_{I+1,j,k}-\frac{c\Delta t}{\Delta x}\left(u^{n}_{I+1,j,k}-u^{n}_{I,j,k}\right) \qquad j=0,\dots,J+1;\quad k=0,\dots,K+1\end{aligned} \tag{24}$$

Relations (22) and (23) can be treated in a similar manner. Reynolds' method is quite effective in reducing reflections from the boundaries. But there is still some residual energy which escapes into the model. It can be eliminated by damping out only the reflected signals as proposed by Israeli and Orszag (1981).

PSEUDO-SPECTRAL AND OTHER SCHEMES

It is widely known that the derivative of a field such as wavefield can be computed very accurately by using the Fourier transform derivative theorem (Bracewell, 1965, p. 117). This idea was intensively pursued by the Seismic Acoustic Lab of the University of Houston for a number of years. It was recognized that except for one-dimensional problems, the operators designed in this fashion would

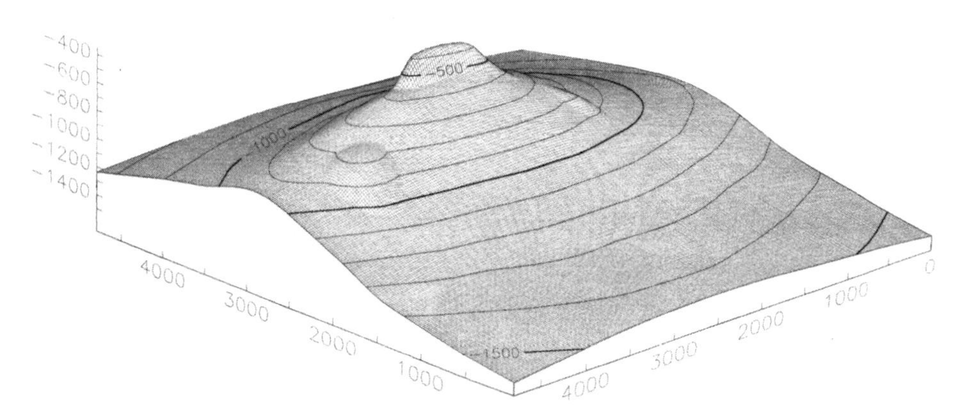

Fig. 2 A perspective plot showing top of the upper interface.

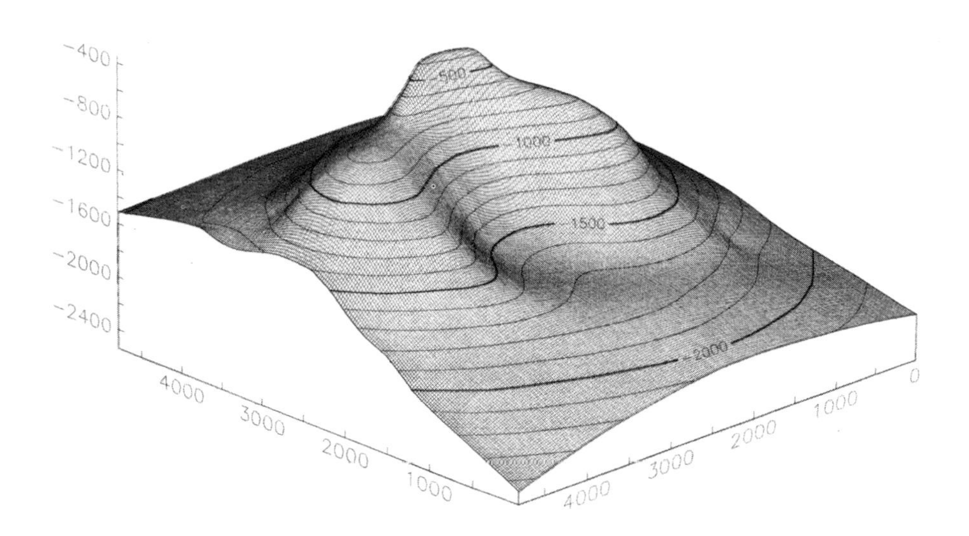

Fig. 3 A perspective plot showing top of the lower interface.

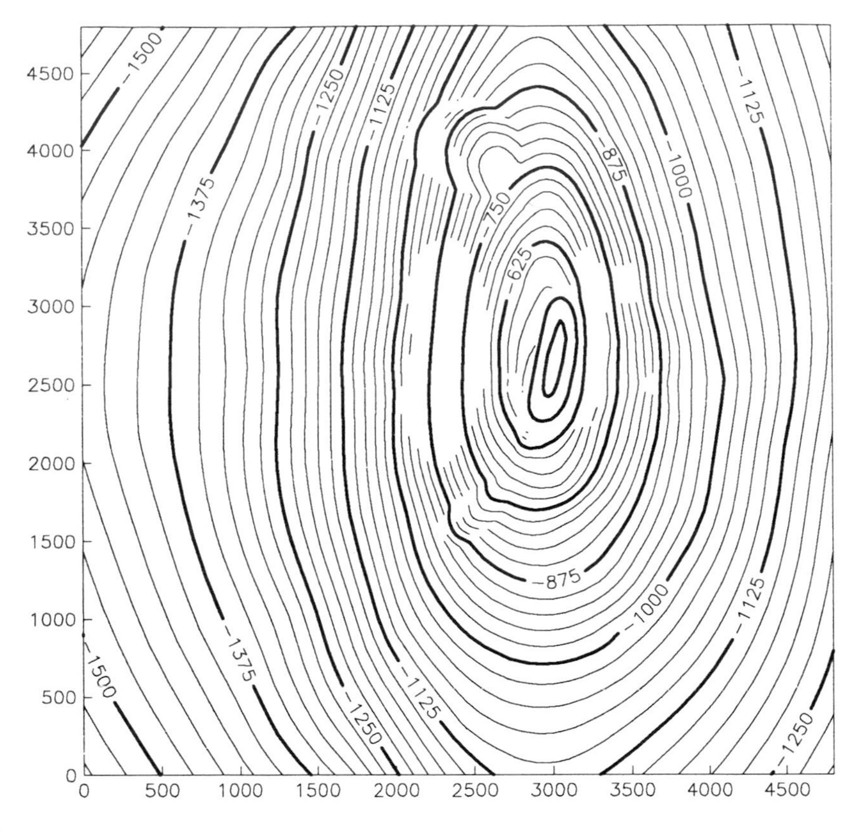

Fig. 4 Contour map of the structure shown in Fig. 2.

be unmanageably large and would require a prohibitive amount of numerical effort. Moreover, such operators cannot be used over a number of grid lines near the boundaries of the model. These difficulties can be easily overcome by introducing an approximation aimed at reducing the size of the operator. Instead of considering the continuous spectrum of the spatial derivative, the accuracy of the operator is restricted so that it would fit in the least-square sense to the continuous spectrum of the spatial derivative over the frequency band of interest. This is the basic idea underlying the so-called pseudo-spectral algorithms. For further details, the reader is referred to Gottlieb and Orszag (1977), Kosloff *et al.* (1984), and Dablain (1986).

The accuracy of a finite-difference operator obtained by using the spectral method must obviously depend on the number of terms retained in the corresponding Fourier series. In this sense, the frequency domain approach for evaluating the derivatives of the wavefield cannot be too different from the widely

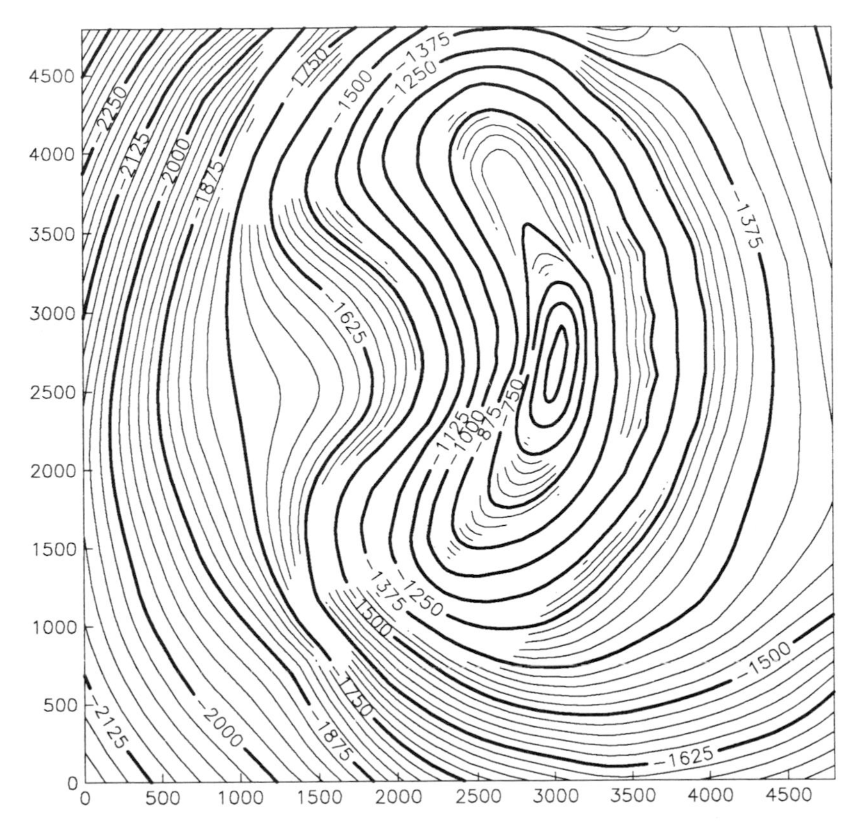

Fig. 5 Contour map of the structure shown in Fig. 3.

known and much simpler procedure of computing such operators by means of the Taylor series, in which the accuracy is likewise determined by the number of terms of the series retained for designing the operator. Besides its inherent simplicity, the latter method is far more flexible and is readily adapted to nonuniform media as well as to nonuniform sampling both in space and time. These considerations raise another question: What else can be done to develop a modeling algorithm which yields more accurate results than the simple fourth-order algorithm described previously. At this stage, it would be helpful to examine the magnitude of the truncation errors more closely. Let the error associated with an algorithm which is m-th order accurate in space and n-th order accurate in time be denoted by $\varepsilon_{m,n}$. Then for (9) and (17), we can write down

$$\varepsilon_{2,2} = O[(\Delta x)^2 + (\Delta y)^2 + (\Delta z)^2 + (c\Delta t)^2] \tag{25}$$

$$\varepsilon_{4,2} = O[(\Delta x)^4 + (\Delta y)^4 + (\Delta z)^4 + (c\Delta t)^2] \tag{26}$$

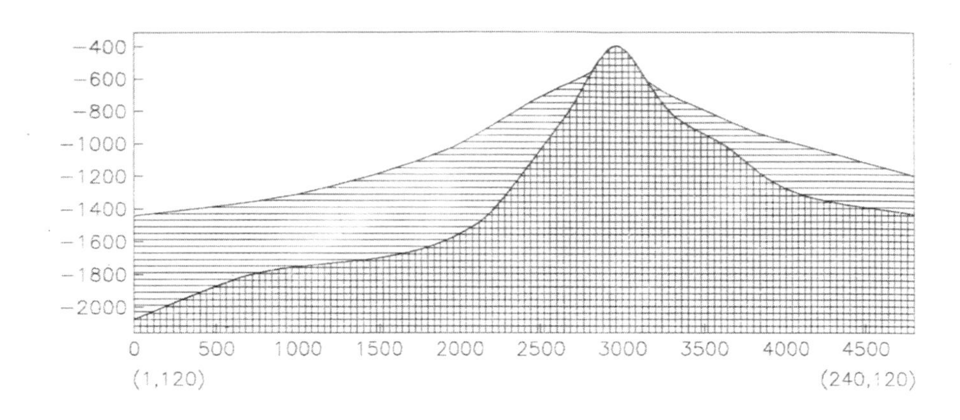

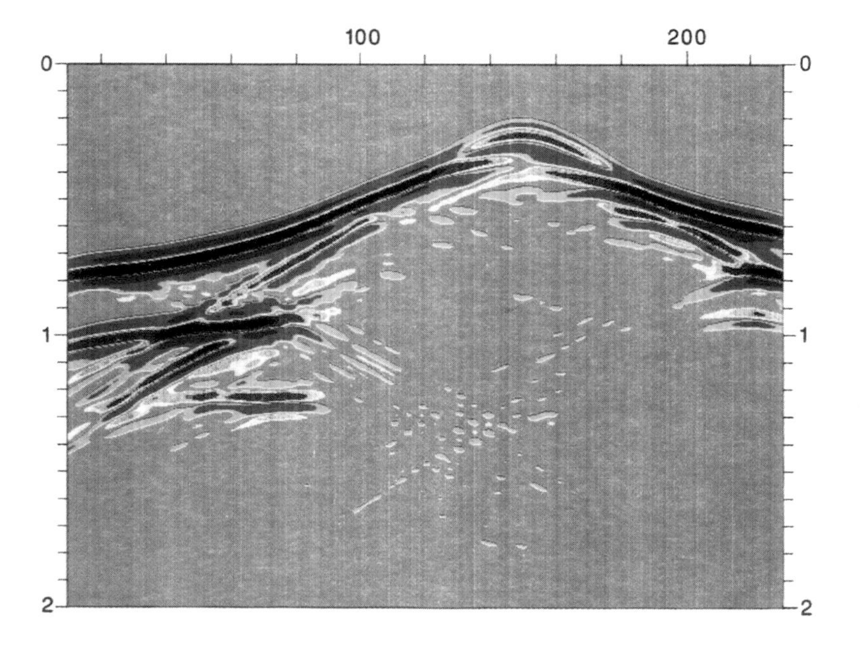

Fig. 6 Geologic and seismic cross-sections along the line EW-120.

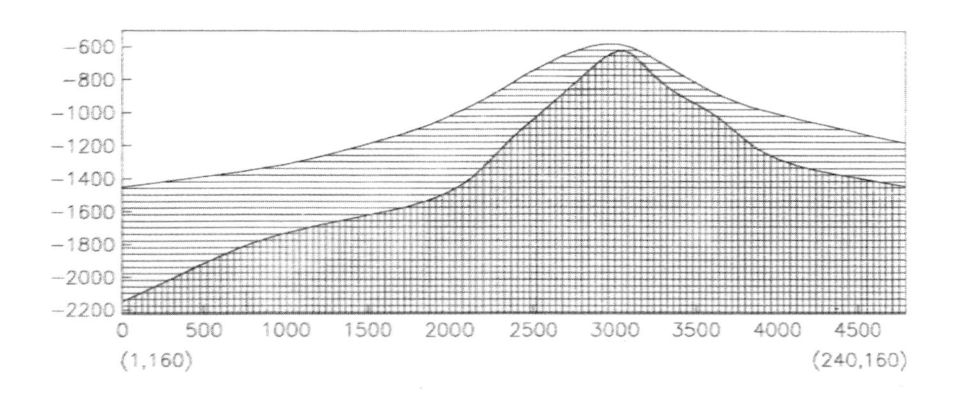

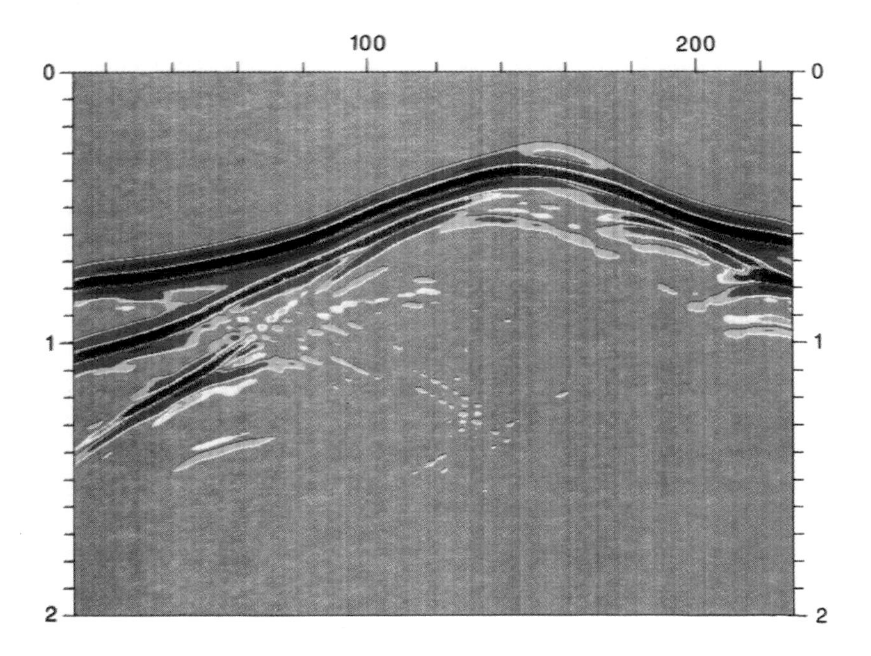

Fig. 7 Geologic and seismic cross-sections along the line EW-160.

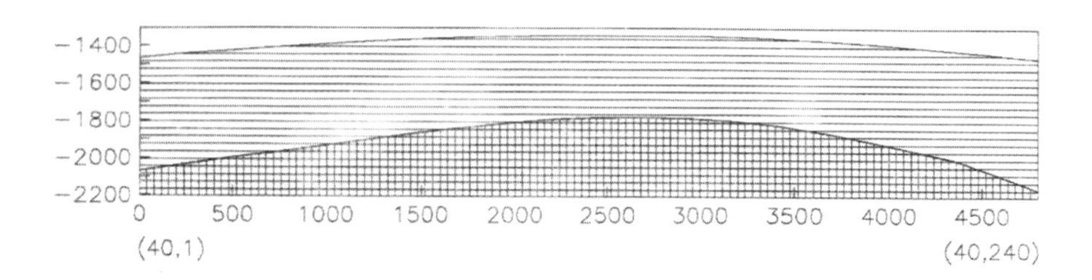

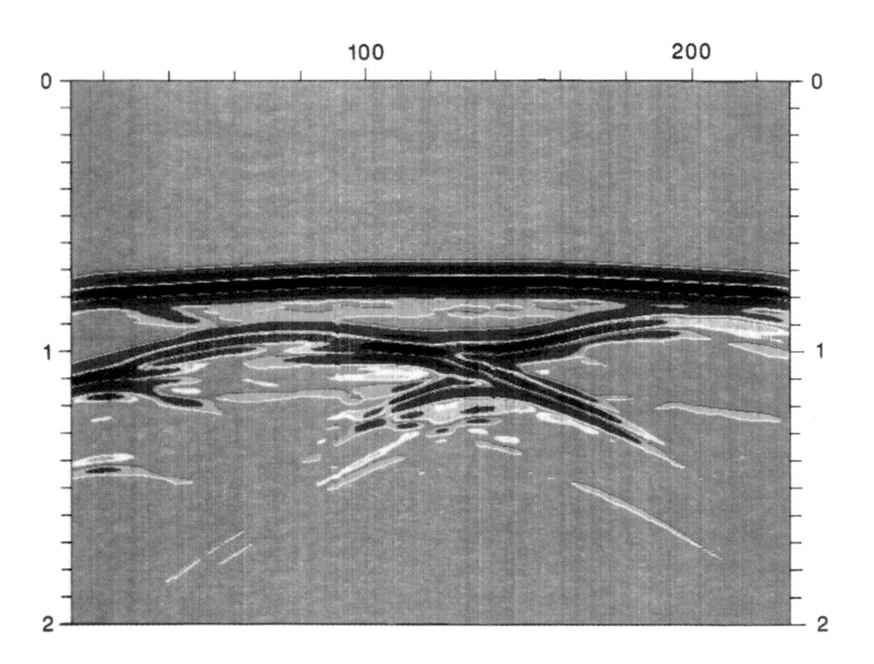

Fig. 8 Geologic and seismic cross-sections along the line NS-40.

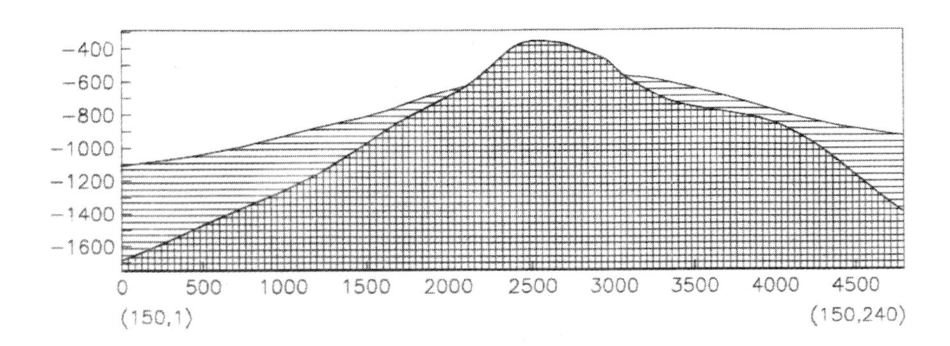

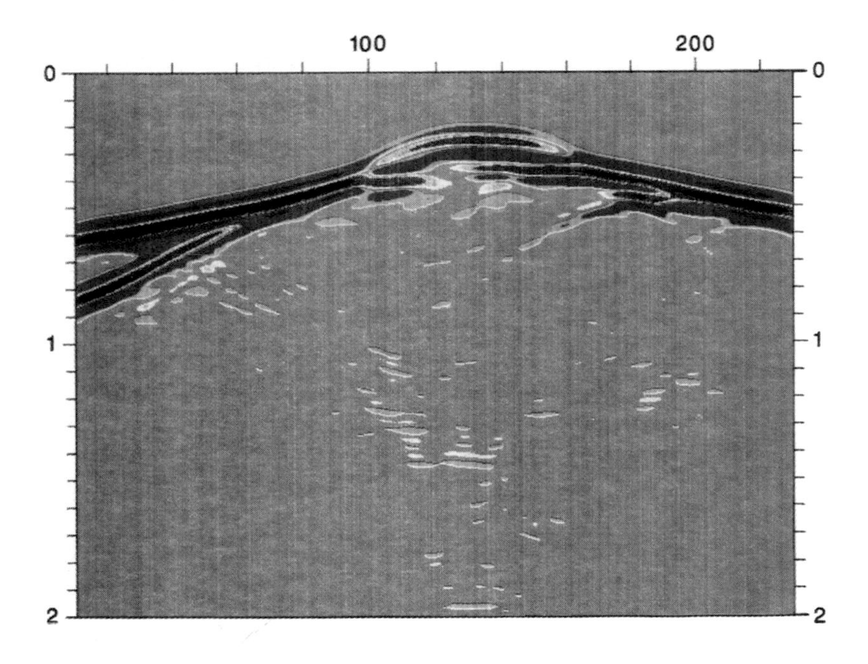

Fig. 9 Geologic and seismic cross-sections along the line NS-150.

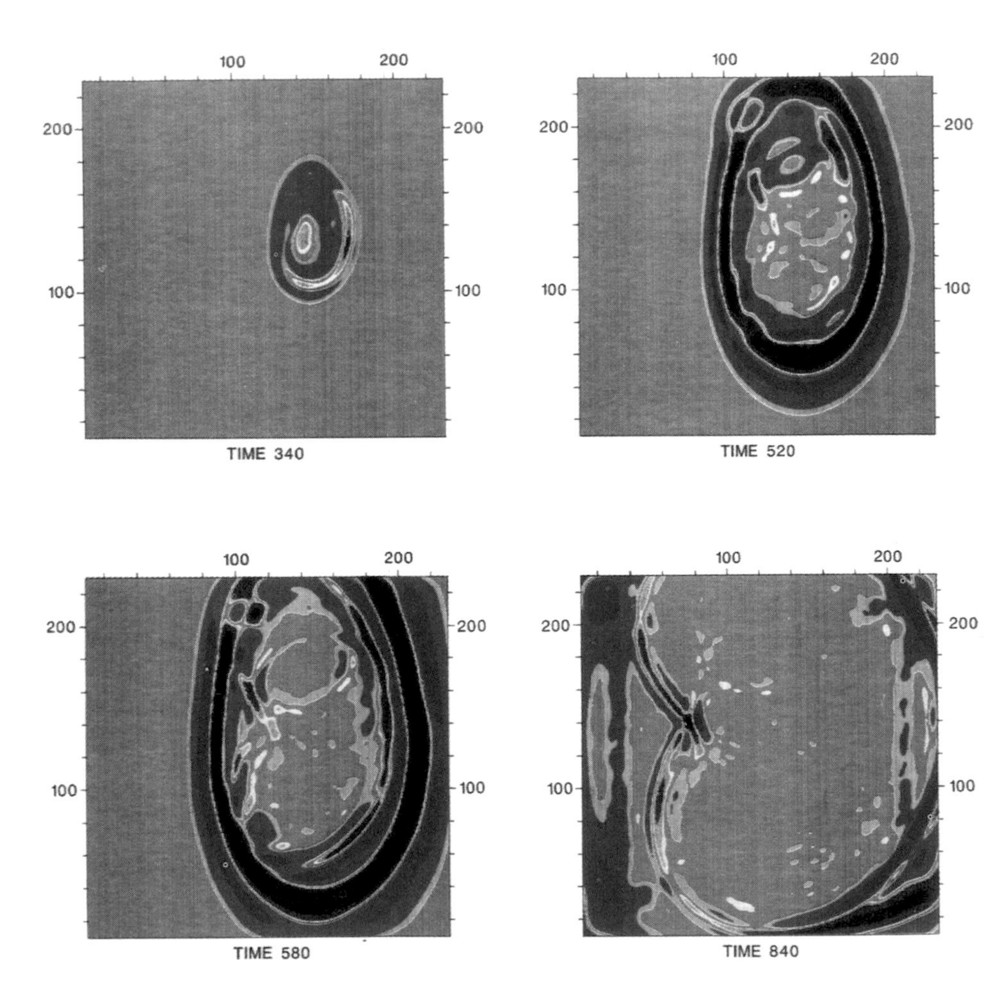

Fig. 10 Seismic time slices for different values of time.

Note that in (26), the component of the truncation error which is associated with the temporal derivative represents the dominant term. Therefore

$$\varepsilon_{4,2} \approx O[(c\Delta t)^2]$$

By the same argument, we can write

$$\varepsilon_{m,2} \approx O[(c\Delta t)^2],\ m=4, 6,...$$

Consequently, the accuracy of results will remain virtually the same if we try to evaluate the spatial derivatives more than fourth-order accurate but retain the expression for the temporal derivative which is only second-order accurate. Ideally one should try to evaluate *both* the spatial and time derivatives to higher and compatible orders of accuracy. Actual numerical tests indicate, however, that any attempts to evaluate the time derivative more than second-order accurate lead to a system which is either unconditionally unstable or yields results of inferior quality. Needless to say that such models are computationally much more intensive and practically unattractive. The rest of the treatment will be based on (17) only.

APPLICATIONS

The most obvious application of finite-difference seismic modeling is to compute a shot record for a given geologic structure. However, for interpretation purposes, a more useful information is the stacked seismic section for such a structure, which can be obtained by computing a large number of shot records, each for a different shot location, followed by conventional data processing. For 2D models, the seismic section can be computed at acceptable costs. In the investigation of 3D problems, one needs to examine a number of seismic sections corresponding to different orientations of the 3D structure. Consequently, the computational task increases by two to three orders of magnitude. One can avoid this cumbersome situation and still obtain reasonably accurate seismic sections by using the method of exploding reflectors (Loewenthal *et al.*, 1976), in which every grid point of the model which belongs to a subsurface interface is assigned a point source of a specified time dependence. All these sources are activated simultaneously and the energy reaching the various grid points along the surface of the model is recorded as a function of time. In a synthetic model, it is accomplished by saving the wavefield data for $k=o$ computed for time steps $n=1, 2,...$ In a practically meaningful modeling problem, it is usually necessary to use a grid size in excess of $250 \times 250 \times 250$ and to evaluate the wavefield for a thousand or more time steps. Such problems can be best solved on a supercomputer preferably with a huge central memory. The following example shows the results of a 3D model obtained by using a Cray 2 machine and interpreted on a seismic workstation.

We shall assume that the ground surface is horizontal and coincides with the plane $z = o$. Figure 2 shows the first interface below the surface; it may be regarded as the top of a sand layer which partially covers a salt dome below it. The dome can be seen subcropping approximately through the middle of the sand; it also subcrops out at two other locations, one to the north and the other to the west of the middle subcrop. The entire surface of the dome which forms the second interface of the model is shown in Figure 3. In both these figures, the vertical axis represents elevation with respect to the ground surface. The counter map of the upper interface is shown in Figure 4; it includes the subcrops. A similar map for the lower interface is shown in Figure 5.

The following parameters were used for setting up the model:

Eastwest extension of the model	= 4800 m
Northsouth extension of the model	= 4800 m
Maximum depth	= 4800 m
Grid spacing $\Delta x = \Delta y = \Delta z$	= 20 m
Grid dimensions of the model	= $240 \times 240 \times 240$
Velocity below the surface	= 3500 m/s
Velocity below the first interface	= 4200 m/s
Velocity of salt	= 4900 m/s
Maximum frequency of the source signal	= 50 Hz
Time sampling interval Δt	= 2 ms
Total number of time steps	= 500

Since the results based on the exploding reflector concept represent one-way travel time, the value of Δt was changed to 4 ms for the purpose of displaying and interpreting the results.

The results to be presented can be regarded as associated with various seismic lines running eastwest or northsouth along the surface boundary ($k = o$) of the model. It would be convenient to define the location of these lines in terms of grid coordinates i and j. Let the NS lines be indicated by different values of i. Then in view of the parameters used for setting up the model, each such line will consist of a total of 240 receiver stations, starting on the south at $j = 1$ and ending on the north at $j = 240$. By a similar argument, there will be 240 EW lines corresponding to different values of j, with each line starting on the west at $i = 1$ and ending on the east at $i = 240$.

Figure 6 (upper frame) shows a vertical cross-section of the 3D structure corresponding to the seismic line EW-120. The vertical scale represents elevation below the surface, whereas the horizontal scale indicates distance along the seismic line. Both scales are in meters. The seismic section for this line is shown in the lower frame. It ranges from trace 11 to trace 230 with each trace being two seconds long. The first and the last 10 traces were slightly distorted due to proximity of the model boundaries and were discarded.

The reflections originating from the two interfaces of the model can be easily identified across the seismic section, which also includes some out-of-plane events. A bow-tie feature can be seen on trace 44 at 1.00 s. Immediately to its right, there is a focussing of out-of-plane energy on traces 40 to 70. Below the bow-tie, a dipping event passes through trace 40 at about 1.1 s. This can only be interpreted as a sideswipe, and it probably originates from the southern "hill" of the dome body (Figure 3). The concentration of energy in this zone is sufficient to generate an interbed multiple arriving on traces 50 to 90 at about 1.25 s. Another interesting feature is readily noticeable on traces 140 to 160 at about 0.4 s; it represents a sectional view of an otherwise three-dimensional diffraction pattern around the peak of the dome.

The results shown in Figure 7 correspond to the seismic line EW-160. Note that this line is parallel to and not too far away from the line EW-120, and the geologic sections of the two lines are also quite similar. But this is not the case with their seismic sections which are significantly different. The seismic section shown in Figure 7 is characterized by a prominent sidewipe. On the west side of the line it can be seen as a steeply dippening event passing through trace 40 at about 1.2 s. Gradually, this out-of-plane event merges into the one above it at traces 100 to 122, followed by a diffraction tail.

Figures 8 and 9 show the results for lines NS-40 and NS-150. The geologic section shown in Figure 8 is almost featureless, except for a mild anticline associated with the lower interface. However, the seismic section is quite involved. An anticlinical event passing through traces 60 to 80 at about 0.95 s can be easily explained. But there is a strong focussing of energy where this event merges into an out-of-plane event. The latter event originates from the northern hill of the dome (Figure 3); it continues on the north side of the section and passes through trace 140 at about 1.0 s. The results shown in Figure 9 are straight forward.

Figure 10 shows 4 horizontal time slices of the wavefield. Each of these slices represents a snapshot of the wavefield reaching the surface of the ground at a specific moment of time. The station positions indicated along the horizontal axis ($i = 11$ to 230) and vertical axis ($j = 11$ to 230) correspond to NS and EW grid lines. The time slice for $t = 0.34$ s shows an early stage of the propagation process when the energy begins to reach the surface. It shows an areal view of the diffraction front emanating from the shallowest outcrop of the dome. In the time slice for $t = 0.52$ s, the reflected energy covers a much broader area, but most of it originates from the upper interface. An eye-shaped event centered at about ($i = 110$, $j = 205$) and a localized wiggle at about ($i = 110$, $j = 80$) are caused by the smaller outcrops of the dome. Most of the energy reaches the surface as a dipping front; this accounts for the apparent broadening of the pulse. Many of these features can also be identified in the time slice for $t = 0.58$ s, but there is also a crescent-shaped event surrounded by the outer event. It represents the salt dome. Note that the pulse associated with this event is much narrower. This is a result of refractive bending of the wavefront as it passes through the upper interface. After undergoing this change in the direction of propagation, the front reaches the surface at a smaller angle of incidence. Geophysicists frequently encounter buried foci in vertical time sections. There is no reason why a similar situation should not take place in a horizontal time section. The time slice for $t = 0.84$ s is an interesting example of a "spatial" buried focus reaching the geometry of the dome below.

CONCLUDING REMARKS

Three-dimensional modeling has been an impossible dream of geophysicists. The introduction of supercomputers has set some of them into a state of dialogue for realizing this dream. This brief investigation is a modest attempt involving the wave-equation analysis of the seismic response of an arbitrarily-shaped 3D structure. The results indicate that a lot of valuable information is lost if we treat the subsurface as a layered medium or a 2D structure. This is particularly true about time slices whch come at no extra cost but can only be obtained from a 3D model. This information is of utmost value to the seismic interpreter enabling him to examine the spatial configuration of the wavefield as it reaches the surface at different times. During the progress of this work, the author received many valuable suggestions and constant encouragement from N. J. Guinzy, M. G. Bloomquist, and S. J. Laster. Permission granted by Mobil Research and Development Corporation to publish the results of this study is appreciated.

REFERENCES

Bracewell, R., 1965, The Fourier Transform and Its Applications, McGraw-Hill Book Co., New York.

Clayton, R. and Engquist, B., 1977, Absorbing boundary conditions for acoustic and elastic equations, Bulletin of the Seismological Society of America, No. 67, pp. 1529–1540.

Dablain, M. A., 1986, The application of high-order differencing to the scalar wave equation, Geophysics, Vol. 51, pp. 54–66.

Gottlieb, D. and Orszag, S. A., 1977, A Numerical Analysis of Spectral Methods: Theory and Applications, Society of Industrial and Applied Mathematics.

Israeli, M. and Orszag, S. A., 1981, Approximation of radiation boundary conditions, Journal of Computational Physics, No. 41, pp. 115–135.

Korn, M. and Stoekel, H., 1982, Reflection and transmission of Love channel waves at coal seam discontinuities with a finite-difference method, Journal of Geophysics, No. 50, pp. 1771–1788.

Kosloff, D., Koren, Z., and Loewenthal, D., Discrete derivative operators for seismic forward modeling and migration, presented in the Annual SEG Meeting, 1984, Atlanta.

Loewenthal, D., Lu, L., Roberson, R., and Sherwood, J., 1976, The wave equation applied to migration, Geophysical Prospecting, Vol. 24, pp. 380–399.

Mufti, I. R., 1985, Seismic modeling in the implicit mode, Geophysical Prospecting, No. 33, pp. 619–656.

Reynolds, A. C., 1978, Boundary conditions for the numerical solution of wave equation propagation problems, Geophysics, Vol. 43, pp. 1099–1110.

Smith, G. D., 1965, Numerical Solution of Partial Differential Equations, Oxford University Press, London.

CHAPTER 11

A FAST BOUNDARY INTEGRAL SOLUTION FOR THE ACOUSTIC RESPONSE OF THREE-DIMENSIONAL AXI-SYMMETRIC SCATTERERS

by
GERARD T. SCHUSTER
Geology and Geophysics Department
University of Utah
Salt Lake City, Utah 84112

ABSTRACT

A Boundary Integral Equation (BIE) method is presented which efficiently computes the harmonic acoustic response of axi-symmetric structures. The key idea is that the BIE's are Fourier transformed via FFT's in the azimuthal variable; this reduces the azimuthal periodic convolutions to simple multiplications. The original three-dimensional problem is transformed into a series of M decoupled two-dimensional problems, where M is the number of azimuthal Fourier components. If N is the number of nodal points along the semi-perimeter of a scatterer, then the harmonic response can be computed with just $O(N^3M)$ algebraic operations. This is far less expensive than solving the original problem which requires $O(N^6)$ algebraic operations per frequency. Moreover, the active memory requirement is reduced from $O(N^4)$ to $O(N^2)$ complex words, and the algorithm is ideally suited to a parallel computer. Examples are given where less than three minutes were required by a Gould computer (4 MIPs) to compute the harmonic response of a scatterer four wavelengths in dimension.

INTRODUCTION

Studying the seismic response of scatterers can reveal valuable information about their internal structure. An important tool in understanding these responses is the forward modeling method. Forward modeling consists of computerized simulations of wave propagation through hypothetical structures; identifying the genesis of simulated arrivals allows them to be used as indicators of the model's internal features. These events can then be correlated with the real seismic data to make inferences about the actual internal structure.

Forward modeling is used in a wide variety of disciplines. In the case of non-destructive testing, internal features of interest might be flaws or cracks in dams, mechanical components, pipes, or building materials. Interaction of seismic waves with these flaws produces diffraction patterns that may be diagnostic of their extent and location. Volcanologists observe that low-frequency (1–5 Hz) volcanic tremors emanate from magma chambers and may reveal information about a chamber's structural and compositional features. Forward modeling of hypothetical chamber models can ultimately provide models consistent with the data. For zoning engineers, it is important to know which areas in a populated basin are most susceptible to basin resonance excited by seismic waves. If the lithologic structure of the basin is known, then forward modeling can aid (Bard and Bouchon, 1980, Trifunac, 1971, Tucker and King, 1984) in predicting these hazardous site locations.

The problem with seismic modeling is that many structures are three dimensional. This means that full wavefield modeling must be performed for three spatial dimensions, a very difficult task even for today's super-computers. Certain algorithms, such as the Pseudo-Spectral method, show much promise but they are still in their development stages. At present, even a $512 \times 512 \times 512$ node model demands many CPU hours and extensive programming efforts on the latest CRAY computer. It is estimated that a new generation of computers must be developed before realistic three dimensional modeling can be performed for large models. In the meantime, many researchers are restricted to 2-D modeling codes based on, for example, wave equation differencing schemes, approximate methods such as ray-tracing, or semi-analytical methods for overly simplified structures.

THREE-DIMENSIONAL MODELING ALGORITHM

A simple, relatively fast and practical method for modeling seismic waves in three-dimensional axi-symmetric structures is presented. The algorithm requires no more than a few hundred lines of computer code and so it can be easily accessible to the non-expert. Its most efficient use is for axi-symmetric scatterers composed of only a few homogeneous regions that can be in contact with one another.

The algorithm formulates the harmonic response of an axi-symmetric structure as a frequency domain boundary integral equation. Rather than solving the set of discretized integral equations by matrix inversion, a Fourier transform in the azimuthal variable is applied. This reduces the periodic convolutions in the azimuthal variable (i.e., azimuthal coordinate surfaces are along the surface of the scatterer) to simple multiplications. The benefit is that the two dimensional surface integrals are transformed into a set of decoupled line integral equations; each can be inverted separately and the composite solution can be inverse Fourier transformed to yield the frequency domain solution. These one-dimensional line

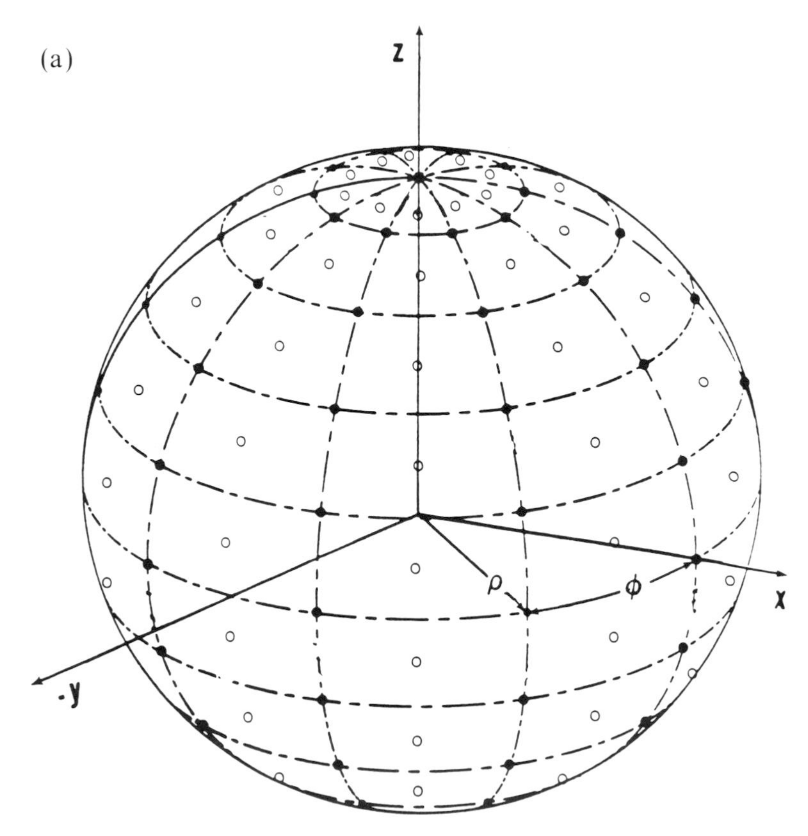

Fig. 1 Sphere where the surface is approximated by quadrilateral patches. An unknown field value is assigned at the center of each patch (open circle) and is assumed to be constant within that patch. Figure (1b) represents the sphere's semi-perimeter in the $y=0$ plane. Azimuthally Fourier transforming the BIE's reduces the two-dimensional integration over the sphere's suface to an integration over the line L_1 in Figure (1b).

Figure continued

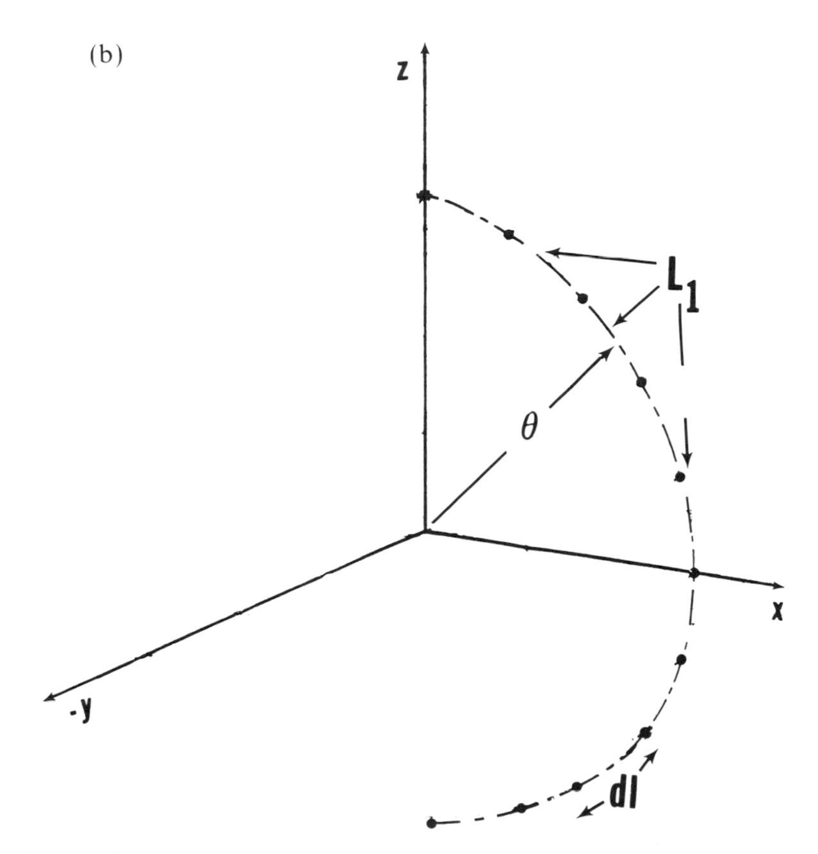

Fig. 1—*(continued)*.

integral equations are far less expensive to solve than the original two-dimensional surface integral equation. It will be understood that frequency corresponds to temporal frequency and wavenumber corresponds to azimuthal frequency.

THEORY

The basic idea is that the boundary integral solution to the Helmholtz equation (or any wave equation in the frequency domain) can be given as

$$[\mathbf{A}]\mathbf{p}=\mathbf{f} \tag{1}$$

where [**A**] is the N^2 by N^2 matrix representing the discretized integral operators in the appendix (equation A.2), **p** is the N^2 by 1 vector containing the unknown field values at each node of the boundary in Figure (1a), and **f** is the N^2 by 1 vector containing the pressure source components. For the spherical model in Figure (1), there are $O(N)$ nodes along the semi-perimeter resulting in, roughly, $O(N^2)$ unknowns on the scatterer's surface. Actually, there are two unknowns at each nodal point, both pressure and its normal derivative, so there should be a total of $O(2N^2)$ unknowns; but for the sake of simplicity we will assume that there is only one unknown per nodal point (such as in the Dirichlet problem). To solve the equations in (1) by Gaussian elimination would cost about $O(N^6)$ algebraic operations per frequency. For N greater than a few hundred, this can be extremely time consuming.

Rather than inverting the N^2 by N^2 matrix in equation (1), we propose to decouple the 3-D problem into a series of 2-D problems. This is possible because: 1) an axi-symmetric scatterer's surface lies along azimuthal coordinate surfaces and 2) the Green's function (i.e., $g=\exp(ikr)/4\pi r$), where $r=(\rho^2+\rho'^2-2\rho\rho'\cos(\phi-\phi')+(z-z')^2)^{1/2}$) is a function of the difference between the source and receiver's azimuthal angles, $\phi-\phi'$. This implies that the integrals in equation (A.2) (or equation 1) are periodic convolutions in the ϕ variable. Fourier transforming equation (1) in the ϕ variable reduces the integration over the surface in Figure (1a) to an integration over the semi-perimeter line in Figure (1b). The three dimensional problem in (ω, ρ, ϕ, z) space is transformed into a series of independent two-dimensional problems in $(\omega, \rho, k_\phi, z)$ space.

Assuming M azimuthal frequencies or wavenumbers, k_ϕ, the azimuthal Fourier transform of equation (1) reduces it to a set of M N by N matrix equations, each denoted by

$$[\tilde{\mathbf{A}}]_m \tilde{\mathbf{p}}_m = \tilde{\mathbf{f}}_m \tag{2}$$

where the tilde denotes the transform pair to equation (1) and N is the number of nodes depicted in Figure (1b). The mth azimuthal Fourier coefficient vector dual to **p** is denoted by $\tilde{\mathbf{p}}_m$. The cost in inverting the N by N matrix in equation (2) is only $O(N^3)$ algebraic operations compared to, roughly, $O(N^6)$ for equation (1). To compute **p**, M matrices similar to that in equation (2) must be inverted and the resulting Fourier coefficient vectors, $\tilde{\mathbf{p}}_m$, inverse Fourier transformed. This is desirable because the computational cost is only $O(MN^3)$ algebraic operations compared to $O(N^6)$ algebraic operations for direct matrix inversion of equation (1). The computational count for the Fourier transformation (via FFT's)

$$\mathbf{p} \approx \sum_{m=0}^{M-1} \tilde{\mathbf{p}}_m e^{im\phi} \tag{3}$$

is neglected because it is much less than the dominant cost of $O(MN^3)$. Note that if the point source is placed along the z-axis then only the $k_\phi = 0$ Fourier coefficient vector $\tilde{\mathbf{p}}_0$, needs to be computed.

Adaptability To Parallel Computers. This method is ideally suited to parallel computers. Each CPU node of a parallel computer can independently compute a Fourier coefficient vector, $\tilde{\mathbf{p}}_m$, by inverting a relatively small N by N matrix. Conceivably, this could reduce the CPU time for axi-symmetric modeling to that of a single two-dimensional model. In addition, the active memory needed to solve each independent problem is reduced from $O(N^4)$ to $O(N^2)$ complex words.

NUMERICAL EXAMPLES

Scattering from two-dimensional cylinders and three-dimensional spheres will now be computed by the fast exact numerical method and compared to the analytical solutions. Such examples will demonstrate the accuracy and convergence characteristics of this method. Scattering from a reef model will also be computed to illustrate a practical use for this method.

Two-Dimensional Scattering From a Cylinder

Figure (2a) depicts the radiation patterns due to plane harmonic waves impinging from the left on a two-dimensional cylinder. The scattered pressure field is computed by the fast exact numerical (dashed lines) method and compared to the analytical solution using an eigenfunction expansion method (undashed lines). There is excellent agreement between the analytical and numerical solutions. In this case the cylinder's internal and external acoustic velocities are, respectively, 7000 and 5000 ft/s, its radius is 100 feet, and the source frequencies are 25 Hz(A), 69 Hz(B), 112.5 Hz(C), 156.2 Hz(D) and 200 Hz(E). (ka = 3.14, 8.6, 14.1, 19, and 25, where ka is the product of the radius with the external wavenumber).

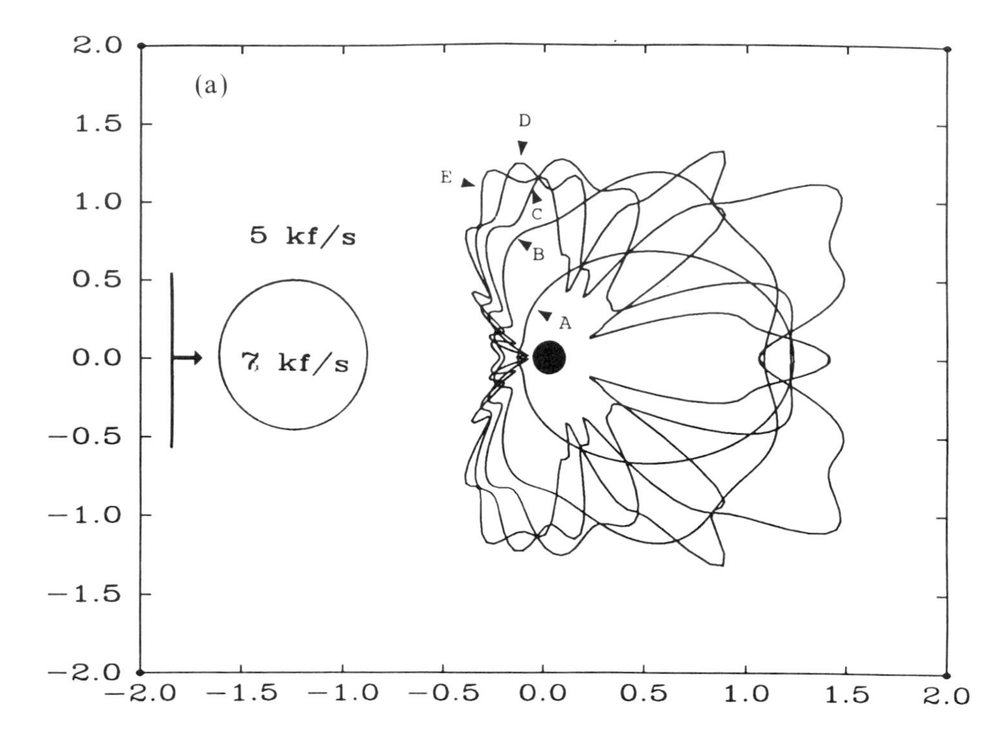

(a)
5 kf/s
7 kf/s
A
B
C
D
E
2.0
1.5
1.0
0.5
0.0
−0.5
−1.0
−1.5
−2.0
−2.0
−1.5
−1.0
−0.5
0.0
0.5
1.0
1.5
2.0

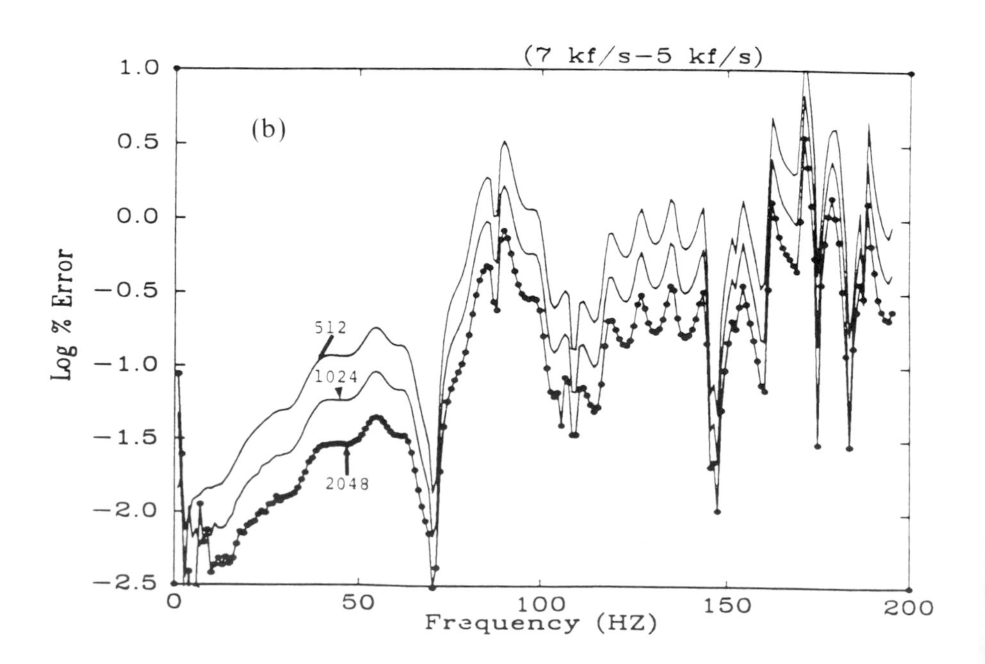

(7 kf/s−5 kf/s)
(b)
Log % Error
512
1024
2048
Frequency (HZ)
1.0
0.5
0.0
−0.5
−1.0
−1.5
−2.0
−2.5
0
50
100
150
200

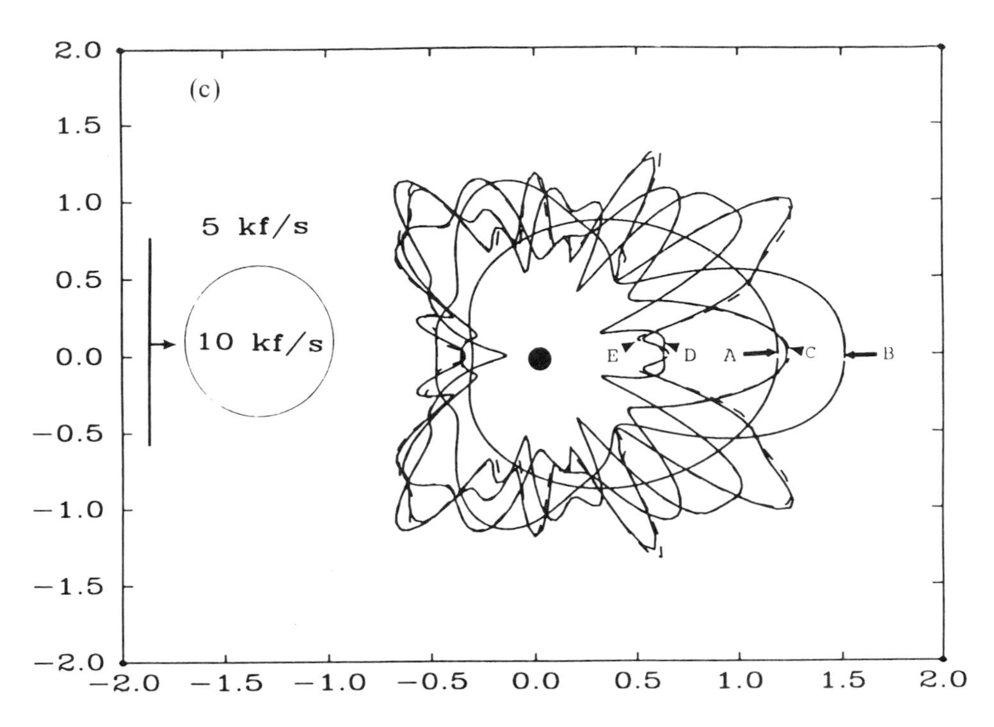

Fig. 2 The scattered pressure field due to a harmonic plane wave incident from the left on a two-dimensional cylinder (radius = 100 ft) is computed by an eigenfunction expansion method (lines) and is compared to the fast exact numerical solution (dashed lines) which use a 2048 point FFT (Figures 2a and 2c). The five radiation patterns correspond to the scattered pressure field at source frequencies of 25 Hz (A), 68.75 Hz (B), 112.5 Hz (C), 156.2 Hz (D), and 200 Hz (E). Figure (2b) depicts the errors for the computed scattered field at the far right hand side of the cylinder for discretizations of 512, 1024, and 2048 points.

The error between the two solution methods for a wide range of frequencies and discretization rates is given in Figure (2b). Higher frequencies demand a finer discretization rate as evidenced by the radiation patterns for the example in Figure (2c). In this case, the discretization of the boundary is kept fixed at 2048 points and the internal (external) velocity is 10,000 ft/s (5000 ft/s). Each frequency response required no more than a few CPU seconds on a Gould computer (4 MIPs) and each could have been computed simultaneously on a parallel computer. This formulation is easily extendible to concentric cylinders, semi-cylinders,

TABLE Ia

Computer program for the fast exact numerical solution (Scheme S.I) for harmonic plane waves incident on a cylinder with a penetrable interface. Numerical results depicted below compare the scattered pressure field (P_s) on the cylinder to the analytic solution computed by an eigenfunction expansion method. In this example, the radius of the cylinder is 100 feet, the interior (exterior) velocity is 7000 ft/sec (5000 ft/sec), and the pressure is measured at the far right hand side of the interface for a plane wave moving to the right.

```
      parameter (iq=512, pi=3.14159265, pi2=2*pi)
      complex g(iq), h(iq), h0(iq), g0(iq), ff(iq), pr(iq)
      read(11,  *) df, ve, vi, ra, nf, ntfr
      do 30 nfr=1, ntfr
         we=pi2* df* nfr/ve
         wi=pi2* df* nfr/vi
           do 10 i=1, nf
              wr=ra* cos((pi2*(i-1))/nf)* we
10         ff(i)=2.*cmplx(cos(wr), sin(wr))
         call fft(nf, ff, -1, 1.0)
         call cself(we, ra, nf, g0, h0, pi)
         call cself(wi, ra, nf, g, h, pi)
           do 20 k=1, nf
20            pr(k)=g(k)*ff(k)/(g(k)* (1.+h0(k))+g0(k)* (1.-h(k)))
         call fft(nf, pr, 1, scale=1./float(nf))
30    continue
      end

      subroutine cself(wi, r, nf, g, h, pi)
      parameter(cnst=pi/(2.*nf), cnst2=pi*r/nf)
      real*4j 0, j1
      complex h(nf), g(nf)
      h(1)=(0., 0.)
      g(1)=-(2.*r/nf)* (alog(cnst2*wi*0.5)-1.+.577216)+cmplx(0., cnst2)
      do 10 i=2, nf
        x=2.*wi*r* sin(pi*float(i-1)/nf)
        call hankel(x, j0, y0, j1, y1)
        h(i)=cmplx(-x*y1*cnst, x*j1* cnst)
10      g(i)=cmplx(- y0*cnst2, j0*cnst2)
      call fft(nf, g, -1, 1.d0)
      call fft(nf, h, -1, 1.d0)
      return
      end
```

Table continued

TABLE Ia (*continued*)

	Scheme S.I. Fast Exact Numerical Solution				Eigenfunction Solution	
Frequency	Real P_{tot}	Cmplx P_{tot}	Real P_{scat}	Cmplx P_{scat}	Real P_{scat}	Cmplx P_{scat}
0.97657	0.98453	0.11659	−0.00795	−0.00583	−0.00805	−0.00577
1.95313	0.94872	0.22050	−0.02131	−0.02248	−0.02143	−0.02239
2.92970	0.89923	0.31147	−0.03376	−0.04843	−0.03389	−0.04830
3.90626	0.83973	0.38943	−0.04219	−0.08197	−0.04235	−0.08180
4.88282	0.77295	0.45428	−0.04464	−0.12153	−0.04482	−0.12134
5.85939	0.70132	0.50592	−0.03963	−0.16564	−0.03985	−0.16542
6.83596	0.62718	0.54448	−0.02599	−0.21273	−0.02625	−0.21250
7.81252	0.55279	0.57034	−0.00278	−0.26113	−0.00308	−0.26088
8.78909	0.48025	0.58427	0.03064	−0.30895	0.03029	−0.30871
9.76565	0.41143	0.58733	0.07454	−0.35422	0.07415	−0.35397
10.74220	0.34792	0.58084	0.12882	−0.39486	0.12840	−0.39463
11.71880	0.29098	0.56634	0.19297	−0.42884	0.19252	−0.42862
12.69530	0.24157	0.54548	0.26611	−0.45422	0.26564	−0.45402
13.67190	0.20027	0.51996	0.34700	−0.46921	0.34651	−0.46902
14.64850	0.16737	0.49152	0.43409	−0.47226	0.43360	−0.47207
15.62500	0.14287	0.46183	0.52556	−0.46204	0.52507	−0.46185
16.60160	0.12646	0.43252	0.61936	−0.43757	0.61889	−0.43733
17.57820	0.11755	0.40502	0.71326	−0.39818	0.71283	−0.39781
18.55470	0.11527	0.38031	0.80481	−0.34394	0.80450	−0.34294
19.53130	0.11913	0.36309	0.89215	−0.27130	0.89145	−0.27274

transient sources, and elastic media. An example of the time-domain response of a cylindrical transient wave impinging upon a semi-cylinder in a half space is given in Figure (3). In this case the cylinder response was computed by the fast exact numerical method and the free surface was taken into account by the method of images.

It is to be noted that the boundary integral solution to the two dimensional cylinder problem is originally cast in the form of one dimensional line integrals that are periodic convolutions in the azimuthal coordinate; hence, application of the azimuthal Fourier transform reduces the line integrals to simple multiplications. No matrix inversions are necessary as the solution only requires simple algebraic division. Appendix A and Schuster and Smith (1988) elaborate upon this method and two simple programs are given in Table I. Scheme S.I refers to the standard BIE formulation (Schuster and Smith, 1985a) and scheme S.IV refers to the formulation based on equation (A.2). These two-dimensional examples are important because they demonstrate the validity of computing the Discrete Fourier Transform

TABLE Ib

Computer program for the fast exact numerical solution (Scheme S.IV) for harmonic plane waves incident on a cylinder with a penetrable interface.

```
      parameter(iq=4096, pi=3.14159265359, pi2=2.*pi)
      complex g(iq), h(iq), f(iq), ff(iq), pr(iq), dpr(iq), df(iq)
    read(11,  *) dff, ve, vi, ra, nf, ntfr
      do 30 nfr=1, ntfr
        wi= pi2*dff*nfr/vi
        we= pi2*dff*nfr/ve
          do 10 i=1, nf
            cs=cos((pi2*(i-1))/nf)
            wr=ra*cs*we
            ff(i)=cmplx(cos(wr), sin(wr))
10          df(i)=cmplx(0.0, we*cs)* ff(i)
        call fft(nf, ff, -1, 1.)
        call fft(nf, df, -1, 1.)
        call cself(wi, we, ra, nf, f, g, h, pi)
          do 20 k=1, nf
            pr(k)=(df(k)* f(k)+ff(k)* (1.-g(k)))
     1      ((1.+g(k))* (1.-g(k))+f(k)* h(k))
20          dpr(k)=(-pr(k)* h(k)+df(k))/(1.-g(k))
        scale=1./float(nf)
        call fft(nf, dpr, 1, scale)
        call fft(nf, pr, 1, scale)
30  continue
      end
c
```

Table continued

TABLE Ib (*continued*)

```
      subroutine cself(wi, we, r, nf, f, g, h, pi)
      real*4 j0i, j0e, j1i, j1e
    complex h(nf), g(nf), f(nf), cnst, a
c ****************************************************************
c compute the asymptotic forms for f(1), g(1), h(1)
c ****************************************************************
      rnf = float(nf)
      cnst = cmplx(0., pi*r*.5/rnf)
      we2 = we**2
      wi2 = wi**2
      f(1) = cnst* cmplx(0., 2.*(alog(wi) - alog(we))/pi)
      g(1) = (0., 0.)
      h(1) = cmplx(0., 1.0)* cnst*
    1( - (wi2 - we2)* .3456 + wi2* alog(wi/2.) - alog(we/2.)* we2)/pi
    2 - .5*r*(alog(2.*pi*r/rnf) - 1.)* (wi2 - we2)/rnf + cnst*.5*(wi2 - we2)
c ****************************************************************
c compute f, g, h for 2 < i < nf
c ****************************************************************
      do 10 i = 2, nf
        an = pi*float(i - 1)/rnf
        s = sin(an)
        s2 = s*s
        rr = 2.*r*s
        x = rr*wi
        call hankel(x, j0i, y0i, j1i, y1i)
        x = rr*we
        call hankel(x, j0e, y0e, j1e, y1e)
        a = wi* cmplx(j1i, y1i) - we* cmplx(j1e, y1e)
        h(i) = cnst*((cos(an*2) + 2.*s2)* a/rr
    1   - s2*(wi2* cmplx(j0i, y0i) - we2* cmplx(j0e, y0e)))
        g(i) =  - cnst*s*a
        f(i) = cnst*(cmplx(j0i, y0i) - cmplx(j0e, y0e))
 10   continue
      call fft(nf, f, - 1, 1.)
      call fft(nf, g, - 1, 1.)
      call fft(nf, h, - 1, 1.)
      return
      end
```

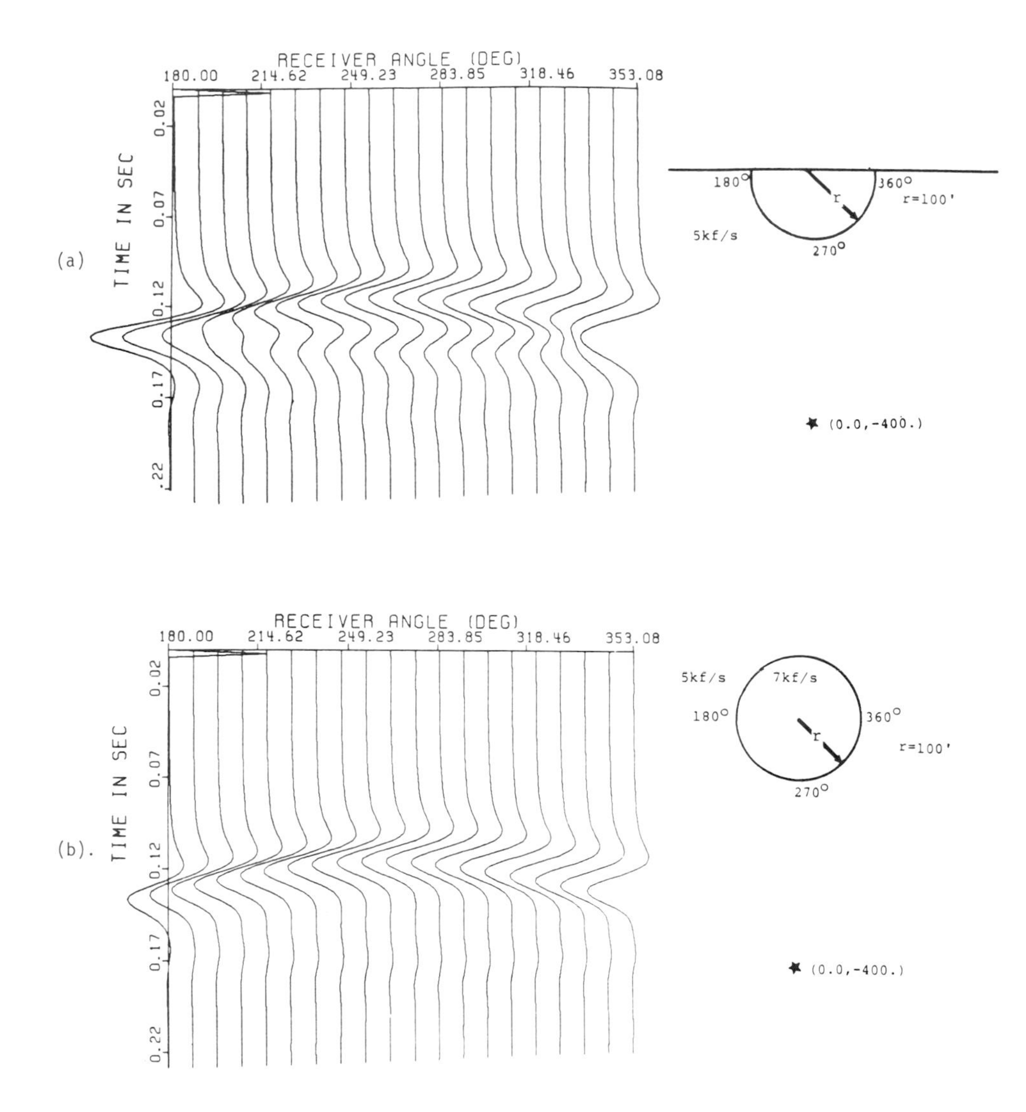

Fig. 3 The transient response of a semi-cylinder intersecting a rigid plane surface ($\partial p/\partial n = 0$ on the plane surface) is computed by the method of images (Fig. 3a). Locating line sources 400 feet above and below the center of the cylinder and computing the response by the fast BIE method insures that the rigid boundary condition is satisfied at the plane surface. The total pressure field at the semi-cylindrical surface (Figure 3a) is compared to that at the surface of a cylinder in Figure (3b).

of integrated Green's functions. Better convergence and accuracy could have been achieved by using a higher order isoparametric representation of both the field variables and boundary geometry. For the examples in Figures (2) and (3), the boundary geometry and pressure field between two nodes were represented, respectively, by a connected series of straight lines and by a constant.

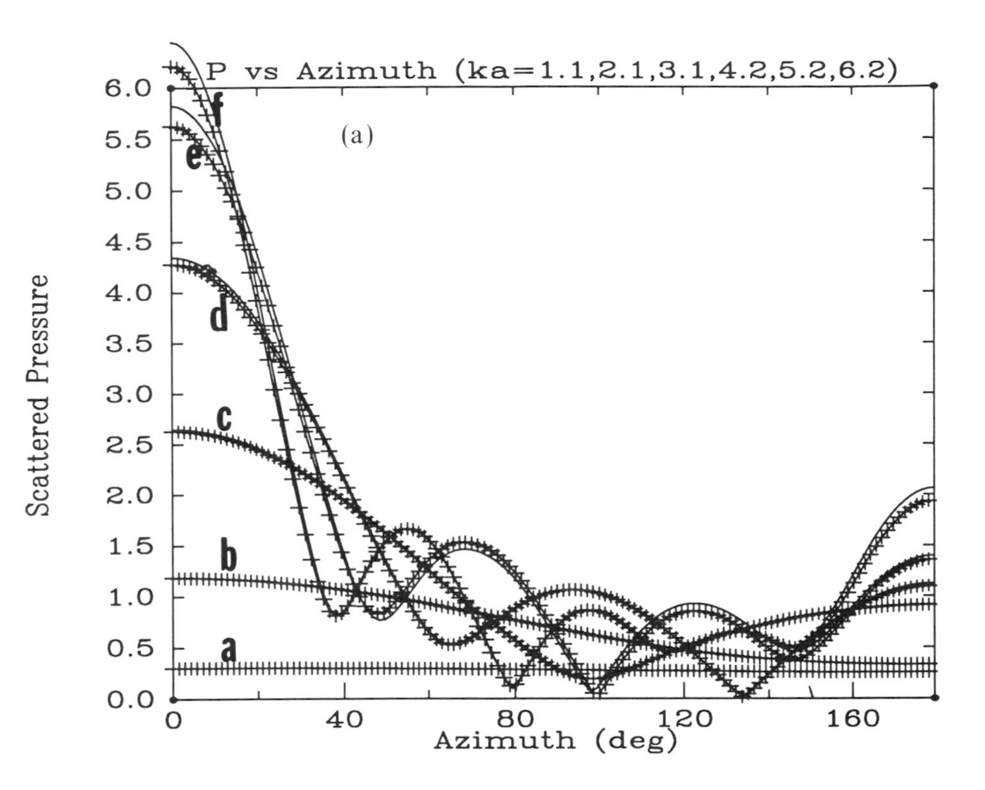

Fig. 4 The plane wave harmonic response of a sphere (radius = 1 km) is computed by the fast BIE method (crosses) and is compared to an eigenfunction expansion solution (lines) for a range of source frequencies. The exterior (interior) velocity of the sphere was taken to be 2.5 (1.5) km/s, the plane wave is moving in the positive x direction in Figure (1) and the field is interrogated at the sphere's surface along the $z = 0$ plane. Higher order oscillations in the radiation patterns correspond to higher source frequencies. Better than 5 per cent accuracy is achieved for $ka < 9$, $N = 35$ and $M = 128$; i.e., accurate results were obtained at 6 to 8 nodes per wavelength. Greater accuracy could have been achieved if the constant interpolation polynomials were replaced by a higher order isoparametric representation of the boundary geometry and field variables.

Figure continued

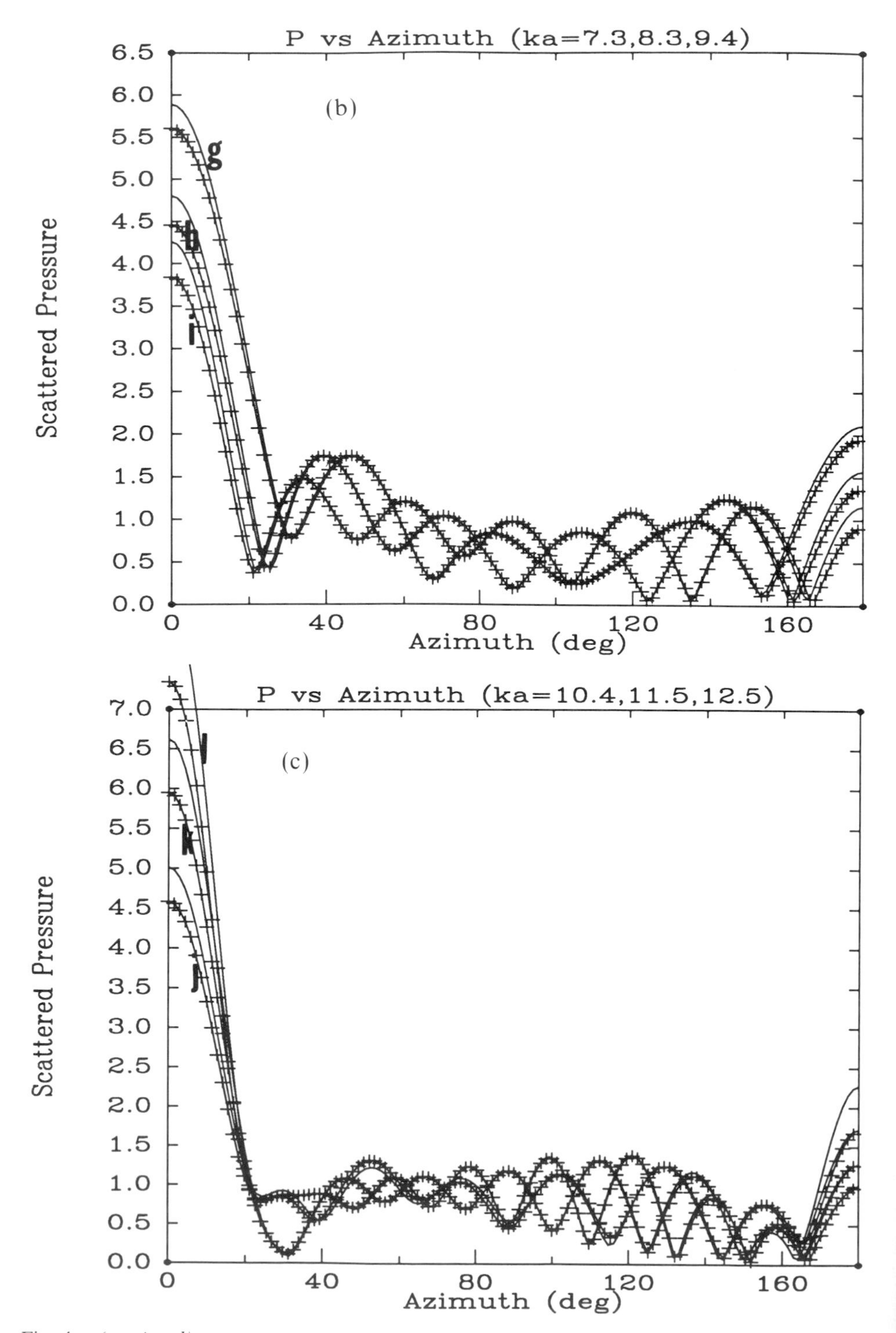

Fig. 4—*(continued)*.

The radiation characteristics due to plane harmonic waves impinging from the left upon a sphere (radius = 1 km) with a penetrable boundary ($v_{ext.} = 2.5$ km/s, $v_{int.} = 1.5$ km/s) are given in Figures (4a–4c). For all of the subsequent numerical examples it will be assumed that the boundaries are penetrable and the plane wave is impinging from the left side of the figure. The numerical solution (crosses) agrees very well with the analytic solution (lines) computed by an eigenfunction expansion method. Disagreement occurs as the source frequency increases and the discretization rate is kept fixed. For this example, $N = 35$ nodes were used to discretize the polar semiperimeter of the sphere (i.e., the line in Figure 1b) and a 256 point FFT was used to compute the azimuthal Discrete Fourier Transform. The accuracy of the numerical solution did not degrade much when a $M = 128$ point FFT was

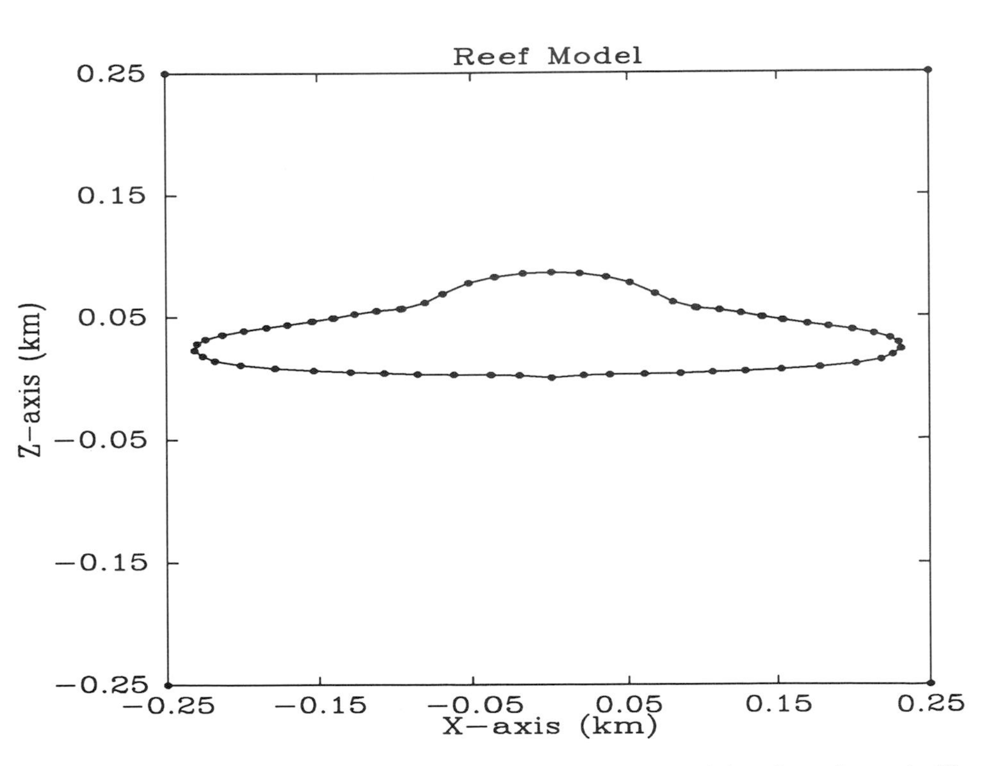

Fig. 5 Axi-symmetric reef model, with the axis of symmetry being about the z-axis. The horizontal (vertical) extent of the reef is about .5(.1) km, and the external (internal) velocity is taken to be 4.(5.) km/s.

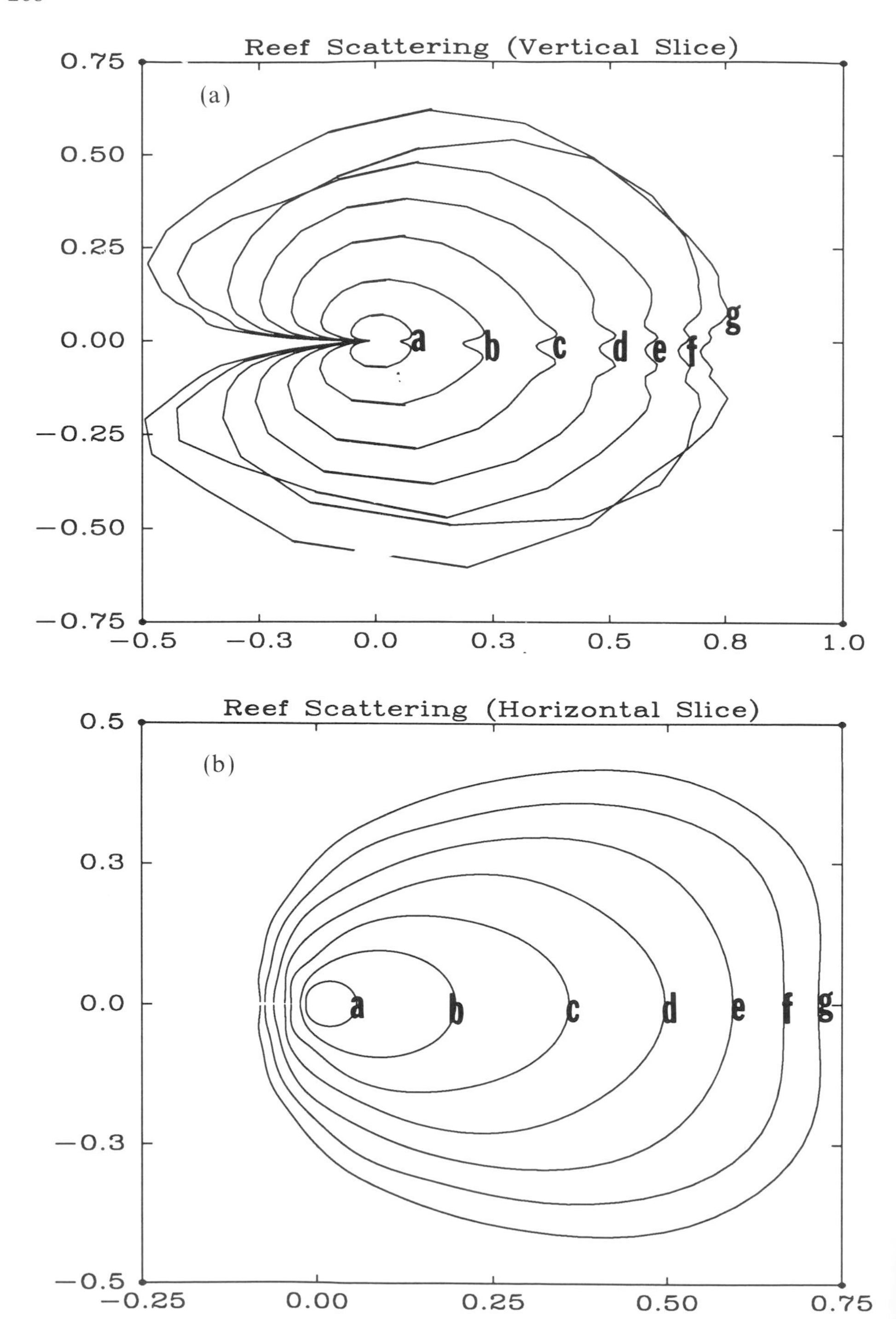
Reef Scattering (Vertical Slice)
(a)
0.75
0.50
0.25
0.00
−0.25
−0.50
−0.75
−0.5
−0.3
0.0
0.3
0.5
0.8
1.0
a
b
c
d
e
f
g
Reef Scattering (Horizontal Slice)
(b)
0.5
0.3
0.0
−0.3
−0.5
−0.25
0.00
0.25
0.50
0.75
a
b
c
d
e
f
g

used in place of the 256 point FFT. Apparently, acceptable accuracy is achieved (ka $<$ 9.4) for a discretization rate greater than 6 or 7 nodes per wavelength.

The computational time per frequency was less than 10 minutes (3 minutes) when the 256 (128) point FFT was used. Since there were 35 nodes, then 70 unknown field values (i.e., pressure and its normal derivative at each nodal point) were solved for at each azimuthal wavenumber. For a parallel computer, each CPU could independently compute a different azimuthal Fourier coefficient vector, $\tilde{\mathbf{p}}_m$, to reduce the computation time to a few seconds per frequency.

To demonstrate the practical utility of this method, the scattering characteristics of a reef model (Figure 5) are computed. In this case, a horizontally moving plane wave is impinging from the left side of the figure and the reef is axisymmetric with an axis of symmetry about the z-axis. The source frequencies are taken to be 5 Hz (a), 10 Hz (b), 15 Hz (c), 20 Hz (d), 25 Hz (e), 30 Hz (f) and 35 Hz (g), and the reef extends about .5 (.1) km in the horizontal (vertical) direction. It is filled with an acoustic material of velocity 5 km/s and the exterior velocity is taken to be 4 km/sec.

Figures (6a) and (6b) display the computed radiation patterns, respectively, in the vertical x-z plane and the horizontal x-y plane, both planes intersecting the center of the reef. This scattering is to be compared to that from a sphere (Figure 7) having the same velocity contrast and a diameter of .5 km. Evidently, the .1 km thinness of the reef tends to induce more backward scattering in the vertical plane than that scattered by the sphere. On the other hand, the reef's radiation pattern in the horizontal plane tends to be predominantly forward scattering, similar to the sphere. This is because the .5 km horizontal extent of the reef is the same as the .5 km diameter of the sphere. Each frequency response required less than 3 minutes on a 4 MIP computer, using 128 point azimuthal FFT's and 32 nodes to discretize the polar semi-perimeter of the reef in Figure (5).

Fig. 6 The plane wave harmonic responses (source frequencies = 5 (A), 10 (B), 15 (C), 20 (D), 25 (E), 30 (F) and 35 (G) Hz.) of the reef in Figure (5) arc computed by the fast BIE method. The plane wave is moving along the positive x-axis and the scattered pressure is interrogated along the reef's perimeter which intersects the $z=0$ plane (Figure 6b) and the $y=0$ plane (Figure 6a). 32 nodes were used to approximate the reef's semi-perimeter and 128 point FFT's were used for the azimuthal Fourier transform. The vertical radiation pattern (Figure 6a) exhibits more backward scattering than that for the horizontal radiation pattern (Figure 6b). Note that scattering increases as the wavelength become smaller than the size of the reef.

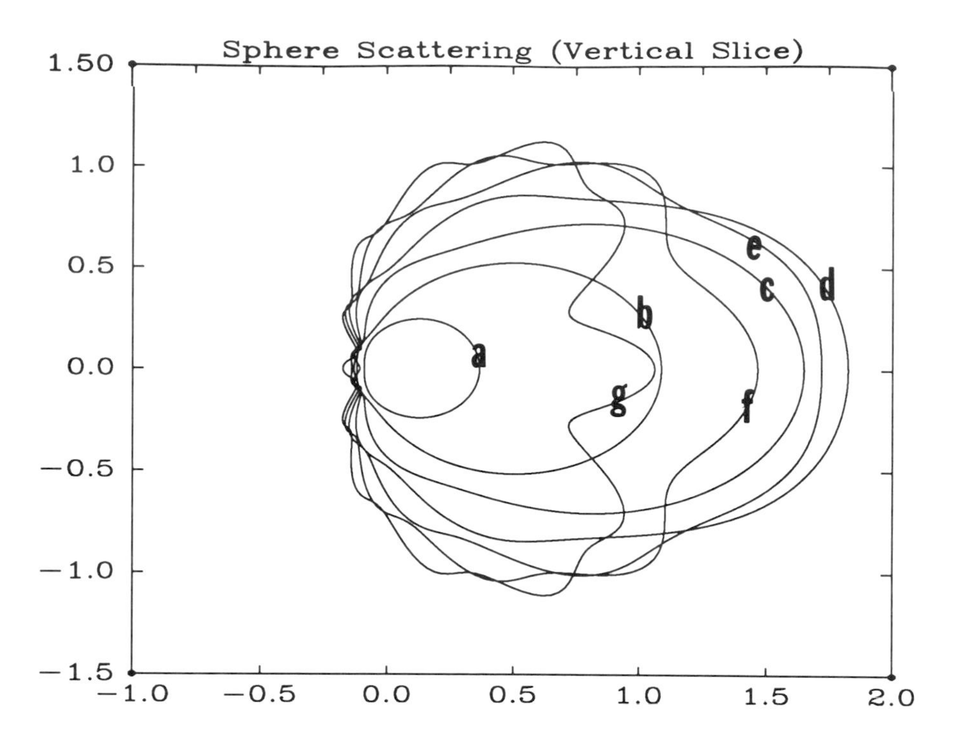

Fig. 7 Same as in Figure 6, except the scatterer is a sphere with a diameter of .5 km.

More extensive analysis on scattering characteristics of three dimensional axi-symmetric reefs, sand lenses, salt domes, and other geologic bodies may provide valuable indicators of their lithology. This fast exact numerical method can easily incorporate attenuation, internal layering within the reef (gas zones), be extendible to elastic wave propagation, and include the effect of external layers via a Generalized Born Series (Schuster, 1985; Schuster and Smith, 1985b.)

SUMMARY

A fast numerical method is presented which efficiently solves for the acoustic response of three dimensional axi-symmetric scatterers. For scatterers with dimensions of 4 to 6 wavelengths, this method required less than 3 minutes of CPU time per frequency on a Gould computer (4 MIPs). Accuracy to better than 5 per cent was achieved when the model was discretized at more than 6 or 7 nodes per wavelength. Greater accuracy could have been achieved if a higher order isoparametric representation had been used for the field variables and boundary geometry. The computer code was not optimized, and so even better efficiency could be achieved with optimization, and especially if an array processor or parallel computer is employed. In fact, the decoupled nature of this algorithm (i.e., the three-dimensional problem decouples into independent two-dimensional problems) makes it optimal for a parallel computer.

Internal layering can easily be incorporated within the scatterers without too much of a cost increase (provided there are only one or two layers). The extension of this method to include attenuation and elastic scattering is straightforward, and should be computationally efficient (on an array processor) provided the number of nodes on a semi-perimeter is less than a few hundred. If external layering is desired then it can be efficiently included by ray tracing and a Generalized Born Series. This method may have a wide variety of uses, including analysis of thin crack radiation patterns, harmonic tremors from magma chambers, resonance and focusing of seismic waves in basins (earthquake hazard studies), oil exploration and non-destructive testing of mechanical components.

The advantages of this method over finite difference methods are: 1) it creates no spurious reflections from the computational boundary; 2) only the semi-perimeter of the scatterer needs to be discretized; 3) it requires only 6 to 8 nodes per wavelength compared to over, sometimes, 20 nodes per wavelength for a 2nd order finite difference scheme; 4) grid dispersion and the computational cost is independent of the receiver offset; 5) the Fortran coding is extremely simple (less than 300 lines of code); 6) and it is easily adaptable to an array processor or a parallel computer. Moreover, a super computer is not needed for reasonable sized models.

The disadvantages of this method include: 1) lack of flexibility to efficiently handle external layering (except by the Generalized Born Series); 2) the computational count is proportional to the cube of the number of nodes along a scatterer's semi-perimeter times the number of azimuthal wavenumbers; and 3) the restriction to homogeneous sub-regions in the model.

APPENDIX

BIE FORMULATION

The boundary integral equation solution (Brebbia, 1978; Kress and Roach, 1978; Schuster and Smith, 1985a) to the Helmholtz equation

$$\nabla^2 p + k^2 p = -f \tag{A.1}$$

for the three dimensional axi-symmetric model in Figure (1a) takes the form

$$\int_{V_0} f \frac{\partial g(k_0 r)}{\partial n_0} dv - \int_{B_1} \left(\frac{\partial p}{\partial n} \left[\frac{\partial g(k_I r)}{\partial n_0} - \frac{\partial g(k_0 r)}{\partial n_0} \right] - p \left[\frac{\partial^2 g(k_I r)}{\partial n_0 \, \partial n_s} - \frac{\partial^2 g(k_0 r)}{\partial n_0 \, \partial n_s} \right] \right) ds = \frac{\partial p(\vec{\mathbf{o}})}{\partial n_0}, \; \vec{\mathbf{o}} \in \mathbf{B}_1 \tag{A.2a}$$

$$\int_{V_0} f g(k_0 r) \, dv - \int_{B_1} \left(\frac{\partial p}{\partial n} [g(k_I r) - g(k_0 r)] - p \left[\frac{\partial g(k_I r)}{\partial n_s} - \frac{\partial g(k_0 r)}{\partial n_s} \right] \right) ds = p(\vec{\mathbf{o}}), \; \vec{\mathbf{o}} \in \mathbf{B}_1 \tag{A.2b}$$

where the normal unit vector at the boundary pointing inward is given by $\hat{n}_0$ and $\hat{n}_s$ at the observer $\vec{\mathbf{o}} = (\rho \cos\phi, \rho \sin\phi, z)$ and source integration $\vec{\mathbf{s}} = (\rho' \cos\phi', \rho' \sin\phi', z')$ locations, respectively. The 1/2 factor does not appear on the right hand sides of equation (A.2) because the jump singularities are subtracted out in the integrands, leading to very weakly singular integrands. Note that $\partial/\partial n_s = \hat{n}_s \cdot \nabla$ and $\partial/\partial n_0 = \hat{n}_o \cdot \nabla$. The scatterer's surface is defined by the set of boundary points, B_1, and $ds = \rho' \, d\phi' \, dl$ where dl is the differential length shown in Figure (1b). The distance between the source integration vector and the observer vector is given by $r = |\vec{\mathbf{o}} - \vec{\mathbf{s}}|$, the free space Green's function is given by $g(k_j r) = \exp(ik_j r)/4\pi r$, where $r = (\rho^2 + \rho'^2 - 2\rho\rho' \cos(\phi - \phi') + (z - z')^2)^{1/2}$ and k_j corresponds to the wavenumber of the jth medium ($j = o, I$). The source is a point source with amplitude f harmonically oscillating at frequency ω in the exterior region V_o, pressure is denoted by p and, without loss of generality, it is assumed that densities are always unity. Equations (A.2a) and (A.2b) were derived from Green's theorem applied to, respectively, the exterior and interior regions in Figure (1a), except that the observer variable $\vec{\mathbf{o}}$ was relocated from the homogeneous region to the scatterer's surface.

The free space Green's function in cylindrical coordinates can be expressed as a Fourier Series (Morse and Feshbach, 1953)

$$g(k_j r) = \sum_{m=-\infty}^{+\infty} \tilde{G}^j_m(\rho, z \mid \rho', z')\, e^{im(\phi - \phi')} \tag{A.3a}$$

where $\tilde{G}^j_m(\rho, z \mid \rho', z')$ is the k_z Fourier transform of the product of a Hankel function and Bessel function of order m, i.e.,

$$\tilde{G}^j_m(\rho, z \mid \rho', z') \approx \int_{-\infty}^{\infty} H_m(\rho\sqrt{k_j^2 - k_z^2})\, J_m(\rho'\sqrt{k_j^2 - k_z^2})\, e^{ik_z(z-z')}\, dk_z \tag{A.3b}$$

for $\rho > \rho'$. The Green's function and its normal derivative are functions of $\phi - \phi'$ (Schuster and Smith, 1988) so that the axi-symmetric surface integrals in equation (A.2) are periodic convolutions in ϕ. Hence, the Fourier transform in ϕ applied to equation (A.2) reduces the periodic convolutions to algebraic multiplications (Oppenheim *et al.*, 1983). Substituting equation (A.3) into equation (A.2) and applying the azimuthal Fourier transform gives

$$\tilde{\mathbf{F}}'_m - \int_{L_1} \left(\frac{\partial \tilde{p}_m}{\partial n_s} \left[\frac{\partial \tilde{G}^I_m}{\partial n_o} - \frac{\partial \tilde{G}^o_m}{\partial n_o} \right] - \tilde{p}_m \left[\frac{\partial^2 \tilde{G}^I_m}{\partial n_o\, \partial n_s} - \frac{\partial^2 \tilde{G}^o_m}{\partial n_o\, \partial n_s} \right] \right) \rho\, dl = \frac{\partial \tilde{p}_m(\vec{\mathbf{o}})}{\partial n_o}, \quad \vec{\mathbf{o}} \in L_1 \tag{A.4a}$$

$$\tilde{\mathbf{F}}_m - \int_{L_1} \left(\frac{\partial \tilde{p}_m}{\partial n_s} [\tilde{G}^I_m - \tilde{G}^o_m] - \tilde{p}_m \left[\frac{\partial \tilde{G}^I_m}{\partial n_s} - \frac{\partial \tilde{G}^o_m}{\partial n_s} \right] \right) \rho\, dl = \tilde{p}_m(\vec{\mathbf{o}}), \quad \vec{\mathbf{o}} \in L_1 \tag{A.4b}$$

where the tilde denotes the transform of the functions in equation (A.2), subscript m denotes the mth azimuthal Fourier coefficient, and $\tilde{\mathbf{F}}_m$ denotes the azimuthal Fourier transform of the incident source field. The two-dimensional surface integrals in equation (A.2) have been reduced to the one-dimensional line integrals in equation (A.4); the line integral is only over the curve L_1 shown in Figure (1b).

NUMERICAL ALGORITHM

Discretizing the curve L_1 into N nodal points with two unknown field values, $(\tilde{p}_m, \partial \tilde{p}_m/\partial n)$, at each node, assuming that the field value on the segment between two nodes is approximated by a constant value (or some piecewise polynomial representation), replacing the integrations by suitable quadratures, and collocating the observer vector, $\tilde{\mathbf{o}}$, at the N nodal points transforms equation (A.4) into a $2N$ by $2N$ matrix with $2N$ unknown field values. Inverting a $2N$ by $2N$ matrix for each of the M k_ϕ wavenumbers requires far fewer computations than inverting the, roughly, N^2 by N^2 matrix associated with equation (A.2).

The problem with equation (A.4) is that $\tilde{G}^j_m$ is computed in equation (A.3b) by an infinite integration of weighted Bessel functions of various orders. The computer coding to compute this is quite complex, involving round-off, over and under flow problems, Horner's rule, singularities when the source and receiver are in close proximity, etc. It can be done (Smith and Schuster, 1985) but the coding is quite tedious.

To avoid this complexity, the strategy employed in Bojarski (1985) and Schuster and Smith (1988) is used; i.e., the integral equations in (A.2) are first approximated by their corresponding quadratures. The azimuthal Fourier transform is then applied (by FFT operations) to transform the equations into a form similar to the discrete form of equation (A.4). The resulting $\tilde{G}^j_m$'s are the Discrete Fourier coefficients of generalized functions (subtracted green's functions) integrated over the quadrilateral patches shown in Figure (1a). They are not computed from the nasty generalized functions in equation (A.3b). Since it can be shown that the integrands in each patch integral of equation (A.2) are integrable in a limiting value sense then their discrete Fourier transforms are quite tractable. Moreover, the computer coding is trivial (see Table I for the two-dimensional cylinder code).

Quadrature of the Integral Equations

The surface of the scatterer in Figure (1) can be approximated by quadrilateral patches, where a nodal point (open circle) is located at the center of each patch. The (m', n') patch is centered at the nodal point having the azimuthal angle $m'2\pi/M$, with a cylindrical radius of ρ, and located a $n'\,\Delta l$ distance along the scatterer's semi-perimeter. The distance $n'\,\Delta l$ is measured along the curve defined by the intersection of the scatterer's surface with the vertical plane oriented at azimuthal angle $m'2\pi/M$. The nodal point associated with $n'=1$ is located in the lowermost patch of the scatterer and the point corresponding to $n'=N$ is located in the scatterer's apex patch.

Designating $B_{m'n'}$ as the elemental area of the patch at (m', n') and designating the subtracted Green's function and first derivative of the subtracted Green's function in equation (A.2b) by G and H respectively, reduces equation (A.2b) to

$$-\sum_{m'=1}^{M}\sum_{n'=1}^{N}\int_{B_{m'n'}}\left(\frac{\partial p}{\partial n}G - pH\right)ds + F(\vec{\mathbf{o}}) = p(\vec{\mathbf{o}}),\ \vec{\mathbf{o}} \in B_1 \tag{A.5}$$

where the integration is over the surface of the $B_{m'n'}$ patch. Approximating p and $\partial p/\partial n$ over each (m', n') patch by the constants $p_{m'n'}$ and $\partial p_{m'n'}/\partial n$, respectively, reduces equation (A.5) to

$$-\sum_{m'=1}^{M}\sum_{n'=1}^{N}\left(\frac{\partial p_{m'n'}}{\partial n}\int_{B_{m'n'}}G\,ds - p_{m'n'}\int_{B_{m'n'}}H\,ds\right) + F_{mn} = p_{mn},$$

$$m \in [1, 2, ..., M], \qquad n \in [1, 2, ..., N] \tag{A.6}$$

where the observer vector on the right hand side of (A.5) is located at the nodal point (m, n) and F_{mn} is the value of the volume source integral interrogated at the nodal point (m, n).

The integration over a patch B_{mn} can be approximated by numerical quadrature in the local coordinate system (η_1, η_2) shown in Figure (A-1). For a

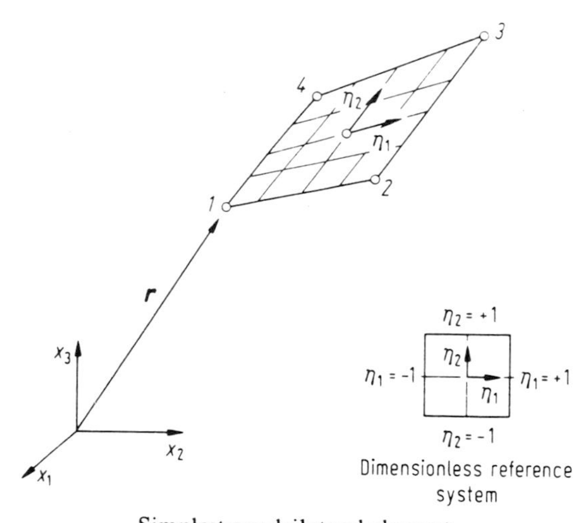

Fig. (A-1) Diagram depicting mapping of quadrilateral patch to local coordinates in (η_1, η_2) (from Brebbia et al., 1984).

four point quadrilateral Gaussian quadrature scheme, the integrals in equation (A.6) can be replaced by

$$\mathbf{G}(m-m', n, n') = \int_{B_{m'n'}} G\,ds = \sum_{i=1}^{2}\sum_{j=1}^{2} W_{ij} G(\eta_{1i}, \eta_{2j})\, J(\eta_{1i}, \eta_{2j}) \tag{A.7a}$$

and

$$\mathbf{H}(m-m', n, n') = \int_{B_{m'n'}} H\,ds = \sum_{i=1}^{2}\sum_{j=1}^{2} W_{ij} H(\eta_{1i}, \eta_{2j})\, J(\eta_{1i}, \eta_{2j}) \tag{A.7b}$$

where the Gaussian weight at the (η_{1i}, η_{2j}) local coordinate of the (m', n') patch is given by W_{ij} and the corresponding values of the subtracted Green's functions and their normal derivatives are given by, respectively, $G(\eta_{1i}, \eta_{2j})$ and $H(\eta_{1i}, \eta_{2j})$. The Jacobian at the points (η_{1i}, η_{2j}) for the mapping from the global coordinates to the local coordinates (η_1, η_2) at the $B_{m'n'}$ patch is given by $J(\eta_{1i}, \eta_{2j})$. By azimuthal symmetry, equations (A.7) are functions of the differences between azimuthal indices, $m-m'$.

Replacing the integrals in equation (A.6) by the bold symbols for quadratures in equation (A.7) gives

$$-\sum_{m'=1}^{M}\sum_{n'=1}^{N} \frac{\partial p_{m'n'}}{\partial n}\mathbf{G}(m-m', n, n') + \sum_{m'=1}^{M}\sum_{n'=1}^{N} p_{m'n'}\,\mathbf{H}(m-m', n, n') + F_{mn} = p_{mn},$$

$$m \in [1, 2, ..., M], \quad n \in [1, 2, ..., N] \tag{A.8}$$

The summations in equation (A.8) are discrete periodic convolutions in the m index. Applying a discrete Fourier transform in the m index to equation (A.8) reduces the discrete periodic convolutions to multiplications (Oppenheim *et al.*, 1983) to give

$$-\sum_{n'=1}^{N} \frac{\partial \tilde{p}_{mn'}}{\partial n}\tilde{\mathbf{G}}_m(n, n') + \sum_{n'=1}^{N} \tilde{p}_{mn'}\tilde{\mathbf{H}}_m(n, n') + \tilde{F}_{mn} = \tilde{p}_{mn},$$

$$m \in [1, 2, ..., M], \quad n \in [1, 2, ..., N] \tag{A.9a}$$

where the tilde denotes the Fourier coefficient associated with the mth azimuthal wavenumber. Steps leading to equation (A.9a) can be applied to equation (A.2a) to give

$$-\sum_{n'=1}^{N} \frac{\partial \tilde{p}_{mn'}}{\partial n}\tilde{\mathbf{G}}'_m(n, n') + \sum_{n'=1}^{N} \tilde{p}_{mn'}\tilde{\mathbf{H}}'_m(n, n') + \tilde{F}'_{mn} = \frac{\partial \tilde{p}_{mn}}{\partial n},$$

$$m \in [1, 2, ..., M], \quad n \in [1, 2, ..., N] \tag{A.9b}$$

where the prime attached to the operators indicates the appropriate derivative in A.2a). Note that the field coefficients in equation (A.8) are coupled to each other with respect to the m index, whereas they are decoupled in equations (A.9).

For a fixed azimuthal wavenumber index m, equations (A.9) form a $2N$ by $2N$ ystem of equations which can be solved for by matrix inversion. These solutions an be inverse Fourier transformed to give the pressure response in the (ω, ρ, ϕ, z) domain.

A good discussion of isoparametric representation of boundaries and field ariables is given in Zienkiewicz (1977), Pina (1984) and Brebbia et al. (1984, Chapter 3 and Appendix A). Isoparametric representations are those in which the order of the polynomial approximation for both the boundary geometry and field values is the same. Quadrature weights appropriate for $1/R$ type singularities are discussed in Critescu and Loubignac (1978) and Appendix A in Brebbia et al. 1984). It was this author's experience that a most convenient way to integrate $1/R$ ingularities (i.e., when the source integration is in the same patch as the observer ariable) was to 1). subtract out from (or smooth) the monopole integrals in quation (A.2) the integral $\int_{\text{semi-disc}} 1/4\pi R\, ds$ over a small semi-disc in the observer's patch, 2). this "smoothed" integrand can now be integrated using regular Gaussian weights, and 3). the $\int_{\text{semi-disc}} 1/4\pi R\, ds$ integral over the semi-disc can now be ntegrated analytically and added back into the monopole integral. An even better esult was achieved when a smoothing polynomial was incorporated into the numerator of the integrand of $\int_{\text{semi-disc}} 1/4\pi R\, ds$. An example might be a linear polynomial in R which attains the value 1 at $R=0$ and becomes 0 at $R=$ radius of the semi-disc.

REFERENCES

Bard, P. and Bouchon, M., 1980, The seismic response of sediment-filled valleys. Part I. the case of incident SH waves: B.S.S.A., 70 No. 4, pp. 1263–1286.

Bojarski, N., 1984, Scattering by a cylinder: A fast exact numerical solution: J. Acoust. Soc. Am., 75, No. 2, pp. 320–323.

Brebbia, C. A., 1978, The boundary element method for engineers: Halsted Press Book (J. Wiley, NY, NY).

Brebbia, C. A., Telles, J. C., and Wrobel, L. C., 1984, Boundary Element Techniques: Springer-Verlag, NY, NY.

Critescu, M., and Loubignac, 1978, Gaussian quadrature formulas for functions with singularities in $1/R$ over triangles and quadrangles: in "Recent Advances in Boundary Element Methods" edited by C. A. Brebbia, Pentech Press.

Kress, R., and Roach, G. F., 1978, Transmission problems for the Helmholtz equation: J. Math. Phys., 19, pp. 1433–1437.

Morse, P. M., and Feshbach, H., 1953, Methods of Theoretical Physics: McGraw-Hill Co., NY, NY.

Oppenheim, A., Willsky, A., and Young, I., 1983, Signals and Systems: Prentice-Hall Co., N.J..

Pina, H. L., 1984, Numerical integration and other computational techniques: in "Boundary Element Techniques in Computer-Aided Engineering" edited by Brebbia, C. A., Martinus Nijhoff Publ., Dordrecht, Netherlands.

Schuster, G. T., 1985, A hybrid BIE + Born series modeling scheme: Generalized Born series: J. Acoust. Soc. Am., 77, No. 3, pp. 865–879.

Schuster, G. T., and Smith., L. C., 1985a, A comparison among four direct boundary integral methods: J. Acoust. Soc. Am., 77, No. 3, pp. 850–864.

Schuster, G. T., and Smith, L. C., 1985b, Modeling scatterers embedded in plane-layered media by a hybrid Haskell-Thomson and boundary integral equation method: J. Acoust. Soc. Am., 78, No. 4, pp. 1387–1394.

Schuster, G. T., and Smith, L. C., 1988, A fast exact numerical solution for the acoustic response of concentric cylinders with penetrable interfaces: (accepted by JASA, subject to minor revision).

Schuster, G. T., 1988, Modeling scatterers embedded in layered media by a hybrid BIE and ray tracing method: (accepted by JASA, subject to minor revision).

Smith, L. C., and Schuster, G. T., 1985, Boreholes with washout zones by a semi-analytic + BIE technique: presented at 55th annual meeting (expanded abstracts), pp. 39–42.

Trifunac, M. D., 1971, Surface motion of a semi-cylindrical alluvial valley for incident plane SH waves: B.S.S.A., 70, pp. 1431–1462.

Tucker, B. E., and King, J. L., 1984, Dependence of sediment filled valley on input amplitude and valley properties: BSSA, 74, No. 1, p. 153.

Zienkiewicz, O. C., 1977, The Finite Element Method: McGraw-Hill Co., NY, NY.

CHAPTER 12 .

SUPERCOMPUTERS IN RESERVOIR SIMULATION: PAST, PRESENT, AND FUTURE

by
THOMAS F. RUSSELL
Computational Mathematics Group
Department of Mathematics
University of Colorado at Denver
1200 Larimer Street, Campus Box 170
Denver, Colorado 80204

1. INTRODUCTION

While the primary emphasis of the volume containing this chapter is supercomputing in seismic processing, it is a reasonable guess that about half of the supercomputing in the petroleum industry is reservoir simulation. To assess how supercomputers have been and are being used in reservoir simulation, we shall give some background on the physical problems being solved, the mathematical models that describe the physics, the numerical models that approximate the mathematical models, and the structure of the computer programs that implement the numerical models. For the most part, we shall find that supercomputers have allowed engineers to extract more information from the same modeling concepts that they were already using, either by faster turnaround of similar studies or by previously unattainable reasonable turnaround of larger studies. To date, there have been few fundamentally new ideas in reservoir modeling due to supercomputers.

We expect this to change in the future. For reasons that we shall outline, the reservoir-simulation problems of the future, specifically those of enhanced oil recovery (EOR), cannot be treated effectively by existing techniques. Different methods are needed, and we see supercomputers as an indispensable tool in the research and development that will bring new methods to practical use. We shall discuss the areas of research that we consider important and how supercomputers can contribute to this research. We also comment on how the capabilities of current technology are likely to be enhanced by more powerful future supercomputers.

An outline of this chapter is as follows. Section 2 provides background on reservoir simulation. We present enough of the physics to clarify later contentions about the kinds of problems that need to be solved. Then we describe the mathematical models of these physical systems and the standard numerical techniques used in the industry. These techniques lead to computer programs consisting of certain modules, and we indicate this structure. In Section 3, we discuss the uses of supercomputers up to the present time. The fairly mature technology of vector computing is considered first, followed by parallel computing, which, at least in reservoir simulation, is in its infancy. We show how vector processing has been used in various types of reservoir models and point out the bottlenecks that could be addressed by future improvements. These improvements are one of the subjects of Section 4, which also deals with directions of fundamental research.

Much of what is said here is based on impressions built up gradually during the author's seven years as a mathematician working on reservoir simulation for Marathon Oil Company, without specific published references in support. Such statements are the opinions of the author, and no claim is made as to the originality of these opinions. Others may, and probably will, disagree; indeed, we see stimulation of discussion as a primary role of this exposition.

2. BACKGROUND ON RESERVOIR SIMULATION

The purpose of reservoir simulation is to optimize profits from a reservoir by finding the best strategy for placement of wells, injection and production flow rates, and choice of injectants. An accurate analysis of a reservoir would require understanding of its geology and resident fluids, together with an accurate model of the flow and interactions of these fluids under the influence of injection and production wells. The flow models involve coupled systems of nonlinear partial differential equations that cannot be solved exactly by analytical means, so simulations must use numerical approximations on computers.

Physics. A petroleum reservoir is a heterogeneous porous medium, whose permeability and porosity vary on a wide range of length scales. In this medium resides a complex collection of chemical fluid species, usually called components. These include water and hydrocarbons, such as methane, ethane, propane, and so on. The components mix to form fluid phases; these are the flowing entities, each of which has its own pressure, density, and viscosity. When multiple phases are present, there is a force of surface tension at the interface between them, leading to a difference between the phase pressures known as capillary pressure. Under most circumstances, there are three phases: a water phase, consisting mostly of water with possibly some dissolved hydrocarbons, a gas phase, with mostly light hydrocarbons and possibly some volatilized heavy hydrocarbons and water vapor, and an oil phase, with mostly heavy hydrocarbons and some dissolved light hydrocarbons and possibly water. The pressure and the amounts of the various components present determine how the components partition themselves among the phases; this mass transfer is extremely important because different phases have very different flow properties.

The convective flow of each phase is usually assumed to be governed by the empirical Darcy's law,

$$\phi u_j = v_j = -\frac{kk_{rj}}{\mu_j}(\nabla p_j - \rho_j g \, \nabla z), \tag{1}$$

where ϕ is the porosity of the rock (ratio of void volume to total volume), j denotes the phase, u is the interstitial velocity (velocity of fluid movement through the pores), v is the Darcy velocity (volume of fluid crossing a unit cross-sectional area in a unit time), k is the permeability of the rock, k_r is the relative permeability of the phase (ratio of flow of this phase under multiphase conditions to flow if it were the only phase present), μ is the viscosity, p is the pressure, ρ is the density, g is the acceleration due to gravity, and z is the depth. In some cases, such as high-velocity flow in fractures, the dependence of the velocity on the pressure gradient is sublinear, and this is often accounted for by adding a term that is quadratic in the velocity to the left-hand side of Equation 1. The relative permeability is normally a strongly nonlinear function of all of the phase saturations S_j, which are the fractions of the total void volume occupied by each phase. Other mechanisms of fluid transport that can be important are molecular diffusion and hydrodynamic dispersion; the latter is mechanical mixing due to the varying tortuosity of different flow paths in a porous medium.

The processes by which oil is extracted from a reservoir are commonly broken down into primary, secondary, and tertiary recovery. We shall discuss these from the point of view of the difficulties in modeling them. In primary recovery, production wells are opened and made to flow, usually at specified rates that change discontinuously from time to time. When a rate changes, a transient pulse is introduced into the reservoir pressure field around the corresponding well. Typically, as time elapses, this pulse dies out and a new smooth pressure field evolves. The pressure controls the fluid flow; phases flow toward production wells according to their respective mobilities k_{rj}/μ_j, but no fluid is injected and no significant displacement occurs. Mathematically, this is a diffuse, parabolic process that presents few problems to standard solution techniques.

Primary recovery produces, on the average, about one third of the original oil in place in a reservoir. In secondary recovery, or waterflooding, some of the remaining oil is produced by injecting water at certain wells in order to displace oil to other wells. This creates a moving displacement front, which is harder for solution methods to represent accurately. However, water continues to be injected for many years or perhaps even decades, so that the process goes on long after the front has broken through at production wells. Except in rare instances, water is thermodynamically inert; temporary errors in its saturation have little if any effect on hydrocarbon phase behavior and the resulting transport of components in phases. Thus, any simulation errors tend to cancel themselves out over time. Ultimate recovery is mainly a function of sweep efficiency (the fraction of the reservoir swept by water), residual oil saturation (the saturation below which oil cannot flow), and capillary pressure, the force that traps oil in pores swept by water. Secondary recovery typically leaves a bit more than half of the original oil behind.

This oil is the target of tertiary recovery, also called enhanced oil recovery (EOR), which is much more difficult to model. EOR processes are usually divided into three categories–miscible, chemical, and thermal (Broome *et al.* 1986). Miscible processes inject a solvent of carbon dioxide, nitrogen, or hydrocarbons to mix with reservoir oil, either upon initial contact ("first-contact miscibility") or, more commonly, after exchanges of hydrocarbon components between the injectant and the oil ("multiple-contact miscibility"). When miscibility is achieved, there is no surface tension between the displacing injectant and the oil (they flow together as one phase), so in principle it is possible to recover all of the oil by displacement. In practice, this does not occur. The injectant is too expensive to use indiscriminately, so only a slug that would fill some small fraction of the reservoir (often 5% to 15%) is injected. This is followed by a cheaper, immiscible fluid that can be injected continuously, such as water. As the miscible slug moves through the reservoir, permeability heterogeneities cause it to prefer certain flow paths and diffusive effects reduce its peak concentration. Because the injectant is normally less viscous than the oil, the displacement front is unstable, and a phenomenon called "viscous fingering" causes the injectant to make finger-like channels into the oil, bypassing much of the oil outside these fingers. Eventually the slug can be expected to break up and lose its effectiveness. Phase behavior (miscibility or the lack thereof) is crucial in these processes, and it is quite sensitive to the pressure and the evolving compositions of the slug and the oil.

Chemical flooding injects a polymer, surfactant, or alkaline solution. The purpose of polymers is to increase the viscosity of injected water so that the mobility ratio of injectant to oil will be reduced. This mitigates viscous fingering and improves the sweep efficiency of the displacement, with less oil being bypassed. A surfactant, which is like a detergent, is intended to lower the surface tension between injectant and oil. Unlike a miscible flood, surface tension is not reduced to zero, but is made low enough so that much more oil can be "washed" out of the porous rock than would be possible without the surfactant. Polymer may be added to the surfactant solution for the sake of mobility control. Alkaline solutions, also with the possible addition of polymer, perform similarly, except that they react with constituents of crude oil to produce surfactants, which then act to lower surface tension. Thc amount of mobility control or surface-tension reduction depends strongly on the evolving chemical concentration and oil composition, and the amount of bypassed or trapped oil in turn depends strongly on mobility and surface tension.

Thermal processes include cyclic ("huff-n-puff") or continuous steam injection and *in situ* combustion. In any of these, the goal is to raise the temperature of the reservoir, lowering the viscosity of the resident oil and changing other properties to improve recovery. In cycling, steam is injected into a well which is then shut in so that the heat will diffuse; the well is then reopened as a producer, and the entire process is repeated. This has the flavor of primary recovery. Continuous injection, or steamflooding, instead displaces oil from an injector to a producer, and can be viewed as waterflooding (secondary recovery) with heat. *In situ* combustion burns a portion of the oil to generate the desired heat. Oxygen ahead of the combustion front reacts with oil to produce coke, which is the actual combustion fuel. The combustion products heat and displace the oil to production wells. The front is narrow, and proper conditions for complex reactions in and around the front must be maintained for the process to work.

Aside from steam injection, all of the EOR processes just described share the property that there is a narrow moving zone in which a crucial variable, either concentration or temperature, varies rapidly. For faithful modeling of the physics, this variable must be approximated accurately in the narrow zone. In miscible flooding, miscibility itself is sensitive to peak solvent concentration, and viscous fingering, which determines sweep efficiency, is governed by the interactions of mobility ratio, heterogeneities, and diffusive effects that depend on the concentration profile of the front as well as its location. Viscous fingering is likewise important in chemical flooding, and the peak concentration of the injected chemical determines the amount of mobility control (for polymer) or surface-tension reduction (for surfactant or alkali) and hence the effectiveness of the process. Peak temperature and concentrations of resulting substances determine whether *in situ* combustion will continue or cease, and how effectively it will reduce oil viscosity.

Mathematical models. Our discussion will be fairly terse; for more details, the reader may consult Peaceman (1977) or some of the papers in Fitzgibbon (1986). We begin with the *compositional model*, which is the most general model of primary and secondary recovery. Ideally, such a model would contain a mass-conservation equation for each chemical component in the fluid system. This is usually not practical, so groups of components are lumped into "pseudocomponents" in such a way that the phase behavior (partitioning into phases, and resulting densities and viscosities of the phases) of the pseudocomponents closely approximates that of the actual components. The mass-conservation equation for pseudocomponent i has the form

$$\frac{\partial}{\partial t}\left(c_i \sum_j \phi \rho_j S_j\right) = -\nabla \cdot \left(\sum_j \rho_j v_j c_{ij}\right) + \sum_j q_j \tilde{c}_{ij}, \tag{2}$$

where j indexes the phases in which the component may appear, c_i is the global concentration (mass of component i divided by total mass), c_{ij} is the local concentration in phase j, q is the injection rate (negative for production) per unit volume, $\tilde{c}_{ij}$ is injected or produced local concentration, and the other symbols were defined after Equation 1. A typical problem may have as many as a dozen pseudocomponents, so that a model will consist of that many coupled partial differential equations of the form of Equation 2. The left-hand side is the accumulation or storage term, and the right-hand side has convection and source terms. This formulation does not account for diffusion or dispersion.

To solve the equations, one chooses a set of primary variables; these commonly consist of one phase pressure (usually p_{oil}), the water-phase saturation S_{water} (because the water phase is usually assumed to contain no hydrocarbons), and the global concentrations $c_2, \ldots, c_n$ of all hydrocarbon components save one. The other global concentration c_1 is obtained from the relationship $c_1 + \cdots + c_n = 1$, and the local phase concentrations depend on the pressure and the global concentrations through phase-equilibrium relationships (variously called flash equations, vapor-liquid equilibrium, or equality of fugacities). The phase equilibria often are derived from an equation of state that also determines the densities and viscosities of the phases. This partitioning of hydrocarbon mass between phases also determines the saturations, which in turn yield the relative permeabilities and capillary pressures. The capillary pressures finally give the pressures in the other phases.

In terms of the pressure, the system in Equation 2 exhibits nonlinear parabolic behavior, because the density on the left-hand side provides a nonlinear first time derivative and the velocity on the right-hand side, which depends on the pressure gradient, leads to a second space derivative. The saturation on the left-hand side and the relative permeability on the right-hand side show that the behavior in terms of saturations is nonlinear hyperbolic (though recent investigations have revealed elliptic properties under certain circumstances (Bell *et al.* 1986)). Similar observations apply to concentrations, based on the global ones on the left-hand side and the local ones on the right-hand side.

In practical calculations, in which a reservoir is discretized with a grid to obtain an approximate solution of the equations, the most time-consuming part of the computation is phase behavior. If there are n hydrocarbon components, this involves the solution of n nonlinear equations in n unknowns in each grid cell. Some applications demand that this be done; examples in primary, secondary, and tertiary recovery, respectively, are gas-condensate fluid systems, in which composition and temperature conditions are such that heavy hydrocarbons are vaporized in the gas phase and condense to form liquids as the pressure is reduced, gas injection, in which oil is displaced by injected gas of composition different from the resident reservoir gas, and carbon dioxide injection, in which the hydrocarbon-CO_2 interaction is sensitive to hydrocarbon composition. However, most studies can be done adequately with a model using simplified phase behavior, which we discuss next.

The *black-oil model* assumes that the hydrocarbon system can be represented by two pseudocomponents, one light and the other heavy; these are often comprised of those hydrocarbons that form gas and oil, respectively, at standard temperature and pressure ("stock-tank conditions"). The model further assumes that fluid densities and viscosities are functions of pressure alone, with no dependence on hydrocarbon composition. The water component exists only in the water phase, the heavy hydrocarbon component exists only in the oil phase, and the light component may exist in the gas phase ("free gas") or dissolve in the oil phase ("solution gas"). The amount of light hydrocarbon that dissolves is given as a function of pressure by the solution gas-oil ratio R_{so}, which is the ratio of light mass to heavy mass in the oil phase. The mass-conservation equations can be formulated in various ways, but the usual formulation is

$$\frac{\partial}{\partial t}(\phi\rho_w S_w) = -\nabla\cdot(\rho_w v_w) + q_w, \tag{3}$$

$$\frac{\partial}{\partial t}(\phi\rho_o S_o) = -\nabla\cdot(\rho_o v_o) + q_o, \tag{4}$$

$$\frac{\partial}{\partial t}(\phi\rho_g S_g + \phi\rho_o S_o R_{so}) = -\nabla\cdot(\rho_g v_g + \rho_o v_o R_{so}) + q_g + R_{so} q_o, \tag{5}$$

for the water, heavy, and light components, respectively, where the subscripts w, o and g denote the water, oil, and gas phases, respectively. Primary variables are often taken to be p_o, S_w, and S_g, with R_{so} substituting for S_g when $S_g = 0$; in the latter case, the oil is "unsaturated", meaning that less light hydrocarbon is dissolved than the function $R_{so}(p_o)$ would indicate. The black-oil model adequately describes most fluid systems that are simulated in practice, and it can model primary and secondary recovery for such systems. A reasonable estimate is that 80% of actual studies use black-oil models.

Other mathematical models in reservoir simulation are more specialized than compositional and black-oil models. *Chemical models* deal with EOR processes that inject chemicals. These can be formulated as either compositional or black-oil systems with additional components corresponding to the injected chemicals. There must be an extra mass-conservation equation for each injectant. An example is the four-component model (Todd and Longstaff 1972, Killough and Kossack 1987) that adds a miscible solvent to the black-oil formulation. Mixing, dispersion, and viscous fingering are treated by empirical "mixing parameters" in such models. Similar in spirit is the addition of polymer to a black-oil model (Russell 1987a), where the polymer component dissolves in the water phase and affects its viscosity. *Thermal models* (Coats 1980, Aziz 1986) treat heat injection and add an energy-conservation equation to the component mass-conservation equations; for *in situ* combustion, reactions must also be incorporated. Like chemical models, these can be designed with either compositional or black-oil assumptions for the hydrocarbons.

A different type of specialization is seen in the *dual-porosity model* (Warren and Root 1963, Gilman and Kazemi 1987). This treats the special geometry of naturally fractured reservoirs, in which most of the fluid flow takes place in the high-permeability fractures, which constitute only a small fraction of the total porosity. The low-permeability rock matrix, with most of the porosity, acts as a source to the fracture network. Especially important in such reservoirs are the mechanisms that control fluid transfer between the matrix and the fractures; capillary pressure, which is negligible in the fractures but significant in the matrix, comes to the forefront. The usual assumption in a dual-porosity model is that the matrix and the fractures can be viewed as continua superimposed on each other, with each variable (pressure, saturation, etc.) having two values at each location. Thus there are twice as many equations as in a single-porosity model. In principle, all of the models mentioned previously could be formulated with dual porosity, but in practice this concept has been applied primarily to black-oil simulators.

Numerical approximations. As we have seen, the system in Equation 2 is expected to be parabolic in terms of the pressure and hyperbolic in terms of saturations and concentrations. For parabolic problems, discretization schemes that are centered in space, such as centered finite differences or standard Galerkin finite elements, work quite well. These schemes must be made implicit in time, because the explicit stability restriction would demand prohibitively small time steps in an explicit method. For hyperbolic problems, centered schemes produce nonphysical oscillations in the results, a difficulty that has traditionally been circumvented by biasing the discretization in the upstream direction of the flow. The explicit stability condition is looser, so explicit as well as implicit time stepping can be considered. This motivates the numerical methods normally used in the petroleum industry, which are centered in space and implicit in time for the pressure, and upstream in space and implicit or explicit in time for the saturations and concentrations. These combinations are usually implemented in the context of the cell-centered finite-difference method, which has conceptual physical appeal.

To make this more concrete, we describe an implicit discretization of Equation 2. Let $m-1$, m, and $m+1$ index three finite-difference cells that are adjacent to each other in the x coordinate direction. Let their dimensions in the x direction be Δx_{m-1}, Δx_m, and Δx_{m+1}, respectively, and let their other dimensions be the common values Δy and Δz. Let $m-\frac{1}{2}$ and $m+\frac{1}{2}$ index the faces between the cells. Let $n-1$ and n index the beginning and end, respectively, of a time step of length Δt. Then the mass conservation of component i in the cell m is expressed by: (mass at step n) $-$ (mass at step $n-1$) $= -$((outward flux at $m+\frac{1}{2}$) $-$ (inward flux at $m-\frac{1}{2}$)) $+$ (injection). In symbols, this is

$$\sum_j \Delta x_m \, \Delta y \, \Delta z((c_i \phi \rho_j S_j)_m^n - (c_i \phi \rho_j S_j)_m^{n-1})$$

$$= \sum_j - \Delta y \, \Delta z \, \Delta t((\rho_j v_j c_{ij})_{m+1/2}^n - (\rho_j v_j c_{ij})_{m-1/2}^n)$$

$$+ \sum_j \Delta x_m \, \Delta y \, \Delta z \, \Delta t \, q_j \tilde{c}_{ij}. \tag{6}$$

Dividing through by $\Delta x_m \, \Delta y \, \Delta z \, \Delta t$, we obtain the cell-centered implicit discretization of Equation 2.

Equation 6 is not a complete description, because values at the faces $m - \frac{1}{2}$ and $m + \frac{1}{2}$ are not directly available; they must be obtained by averaging nearby node values. As seen by Equation 1, v requires evaluation of k, k_r, μ, and ∇p (neglecting gravity for simplicity of exposition). The pressure gradient is approximated by a centered difference $(\partial p/\partial x)_{m-1/2} \approx (p_m - p_{m-1})/\frac{1}{2}(\Delta x_{m-1} + \Delta x_m)$, with an analogous formula at $m + \frac{1}{2}$; this leads to centered differences for the pressure. For k_r, which is a strong nonlinear function of saturation, the evaluation is done at the upstream cell; at face $m - \frac{1}{2}$, this is $m - 1$ if the x component of v is positive, m if it is negative. The same procedure is followed for c, yielding the desired upstream differencces in terms of saturations and concentrations. In black-oil models, where μ and ρ are functions of pressure, they are usually evaluated at faces by arithmetic averages, reflecting the appropriateness of centered differences for pressure. In compositional models, these may be strong functions of concentrations, and upstream evaluation is often used. The permeability k is harmonically averaged, since the average is to represent conductivities in series.

Boundary and initial conditions must be added to Equation 6 (and to Equation 2) to close the system. The usual boundary condition is "no flow," meaning that the flux across a cell face at the outer boundary of the reservoir is zero. This is implemented in Equation 6 simply by suppressing fluxes at boundary faces. Initial conditions specifying pressures, saturations, and concentrations at the outset of a simulation are obtained by assuming that all flow mechanisms (capillary, gravity, and viscous Darcy forces) and mass transfers are in equilibrium, so that nothing changes if no wells are open; we shall not discuss this further here.

The implicit scheme of Equation 6 leads, at each time step, to a set of nonlinear discrete equations to be solved simultaneously. The number of equations (or unknowns) is the product of the number of components and the number of finite-difference cells. This can be impractically large in many cases, so alternative methods that are explicit in saturations and concentrations have been sought. These methods are generally called IMPES, for IMplicit Pressure, Explicit Saturation. IMPES methods take many forms (see, e.g., Peaceman 1977, Aziz and Settari 1979, Thomas 1981), all of which construct a pressure equation as a combination of the component mass-conservation equations. The idea is that all dependences on saturations and concentrations in the convection terms of Equation 6 are evaluated at time level $n-1$ instead of n, and that some linear combination of the equations will be free of time derivatives of saturations and concentrations. This combination, in which the pressure is the only variable appearing at time level n, is solved for the pressure simultaneously at each grid cell. Then the conservation equations for all but one component (where one of the original equations has been replaced by the pressure equation) can be solved explicitly for the saturations and concentrations; these variables appear at level n only in the accumulation terms, which do not couple different grid cells to each other. The pressure equation and the saturation and concentration equations are all nonlinear because the coefficients depend on these variables, so a linearization technique is needed to solve them; Newton's method is most often used. The formulation of the pressure equation can be done at the differential-equation level before any discretization, at the nonlinear discrete level, or at the linear discrete level after linearization.

The number of simultaneous nonlinear equations solved by an IMPES procedure at each time step is the number of finite-difference cells, not multiplied by the number of components. Thus, an IMPES time step is much less costly than an implicit step. However, the explicit saturations and concentrations impose Courant-Friedrichs-Lewy stability restrictions essentially of the form $\Delta t \leqslant u\,\Delta x$ (the actual conditions are more complicated; an analysis of the black-oil case is in Russell (1987b)). This can force IMPES models to take smaller time steps than implicit models. In applications involving high flow velocities, such as near-well analysis of gas and water coning into a producing oil zone, flow in natural or hydraulic fractures, and gas percolation (in which lowered pressure causes a gas bubble to form, which then moves rapidly upward because of its low density and viscosity), stable IMPES time steps can be prohibitively small. Hence the industry uses both implicit and IMPES models, depending on the problems to be solved.

The upstream difference method, on which this whole structure is based, has some well-known drawbacks (see, e.g., Peaceman 1966, Watts and Silliman 1980, Shubin and Bell 1984). It yields a diffusive leading truncation-error term that, in two or more dimensions, varies as the numerical coordinates are rotated. On coarse grids of practical size, the resulting error known as "numerical dispersion" produces displacement fronts that are spread out more than physics would dictate. The rotation sensitivity, called the "grid-orientation effect," causes fronts to be geometrically misplaced. This effect is especially pronounced when the displacing fluid is less viscous than the displaced fluid, so that viscous fingering is important. As we noted previously, these types of errors will not be disastrous in simulations of primary or secondary recovery because the physics will not be seriously misrepresented. Unfortunately, with the exception of steam injection, this does not extend to simulations of EOR processes. Smeared or misplaced fronts will lead to irreversible errors, such as declaring a flood to be immiscible when it is in fact miscible. We shall point out some suggested improvements in this unsatisfactory state of affairs in our discussion of research directions.

Computational modules. The mathematical and numerical models just described give rise to a set of significant computational tasks. In the context of supercomputing, each of these has its own peculiarities. We categorize the tasks as phase-behavior calculation, table lookup, evaluation of accumulation and convection coefficients, calculation of well rates, and linear-equation solution. A time step in a production code normally involves the solution of nonlinear equations by Newton's method, and each iteration of Newton's method consists primarily of these tasks.

Phase-behavior calculation is negligible in black-oil models but paramount in compositional simulation. In the latter context, it generally consumes more than half of the computing time (Young 1987). Consider the usual situation, with n hydrocarbon components partitioning themselves among two hydrocarbon phases. In each grid cell, the gas-phase mass fraction V (mass of hydrocarbon gas divided by total hydrocarbon mass) and the local concentrations $c_{2g}, ..., c_{ng}$ can be determined from the pressure and the global concentrations by n coupled nonlinear phase-equilibrium equations. Then the relationship $c_{1g} + \cdots + c_{ng} = 1$ fixes c_{1g}, and $c_{io}(1 - V) + c_{ig} V = c_i$ yields $c_{1o}, ..., c_{no}$. The calculation in one cell is independent of all other cells. However, cells in which both hydrocarbon phases are present are handled by a solution algorithm different from that for cells with a single gas or oil phase. This has important implications, which we discuss later, for supercomputing.

Table lookup is the process by which, for example, R_{so} is evaluated as a function of p or k_{rg} is evaluated as a function of S_g. Pressure-dependent functions are usually given in simulators by a table of values at evenly spaced pressures. Saturation tables may, in general, be unevenly spaced. In either case, the function is evaluated for a given pressure or saturation by finding the table interval in which the given value lies and then interpolating between the endpoints of the interval. Linear interpolation is most often used. The interval may be found by a simple division when the table is evenly spaced, while an unevenly spaced table requires a binary search (test the midpoint of the table to see which half contains the value, then test the midpoint of that half to find the quarter containing the value, and so on). The logical path of such a search depends on where the value lies in the table. The subsequent linear interpolation uses array subscripts that depend on the location in the table. Like phase behavior, table lookup at one cell is independent of all other cells.

Accumulation and convection coefficients are evaluated by straightforward arithmetic once the results of phase-behavior calculations and table lookups are available. This can be seen by a glance at Equation 6. With no-flow boundary conditions, one can take $v=0$ at boundary faces and use the same formulas for both boundary and interior cells. Coefficient generation can approach half of the total computation in black-oil models, but it is much less important in compositional simulators.

Calculation of well rates shares some properties with phase behavior and table lookups. A well can be under various forms of control, such as bottom-hole pressure, surface pressure, injection rate, total production rate, or production rate of one phase. Each possibility requires its own logic. Also, individual wells can be independent of one another. However, there has been a trend in recent years toward more comprehensive well-management routines that allow controls such as target rates to apply to a group of wells or an entire field (Holmes 1983); in this situation, wells are not independent. A well is usually completed in several layers of the computational grid, and a rate must be assigned to each layer. This can be done explicitly, in the sense that the allocation among layers depends only on the state of the reservoir at the end of the preceding time step, or implicitly, in which case the well contributes to the system of equations to be solved for the new time step. Each implicit rate-controlled well gives rise to a rate equation that must be solved simultaneously with the cell mass balances such as Equation 6; the corresponding extra unknown is the bottom-hole pressure of the well (Trimble and McDonald 1976). The equation couples all cells in which the well is completed, disrupting the regular pattern of couplings in Equation 6.

The four modules in the preceding paragraphs inevitably lead to a large, sparse system of linear equations. In black-oil models, solution of this system most often takes the majority of the computer time (Killough and Levesque 1982). It is less important in compositional simulations, which are dominated by phase calculations. An IMPES model yields one equation and one unknown per grid cell, while for an implicit black-oil simulator this number is three. The Jacobian of Newton's method can be viewed as a band matrix, where the number of nonzero bands is the number of cells with which a given cell interacts; in three dimensions, this is usually seven (the cell itself, plus two neighbors in each direction). For implicit black oil, each entry is a 3-by-3 submatrix whose rows and columns correspond to components and primary variables, respectively; in the IMPES case, each entry is simply a number. An IMPES matrix is diagonally dominant and nearly symmetric, while an implicit matrix is highly nonsymmetric and may not be diagonally dominant (Wallis *et al.* 1985). Further complicating matters is the possibility of implicit well equations for bottom-hole pressures; these are usually put at the beginning or the end of the grid-cell equations, resulting in a "bordered Jacobian" that has a regular structure except for irregular couplings at its edges. Direct Gaussian elimination is preferable for small problems of no more than a few hundred cells, but considerations of storage and computer time dictate that larger systems be solved by iterative methods.

3. PAST AND PRESENT USE OF SUPERCOMPUTERS

Reservoir simulation has been a major application of the current generation of supercomputers almost from the beginning. As early as February 1979, papers dealing with reservoir simulation on array processors (Killough 1979, Woo 1979) and vector machines (Nolen *et al.* 1979) appeared in the literature. Most major companies now have vector machines, so that the use of these computers has grown more sophisticated over the years, while attached array processors have received little attention. Parallel hardware seems sure to have a large impact in the future, but its use has only recently begun. We now consider how these types of machines have been used.

Vector machines. Most vectorized reservoir simulation has been carried out on Cray-1 or Cray X-MP machines, with some use of the Cyber 205, the IBM 3090 vector facility, and FPS attached array processors. We view this software technology as fairly mature at present, although some incremental improvements are no doubt on the horizon. We review some vector implementations of the computational modules discussed previously.

The greatest effort has been directed toward vectorization of linear-equation solvers. This began in 1979 with the direct method of sparse Gaussian elimination on array processors (Woo 1979) and the iterative method of line successive over-relaxation (LSOR) on a vector machine (Nolen *et al.* 1979). The latter reference demonstrated the usefulness of red-black checkerboard ordering in vectorization of

LSOR. The most popular direct method today is Gaussian elimination with alternate-diagonal (D4) ordering (Price and Coats 1974); large bandwidths make efficient vectorization straightforward, and a typical vector performance is a factor of 5 over scalar speed (Behie and Forsyth 1983). Sparse elimination has also been considered, with speedups of 2 to 2.5 (Sherman 1985). However, the trend is toward larger problems that must be treated iteratively, so iterative methods are our main focus here.

The most popular of the older generation of iterative methods were LSOR and the strongly implicit procedure (SIP). Diagonal planes can be used to vectorize SIP, with a speedup of about 4 over scalar performance in Fortran (Killough and Levesque 1982); further improvements were possible with Cray assembly language (CAL). Red-black LSOR achieves factors of 5 to 10 in Fortran (Young 1987). LSOR can use longer vectors than SIP (half of the product of the horizontal dimensions, as opposed to half of the larger horizontal dimension), so its speedup would be expected to be greater, but SIP is a somewhat more reliable method in the sense that it is less likely to fail to converge to an answer. Both of these methods have been largely supplanted by a new family of procedures exhibiting much greater reliability.

The newer methods are based on the conjugate-gradient algorithm for symmetric positive-definite matrices and on nonsymmetric generalizations such as Orthomin (Vinsome 1976). At each iteration, these schemes minimize a quadratic functional of the residual over a subspace of the solution space. The rate at which these iterations converge to the desired solution (which makes the quadratic functional zero) depends on how close the matrix is to the identity matrix. A technique known as *preconditioning* can make the matrix close to the identity as follows. Let the problem to be solved be $Ax=b$, and let P be a matrix that approximates A and has the property that $Py=c$ is much easier to solve than $Ax=b$. Then apply the iterative method to the problem $(AP^{-1})(Px)=b$. The matrix AP^{-1} is close to the identity, and it turns out that each iteration involves the same computations as the unpreconditioned scheme, except that a system of the form $Py=c$ must be solved in addition. The hope is that preconditioning will obtain convergence in many fewer iterations, each of which is slightly more expensive, so that the entire procedure is much faster. The key is the choice of P.

Probably the most prevalent preconditioners are generated by applying Gaussian elimination to A and neglecting most of the fill-in terms; these are called incomplete LU (ILU) decompositions. This was first tried in the context of IMPES models, where the pressure matrix is nearly symmetric, with the conjugate-gradient algorithm (Watts 1981). Subsequently, the concept was extended to implicit models with Orthomin (Behie and Vinsome 1982). It was soon noticed that these procedures could benefit by a preprocessing step in which the unknowns of one checkerboard color eliminated the other color (Tan and Letkeman 1982); this reduction is possible because the cells of one color are directly connected only to the other color (Equation 6). The reduced system has half as many unknowns and represents the first step of exact Gaussian elimination with red-black or D4 ordering, so each iteration is cheaper and good convergence can be expected. Comparisons (Behie 1985, Eisenstat *et al.* 1985) have found reduced-system ILU/Orthomin with D4 ordering to be extremely reliable and probably also the fastest iterative method in terms of scalar work. The method extends readily to bordered matrices with well equations (Behie and Forsyth 1983). The author coded such a method for Marathon Oil Company, and it has received heavy use for three years; to our knowledge, no simulation has aborted because of failure of the solver to converge. Unfortunately, the method is not ideal for supercomputers. The ILU decomposition is highly recursive, inhibiting either vectorization or parallelization, and the solution of $Py = c$ involves forward and backward substitutions that are also recursive and allow only short vector lengths spanning small numbers of nonzero bands. Fortran code runs only 1.5 to 2 times faster in vector mode than scalar mode (Behie and Forsyth 1983).

A generalization of the just-described "point" incomplete factorizations is to eliminate blocks of unknowns simultaneously, leading to "block" incomplete factorizations. One such procedure that has been extensively used in reservoir simulation is nested factorization (Appleyard and Cheshire 1983), and a more general framework for these methods has also been proposed (Meijerink 1983). These methods do not outperform the best point methods in scalar work, but they can be competitive and they require less storage (Behie 1985, Eisenstat *et al.* 1985). When the blocks correspond to lines of the reservoir grid, the recursive work of these algorithms is in the solution of tridiagonal systems, unlike point methods that solve systems with more bands. For tridiagonal systems, vectorization or parallelization can be attempted via cyclic reduction (Meijerink 1983); to our knowledge, this has not yet been done. For nested factorization, 60% of the work is readily vectorizable, but the authors did not try to vectorize the remaining 40% that involved tridiagonal systems (Appleyard and Cheshire 1983).

We now give an idea of the computer resources required by these methods. A "work unit" will be one multiplication or division per grid cell, and a "storage unit" will be one word per cell. Let m be the number of unknowns per cell (1 for IMPES,

3 for implicit black oil, etc.), and let k be the number of previous iterations used by Orthomin. Typical values of k are 5 and 10 for difficult IMPES and implicit problems, respectively (Behie and Forsyth 1983). Each method incurs a one-time setup cost at the outset, followed by repeated iteration costs. We compare the most reliable point and block preconditioners, namely reduced-system ILU with D4 ordering and no fill-in beyond the bands of the reduced matrix itself (RS/ILU(0)) and nested factorization (NF). For three-dimensional problems, the costs are as follows:

	RS/ILU(0)	NF
Storage units	$16.5m^2+(k+3)m$	$9m^2+(2k+4)m$
Work units for setup	$51m^3+6.5m^2$	$16m^3+25m^2$
Work units per iteration	$16m^2+\dfrac{k+3}{2}m$	$25m^2+(k+3)m$

For an IMPES problem ($m=1$, $k=5$), RS/ILU(0) requires 24.5 storage units, 57.5 setup work units, and 20 work units per iteration, compared to 23, 41, and 33 for NF. In the implicit black-oil case ($m=3$, $k=10$), RS/ILU(0) needs 187.5 storage units, 1435.5 setup work units, and 163.5 work units per iteration, while NF uses 153, 657, and 264. NF has a moderate advantage over RS/ILU(0) in terms of storage and a significant advantage in setup cost, but RS/ILU(0) costs less per iteration and is less sensitive to choices such as the ordering of the coordinate directions (Behie 1985, Eisenstat *et al.* 1985). We believe that RS/ILU(0) is the method most often used in practice for difficult linear systems. A 20,000-cell IMPES compositional study (Young 1987) or a 20,000-cell implicit black-oil run would roughly consume the maximum core storage of 8 million words generally available in the petroleum industry (for example, on a Cray X-MP/48), and the resulting linear system could be expected to be solved in a number of iterations of the order of 20. For RS/ILU(0), this leads to an estimate of 500 work units, or 10 million operations, in the IMPES case, and 5000 work units, or 100 million operations, in the implicit case. With a typical execution rate of 10 megaflops (10 million floating-point operations per second) on one of the usual machines, this corresponds to 1 second for an IMPES linear solution and 10 seconds for an implicit linear solution. A lengthy simulation with 1200 time steps and 3 Newton iterations per time step would then require one hour in the IMPES case and 10 hours in the implicit case.

For the problems considered, the existing limitations of storage and computer time are in reasonable balance. This is not the case for implicit thermal problems, for which a typical value of m is 6; here, the problems that fit into storage have prohibitive work counts as the m^3 setup terms increase in importance. When the Cray-2, which has 32 times the storage of the Cray X-MP/48 and perhaps twice the processing speed, is taken into account, it becomes clear that the hardest reservoir-simulation problems are work-limited. Vectorization of block preconditioners could help to redress the imbalance, but some other ideas that have been suggested, and which we now discuss, could also contribute.

The first idea involves the imposition of constraints on the residual of the linear system. In reservoir simulation, this dates back almost 20 years (Watts 1971), when it was suggested that line SOR would converge faster if the residuals of lines or planes of the grid were made to sum to zero. Exact conservation of mass, or making the residuals of the entire grid sum to zero, has been recommended for nested factorization (Appleyard and Cheshire 1983). Recently, the possibility of conjugate-gradient-like methods with simple, easily vectorizable preconditioners (such as line Jacobi relaxation) in conjunction with extensive, adaptive constraints has been proposed (Wallis *et al.* 1985). The constraints, which could be physically motivated, would combine with the vectorizable preconditioner to make the iterative method converge rapidly. This has been implemented with considerable success (Wallis *et al.* 1985).

Another idea that is prominent outside but not inside the reservoir-simulation community is the multigrid method. It uses vectorizable relaxation methods, such as Gauss-Seidel relaxation with red-black ordering, on grid scales of varying fineness to damp out quickly the entire range of error frequencies in the residual. For reservoir simulation, this procedure has been applied successfully to IMPES pressure equations in two dimensions (Behie and Forsyth 1982), but it has not been extended to implicit systems or to three dimensions. Multigrid has the theoretical advantage that its work count is proportional to the number of unknowns, unlike the methods discussed above, which can be expected to take more iterations as the number of unknowns increases. There is no obvious reason why it could not be as successful in reservoir simulation as it has been in many other applications.

A third innovative idea is domain decomposition, which will be discussed in the context of parallel computation.

The computational modules other than linear-equation solvers have received considerably less attention. Of these, phase-behavior calculation has held the most interest because of its critical role in compositional models. Several approaches have been tried, but the clear winner that should become an industry standard is based on reordering of grid fluid-property arrays by grouping cells of the same fluid type (Young 1987). Because few cells change type in a typical time step, the reordering can be carried out with little work, and the time-consuming flash calculations can be vectorized with vectors of length equal to the large number of cells with a given fluid type; cells with the same type follow the same logic. For a large problem (over 3,000 cells), phase behavior was calculated in vector mode on a Cray X-MP at 10 times scalar speed (Young 1987), and the scalar speed was already superior to previous published scalar results. For test problems dominated by phase behavior, in which the linear systems were easy to solve so that LSOR would work well, the simulator incorporating this reordering outperformed others by a wide margin (Killough and Kossack 1987, Young 1987). When the linear system is easy, it is estimated that this IMPES simulator could solve a problem with 9 components, 20,000 cells, and 1,000 time steps in 40 minutes on a Cray X-MP (Young 1987). With fluid properties needing 250 to 300 words of storage per cell, such a problem would approximately fill the normally available 8 million words.

Vectorization of table lookups has evolved with changes in hardware, specifically the vectorizability of indirect addressing. In the past, assembly language was needed to vectorize lookups because of the array index that depended randomly on the grid cell; Fortran code would run in scalar mode, and assembly-language code could be 10 times faster (Killough and Levesque 1982). More recently, vectorizable indirect addressing has allowed vectorizable Fortran evenly spaced lookups (Bolling 1987), though the searching logic for unevenly spaced lookups still precludes their vectorization in Fortran. Computation of accumulation and convection coefficients vectorizes over an entire grid in a straightforward manner, assuming that sufficient storage is available; vector speeds of 10 times scalar speed have been observed (Young 1987). Well calculations appear to be unvectorizable because of the unique logic for each well; in a highly vectorized code, they have taken 4% of the execution time in scalar mode and 28% in vector mode, because all other segments were vectorized (Young 1987).

Parallel machines. Many observers believe that current vector processors are approaching a physical limit, imposed by the speed of light, in processing power. The way to obtain greater megaflop rates appears to be simultaneous execution of multiple processors on different parts of a problem. Such parallel computers are classified as SIMD (single instruction, multiple data) or MIMD (multiple instruction, multiple data); SIMD machines require algorithms similar to those for vector hardware, while MIMD machines allow different processors to perform completely different tasks at the same time. Another classification of parallel computers involves the type of memory; shared-memory machines have common storage accessible to all processors, while each processor in a distributed-memory machine has its own memory and data must be passed from one processor to another. A shared-memory machine typically has 4 or 8 or perhaps 16 processors, but a distributed-memory architecture allows for hundreds or thousands of processors. These are often connected like the edges of a hypercube, because this nearly achieves the optimal goal of the shortest communication paths for the fewest connections.

To our knowledge, the first use of parallel processing in reservoir simulation was on the ICL Distributed Array Processor (DAP), a distributed-memory SIMD machine with 4,096 processors (Scott *et al.* 1982). The investigators implemented an implicit black-oil simulator, using line Gauss-Seidel relaxation to solve linear equations. The tridiagonal systems of line Gauss-Seidel were solved in parallel by the cyclic-reduction algorithm. Subsequent work, all of which is quite recent, has used shared-memory MIMD machines. One paper (Scott *et al.* 1987) studied various forms of SOR on the Denelcor HEP and found that red-black ordering and assignment of a submatrix to each processor were two effective alternatives. The authors also proposed asynchronous updating of primary variables in an IMPES simulator to converge to an implicit solution; they did not implement this idea, but it is known (Russell 1987b) that it is subject to the same stability limitations as the usual IMPES method. The Cray X-MP/48, which has 4 vector processors acting in parallel, was used in another study (Chien *et al.* 1987) that addressed the computational modules other than linear-equation solvers. All modules except well calculations were vectorized, and all modules were parallelized. With 4 processors, speedups of up to 3.3 over single-processor vector speed were achieved. Because of the small number of processors, cells with different fluid types were not processed simultaneously in phase-behavior calculations, though the authors noted that this could be done.

The most intriguing parallel concept considered so far is domain decomposition (Killough and Wheeler 1987). This involves partitioning of the computational grid into three-dimensional subdomains that are handled in parallel. The overall algorithm is quite complicated, requiring approximate ILU subdomain solutions that are used in obtaining solutions on the faces of the subdomains; these in turn yield improved subdomain solutions by backsubstitution. This entire scheme then functions as a preconditioner for the Orthomin method. Line corrections such as those employed with LSOR (Watts 1971) can be incorporated. On Cray X-MP and IBM 3090 machines with vector and parallel capabilities, the method vectorized and parallelized well and was promising in comparison with other methods for some difficult IMPES test problems.

We are not aware of any reservoir-simulation work on distributed-memory MIMD machines. This is probably due to the fact that such computers usually require extensive reprogramming, and the commercially available machines of today do not outperform the most powerful vector computers.

Applications and bottlenecks. We summarize briefly the types of applications in which supercomputers are important and the principal barriers to improved performance in these applications. For black-oil models, the typical study is a large full-field simulation with possibly tens of thousands of cells and hundreds of wells separated by a few cells. In either an IMPES or an implicit model, the computation is dominated by linear-equation solution, and unfortunately the best methods in common use involve recursive preconditioners that do not vectorize or parallelize well. Well calculations do not vectorize, and these can become burdensome in certain cases. Dual-porosity models are similar to implicit black-oil models because the equations for the rock matrix are eliminated in a preprocessing step, leaving a system with the same structure as implicit black oil. However, table lookup, phase behavior, and coefficient evaluation must be performed in both the fractures and the matrix, so the emphasis on linear-equation solution is somewhat less and storage limitations become more important; grids will not have as many cells. Thermal models, which are usually implicit, return the emphasis to linear equations because there are more unknowns per cell, as in dual-porosity models, but none are eliminated by preprocessing. Grids are limited to a few thousand cells by the computational burden, and the need for preconditioners that take maximum advantage of supercomputers is acute. Until the recent work on reordering of arrays (Young 1987), compositional models were limited to a few thousand cells by the poorly vectorized phase-behavior computations; this should now be changed and models as large as black-oil studies should be feasible, with memory being an important limitation in the absence of a machine like the Cray-2. Because of numerical dispersion, we do not feel that existing chemical models represent the relevant physics adequately, so the bottleneck to useful results here lies in discretization methods rather than efficient use of supercomputers.

4. FUTURE USE OF SUPERCOMPUTERS

A review of the preceding section shows that supercomputers have spawned clever ideas for speeding up certain computational modules. They have not, however, caused fundamental changes in the types of reservoir-simulation problems attacked or in how these problems are solved; the mathematical models and numerical approximations have remained essentially the same since well before the present generation of machines. We expect future developments to fall into two categories: fairly routine extensions of current techniques, which will allow existing models to solve larger problems, and research that will lead to new methods capable of solving problems beyond the reach of today's methods, such as physically realistic simulation of most EOR processes.

Extensions of current uses. As vector computers become more powerful, it can be expected that engineers will seek greater accuracy with existing codes by using finer grids and smaller time steps. The forthcoming Fortran 8X standard may lead to vector constructs that will be handled more efficiently by hardware and compiler-generated machine language. However, these developments are not likely to improve processing power by many orders of magnitude because of the speed-of-light barrier. This limitation should eventually give rise to widespread use of parallel machines.

Parallelism offers some obvious benefits to computational modules in existing simulators. Phase-behavior calculations, table lookups, and well-rate computations (for individual wells not under collective control) share the property that one calculation is independent of all others, but that the logical paths of different calculations may be different. The logic inhibits vectorization, but it presents no difficulty on MIMD parallel machines. We see these modules as incentives for model developers to investigate massively parallel distributed-memory MIMD machines when these become competitive with leading vector processors.

In linear-equation solvers, it appears that most nonvectorizability is caused by recursiveness rather than logic, so that parallelism by itself will not yield obvious improvements; the vectorizability as well as parallelizability of residual constraints and the multigrid method could have a strong impact. Domain decomposition, however, could take significant advantage of massive parallelism. Another technique of interest here is the adaptive implicit method, or AIM (Thomas and Thurnau 1982). In this procedure, saturation and concentration dependences in Equation 6 are evaluated at the implicit time level n in some cells and explicitly at level $n-1$ in others. The idea is that implicitness is often needed only in a small part of the grid, so that it is wasteful to impose it everywhere. Each saturation or concentration has a region of implicitness that can change as a simulation proceeds. This leads to linear systems with a variable number of equations per grid cell, making vectorization extremely difficult, and we believe that this is the reason why AIM has not been more widely used. Such systems should parallelize as well as any, so AIM could become more attractive in the future. A black-oil stability analysis, extendable to compositional models, shows how to obtain optimal efficiency with AIM Russell 1987b).

Research. We begin with a brief review of the state of the art for various mathematical models, indicating suggested areas of improvement. We find that the accuracy of black-oil models is generally satisfactory, with the possible exception of near-well behavior. Full-field simulations must represent this behavior by a fairly crude approximation that assumes a local pressure field based on steady-state single-phase two-dimensional flow (Peaceman 1978). A recent comparison of several industry simulators (Killough and Kossack 1987) found that differing models of injectivity, or the ability of fluids to flow from wells into rock layers, led to significant differences in results. Research in local grid refinement (Heinemann *et al.* 1983, Pedrosa and Aziz 1985) needs to be continued to address this problem; refinement will be even more important in EOR modeling, as we discuss below. The speed of black-oil models is also relatively good, with vectorizable and parallelizable preconditioners for linear-equation solution being the leading concern; a theory explaining the observed convergence with conjugate-gradient-like methods would be desirable. Reliability is an issue at the nonlinear level because Newton's method sometimes fails to converge, forcing simulators to resort to *ad hoc* remedies such as reducing the time step or multiplying primary-variable changes by a damping factor. Worse, the solution of the nonlinear equations for a time step may not be unique; we have seen different *ad hoc* strategies for the same equations converge to different solutions. Perhaps Newton's method is not the best approach, and a continuation method or a nonlinear version of the multigrid method would be better. A continuation method would be founded on the belief that reservoir physics does not lead to true bifurcations of physical solutions, so that there is a unique physical path to be followed. In multigrid, the solution on a coarse grid provides a good initial guess for the iteration to a solution on a finer grid; any improvement over the standard initial guess, which is the solution at the previous time step, would be welcome.

The status of compositional models is similar to black oil but a bit more complicated. The secondary-recovery application of gas injection is more vulnerable to grid-orientation sensitivity than waterflooding is, because the low viscosity of gas makes the mobility ratio of injected gas to resident oil very high. This can be handled with reasonable accuracy by existing nine-point finite-difference methods, which connect grid cells diagonally in horizontal planes along with the usual five-point connections. This reduces the variation of the diffusive leading truncation-error term of upwind differencing as the coordinates are rotated (Yanosik, and McCracken 1979, Russell and Wheeler 1984, Shubin and Bell 1984). Tertiary recovery (e.g., CO_2 flooding) is discussed below. The other aspect not found in black-oil models is phase behavior; we have covered the speed of these computations previously but have not discussed their reliability and their effect on nonlinear convergence of Newton's method. Phases can become nearly indistinguishable as a fluid system approaches a critical point, so that the phase-equilibrium equations become nearly singular and the gas-phase mass fraction V becomes very sensitive to changes in the global concentrations c_i. Then V affects saturations and relative permeabilities strongly. Thus, the phase equations themselves are difficult to solve, and they create strong nonlinearities that make the global flow equations difficult to solve. Yet these difficulties are in some sense artificial, because saturations of nearly indistinguishable phases should not be important; their importance is a defect of the mathematical model. Perhaps one should define a special set of phase variables that are meaningful at a critical point and solve for near-critical behavior by perturbations of these variables.

In dual-porosity models, the crucial addition to the black-oil formulation is matrix-fracture interchange of fluids. A rigorous treatment of capillary imbibition and drainage should model boundary conditions at the interface. The usual finite-difference methods cannot do this, but an alternative finite-element model has been suggested (Douglas *et al.* 1987). This model would require a massively parallel computer to be implemented efficiently, because it involves the solution of a large number of independent boundary-value problems on small matrix subdomains. When such a machine becomes available, a combination of the new model with the physical sophistication of the most advanced finite-difference models (Gilman and Kazemi 1987) could achieve physical realism for the first time.

Thermal models, when applied to steamflooding, are in a state similar to compositional models applied to gas injection; grid orientation is important, but existing nine-point upwind difference methods are satisfactory. The resulting linear systems were once quite difficult to solve reliably with iterative methods, but preconditioned conjugate-gradient-like techniques have changed that. Thermal models applied to *in situ* combustion and chemical models require new discretization methods, which we address next.

Many alternatives to upwind differences, aiming at the accuracy needed for EOR modeling, have been proposed (Carr and Christie 1983, Ewing *et al.* 1983, Ewing and Heinemann 1983, Taggart and Pinczewski 1985, Bell and Shubin 1985).

The common theme is elimination of numerical dispersion, so that the sharp fronts and viscous fingering characteristic of EOR can be modeled in a way that preserves the physics. It is not our purpose here to compare these methods, but rather to identify significant practical contributions that they could make with the help of supercomputers. The methods have been tested successfully on model problems, but to the best of our knowledge none have been extended to practical EOR simulation.

The obvious benefit of better methods, given sufficient data and physical understanding of the problem at hand, is physically realistic simulation of EOR. Unfortunately, life is not that simple; reservoir simulation is data-poor and somewhat lacking in physical understanding. Rock heterogeneities, which manifest themselves in permeability and porosity coefficients, exist on all length scales in real reservoirs. These quantities are typically measured at wells by using rock samples and log data, and at short distances from wells by analyzing transient pressure behavior. Elsewhere, they are guessed at or obtained by a trial-and-error history-matching procedure, in which an engineer attempts to find values that cause a simulator to match observed well histories. History matching is much more art than science, and its results are far from unique. Fluid properties such as viscosity, surface tension, relative permeability, and dispersivity can be measured in laboratory rock samples, but it is not at all clear how to translate these values from the laboratory to the field length scales used in simulators. Thus, most of the data submitted to a simulator are questionable at best.

A prime issue in physical understanding is the relationship between microscopic and macroscopic flow–how can laboratory-scale physics be averaged up to simulator length scales? In particular, how can small-scale heterogeneities, smaller than the size of a grid cell, be represented? We have some notion of how viscous fingering operates at laboratory scale, but what happens at field scale as a function of heterogeneity, dispersivity, and mobility ratio? Can fingering be viewed as a macroscopically deterministic process through averaging of chaotic microscopic behavior, or do microscopic perturbations have macroscopic implications so that it must be treated probabilistically? In history matching, how sensitive is the flow to the unknowns; i.e., how accurate or voluminous must our data be before we can calculate reasonable approximations to flow?

The important role of better discretization methods, in our view, is in the attack on these types of questions for simulation of EOR. If these questions are to be answered at all, we believe that the answers will come from large-scale simulations on supercomputers using accurate methods. We see some use for theories such as homogenization in the averaging process, but the instabilities of viscous fingering are such that the averaging of rock properties will probably depend on fluid properties; theories do not account for this, so empirical work will play an indispensable part. We envision a bootstrapping process, in which simulators extend understanding from laboratory scale to somewhat longer scales in turn, repeating until field scale is reached. It may turn out that new concepts derived from this process will require that the mathematical model of reservoir flow be reformulated for EOR. In any case, better methods are a necessity, because upwind differences mask the very physics that this program hopes to illuminate. Local grid refinement is essential to achieve sharp resolution of moving fronts; even with local refinement, the problems are difficult enough that this work could fill any foreseeable supercomputer to capacity.

Another use of better methods on supercomputers, which could lead to more practical gains in the short term, is in the context of simulations of full oil fields with very coarse grids; large black-oil studies, or compositional studies with the vectorizable phase-behavior calculations based on array reordering, would be typical. Current limitations in supercomputers are such that a full three-dimensional simulation of a large field can allow only two or three grid cells between wells, if that. Better methods are unlikely to yield much direct improvement on such coarse grids. However, they could be used indirectly in order to obtain coarse-grid data. Accurate local simulations could be performed with fine grids and compared to local coarse-grid results; when a match is obtained, the corresponding coarse-grid data could be put into the full-field model. This would put the process of averaging fine-grid data to a coarse grid on a sound footing. Again, this process could fill any foreseeable supercomputer to capacity.

REFERENCES

Abbreviations

JPT = *Journal of Petroleum Technology*
SPEJ = *Society of Petroleum Engineers Journal*
SPERE = *SPE Reservoir Engineering*
SPE4 = *Proceedings of the Fourth SPE Symposium on Numerical Simulation of Reservoir Performance*, Society of Petroleum Engineers, Dallas, 1976

SPE5 = *Proceedings of the Fifth SPE Symposium on Reservoir Simulation*, Society of Petroleum Engineers, Dallas, 1979

SPE6 = *Proceedings of the Sixth SPE Symposium on Reservoir Simulation*, Society of Petroleum Engineers, Dallas, 1982

SPE7 = *Proceedings of the Seventh SPE Symposium on Reservoir Simulation*, Society of Petroleum Engineers, Dallas, 1983

SPE8 = *Proceedings of the Eighth SPE Symposium on Reservoir Simulation*, Society of Petroleum Engineers, Dallas, 1985

SPE9 = *Proceedings of the Ninth SPE Symposium on Reservoir Simulation*, Society of Petroleum Engineers, Dallas, 1987

Appleyard, J. R., and Cheshire, I. M., 1983. Nested Factorization. *SPE7*, pp. 315–324.

Aziz, K., 1986. Simulation of Thermal Recovery Processes. In *Mathematical and Computational Methods in Seismic Exploration and Reservoir Modeling*, W. E. Fitzgibbon, ed., Society for Industrial and Applied Mathematics, Philadelphia.

Aziz, K., and Settari, A., 1979. *Petroleum Reservoir Simulation*. Applied Science Publishers, London.

Behie, A., 1985. Comparison of Nested Factorization, Constrained Pressure Residual, and Incomplete Factorization Preconditionings. *SPE8*, pp. 355–373.

Behie, A., and Forsyth, P. A., 1982. Multi-Grid Solution of the Pressure Equation in Reservoir Simulation. *SPE6*, pp. 29–42; *SPEJ*, 23 (1983), 623–632.

Behie, A., and Forsyth, P. A., 1983. Practical Considerations for Incomplete Factorization Methods in Reservoir Simulation. *SPE7*, pp. 305–314.

Behie, A., and Vinsome, P. K. W., 1982. Block Iterative Methods for Fully Implicit Reservoir Simulation. *SPEJ*, 22, 658–668.

Bell, J. B., and Shubin, G. R., 1985. Higher-Order Godunov Methods for Reducing Numerical Dispersion in Reservoir Simulation. *SPE8*, pp. 179–190.

Bell, J. B., Trangenstein, J. A., and Shubin, G. R., 1986. Conservation Laws of Mixed Type Describing Three-Phase Flow in Porous Media. *SIAM J. Appl. Math.*, 46, 1000–1017.

Bolling, J. D., 1987. Development and Application of a Limited-Compositional, Miscible Flood Reservoir Simulator. *SPE9*, pp. 21–37.

Broome, J. H., Bohannon, J. M., and Stewart, W. C., 1986. The 1984 National Petroleum Council Study on EOR: An Overview. *JPT*, 38, 869–874.

Carr, A. H., and Christie, M. A., 1983. Controlling Numerical Diffusion in Reservoir Simulation Using Flux Corrected Transport. *SPE7*, pp. 25–32.

Chien, M. C. H., Wasserman, M. L., Yardumian, H. E., Chung, E. Y., Nguyen, T., and Larson, J., 1987. The Use of Vectorization and Parallel Processing for Reservoir Simulation. *SPE9*, pp. 329–341.

Coats, K. H., 1980. In-Situ Combustion Model. *SPEJ*, 20, 533–554.

Douglas, J., Jr., Paes Leme, P. J., Arbogast, T., and Schmitt, T., 1987. Simulation of Flow in Naturally Fractured Reservoirs. *SPE9*, pp. 271–279.

Eisenstat, S. C., Elman, H. C., and Schultz, M. H., 1985. Block-Preconditioned Conjugate-Gradient-Like Methods for Numerical Reservoir Simulation. *SPE8*, pp. 397–405.

Ewing, R. E., and Heinemann, R. F., 1983. Incorporation of Mixed Finite Element Methods in Compositional Simulation for Reduction of Numerical Dispersion. *SPE7*, pp. 341–347.

Ewing, R. E., Russell, T. F., and Wheeler, M. F., 1983. Simulation of Miscible Displacement Using Mixed Methods and a Modified Method of Characteristics. *SPE7*, pp. 71–81.

Fitzgibbon, W. E., ed., 1986. *Mathematical and Computational Methods in Seismic Exploration and Reservoir Modeling*. Society for Industrial and Applied Mathematics, Philadelphia.

Gilman, J. R., and Kazemi, H., 1987. Improved Calculations for Viscous and Gravity Displacement in Matrix Blocks in Dual-Porosity Simulators. *SPE9*, pp. 193–208; *JPT*, 40 (1988).

Heinemann, Z. E., Gerken, G., and von Hantelmann, G., 1983. Using Local Grid Refinement in a Multiple-Application Reservoir Simulator. *SPE7*, pp. 205–218.

Holmes, J. A., 1983. Enhancements to the Strongly Coupled, Fully Implicit Well Model: Wellbore Crossflow Modeling and Collective Well Control. *SPE7*, pp. 255–266.

Killough, J. E., 1979. The Use of Array Processors in Reservoir Simulation. *SPE5*.

Killough, J. E., and Kossack, C. A., 1987. Fifth Comparative Solution Project: Evaluation of Miscible Flood Simulators. *SPE9*, pp. 55–73.

Killough, J. E., and Levesque, J. M. 1982. Reservoir Simulation and the In-House Vector Processor: Experience for the First Year. *SPE6*, pp. 481–487.

Killough, J. E., and Wheeler, M. F., 1987. Parallel Iterative Linear Equation Solvers: An Investigation of Domain Decomposition Algorithms for Reservoir Simulation. *SPE9*, pp. 293–312.

Meijerink, J. A., 1983. Iterative Methods for the Solution of Linear Equations Based on Incomplete Block Factorization of the Matrix. *SPE7*, pp. 297–304.

Nolen, J. S., Kuba, D. W., and Kascic, M. J., Jr., 1979. Application of Vector

Processors to the Solution of Finite Difference Equations. *SPE5*; *SPEJ*, 21 (1981), 447–453.

Peaceman, D. W., 1966. Improved Treatment of Dispersion in Numerical Calculation of Multidimensional Miscible Displacements. *SPEJ*, 6, 213–216.

Peaceman, D. W., 1977. *Fundamentals of Numerical Reservoir Simulation*. Elsevier, Amsterdam.

Peaceman, D. W., 1978. Interpretation of Well-Block Pressures in Numerical Reservoir Simulation. *SPEJ*, 18, 183–194.

Pedrosa, O. A., Jr., and Aziz, K., 1985. Use of Hybrid Grid in Reservoir Simulation. *SPE8*, pp. 99–111; *SPERE*, 1 (1986), 611–621.

Price, H. S., and Coats, K. H., 1974. Direct Methods in Reservoir Simulation. *SPEJ*, 14, 295–308.

Russell, T. F., 1987a. Mathematical Formulation of the Polymer-Flooding Option for the New Black Oil Simulator (NBOS). *Technical Report* 87–7*R*, Marathon Oil Company, Littleton, CO.

Russell, T. F., 1987b. Adaptive Implicit Finite Difference Methods: Structure, Stability Analysis, and Switching Criteria, *Technical Report* 87-6*R*, Marathon Oil Company, Littleton, CO.

Russell, T. F., and Wheeler, M. F., 1984. Finite Element and Finite Difference Methods for Continuous Flows in Porous Media. In *The Mathematics of Reservoir Simulation*, R. E. Ewing, ed., Society for Industrial and Applied Mathematics, Philadelphia, pp. 35–106.

Scott, A. J., Sutton, B. R., Dunn, J., Minto, P. W., Thomas, C. L., Habib, S., Oakley, C. A., and Krzeczkowski, A. J., 1982. An Implementation of a Fully Implicit Reservoir Simulator on an ICL Distributed Array Processor. *SPE6*, pp. 523–533.

Scott, S. L., Wainwright, R. L., Raghavan, R., and Demuth, H., 1987. Application of Parallel (MIMD) Computers to Reservoir Simulation. *SPE9*, pp. 281–292.

Sherman, A. H., 1985. Sparse Gaussian Elimination for Complex Reservoir Models. *SPE8*, pp. 407–413.

Shubin, G. R., and Bell, J. B., 1984. An Analysis of the Grid Orientation Effect in Numerical Simulation of Miscible Displacement. *Comp. Meth. Appl. Mech. Eng.*, 47, 47–71.

Taggart, I. J., and Pinczewski, W. V., 1985. The Use of Higher-Order Differencing Techniques in Reservoir Simulation. *SPE8*, pp. 163–177; *SPERE*, 2 (1987), 360–372.

Tan, T. B. S., and Letkeman, J. P., 1982. Application of D4 Ordering and

Minimization in an Effective Partial Matrix Inverse Iterative Method. *SPE6*, pp. 43–58.

Thomas, G. W., 1981. *Principles of Hydrocarbon Reservoir Simulation.* IHRDC, Boston.

Thomas, G. W., and Thurnau, D. H., 1982. The Mathematical Basis of the Adaptive Implicit Method. *SPE6*, pp. 69–73; *SPEJ*, 23 (1983), 759–768.

Todd, M. R., and Longstaff, W. J., 1972. The Development, Testing, and Application of a Numerical Simulator for Predicting Miscible Flood Performance. *JPT*, 24, 874–882.

Trimble, R. H., and McDonald, A. E., 1976. A Strongly Coupled, Fully Implicit, Three Dimensional, Three Phase Well Coning Model. *SPE4*; *SPEJ*, 21 (1981), 454–458.

Vinsome, P. K. W., 1976. ORTHOMIN, An Iterative Method for Solving Sparse Sets of Simultaneous Linear Equations. *SPE4.*

Wallis, J. R., Kendall, R. P., and Little, T. E., 1985. Constrained Residual Acceleration of Conjugate Residual Methods. *SPE8*, pp. 415–428.

Warren, J. E., and Root, P. J., 1963. The Behavior of Naturally Fractured Reservoirs. *SPEJ*, 3, 245–255.

Watts, J. W., 1971. An Iterative Matrix Solution Method Suitable for Anisotropic Problems. *SPEJ*, 11, 47–51.

Watts, J. W., 1981. A Conjugate Gradient-Truncated Direct Method for the Iterative Solution of the Reservoir Simulation Pressure Equation. *SPEJ*, 21, 345–353.

Watts, J. W., and Silliman, W. J., 1980. Numerical Dispersion and the Origins of the Grid Orientation Effect–A Summary. Paper 96C, presented at the 73rd Annual Meeting of AIChE, Chicago.

Woo, P. T., 1979. Application of Array Processor to Sparse Elimination. *SPE5.*

Yanosik, J. L., and McCracken, T. A., 1979. A Nine-Point, Finite-Difference Reservoir Simulator for Realistic Prediction of Adverse Mobility Ratio Displacements. *SPEJ*, 19, 253–262.

Young, L. C., 1987. Equation of State Compositional Modeling on Vector Processors. *SPE9*, pp. 313–320.

INDEX